SCIENCE UNDER SIEGE

SCIENCE UNDER SIEGE

DEFENDING SCIENCE, EXPOSING PSEUDOSCIENCE

EDITED BY KENDRICK FRAZIER

Prometheus Books

59 John Glenn Drive
Amherst, New York 14228–2119

Published 2009 by Prometheus Books

Inquiries should be addressed to
Prometheus Books
59 John Glenn Drive
Amherst, New York 14228–2119
VOICE: 716–691–0133, ext. 210
FAX: 716–691–0137
WWW.PROMETHEUSBOOKS.COM

13 12 11 10 09   5 4 3 2 1

Library of Congress Cataloging-in-Publication Data

Science under siege : defending science, exposing pseudoscience / edited by Kendrick Frazier.
    p. cm.
  ISBN 978–1–59102–715–7 (pbk. : alk. paper)
    1. Science—Miscellanea. 2. Pseudoscience. I. Frazier, Kendrick.

Q173.S4265 2009
500—dc22

                  2009005800

Printed in the United States of America on acid-free paper

CONTENTS

II. CRITICAL INQUIRY AND PUBLIC CONTROVERSIES

III. UNDERSTANDING PSEUDOSCIENCE, INVESTIGATING CLAIMS

INTRODUCTION

Science, reason, and rationality have, by nearly any recent measure, suffered serious diminishment in the public sphere. When asked, people say they like and admire science (and I think they do), but the evidence also shows that they don't know much about it and they poorly understand and appreciate the methods science uses to pursue the truth about nature. This dual conclusion has been repeatedly confirmed in the broad Science and Engineering Indicators surveys carried out biennially for the National Science Board / National Science Foundation, most recently in 2008. The same surveys, and others by private polling agencies, have consistently found that belief in paranormal phenomena, including extrasensory perception and alien visitations, is fairly widespread. Disbelief in evolution is notoriously widespread in America, so much so that among thirty-four nations of the world the United States ranks second from last (only to Turkey) in acceptance of evolution. And, worse, many findings of science are under attack from a wide variety of social, political, and cultural forces. These forces selectively dismiss, deny, denigrate, and distort legitimate results of scientific research they construe as unwelcome or uncomfortable.

Science engages in robust internal debate about all new findings and is always open to new evidence. Discovering new evidence about nature is what science is all about. But debates in the public that concern science rarely involve the actual scientific content and frequently caricaturize and stereotype both scientists and the scientific process. In the public arena partisans increasingly misuse or misrepresent the science. This distorts the democratic process and leads to poorly informed decision making. And that is dangerous in our increasingly complex and challenging world.

Some of these problems are due to antiscience sentiments, some to strong ideological or cultural agendas that seemingly require ignoring or dismissing unwanted scientific results. Some appear more to be a troubling *indifference* to science. This is true even at the highest levels of leadership. Last year's exhausting presidential campaigns in the United States saw the candidates in endless debates and media appearances. Yet crucial issues involving science and technology were scarcely mentioned. When a debate on issues of science and technology was formally proposed to the candidates, they ignored the request, despite the fact that science and technology profoundly shape the future. And despite the fact that virtually all the nation's scientific leaders, including dozens of Nobel laureates and every major scientific society, including the National Academy of Sciences and the American Association for the Advancement of Science, had signed the requesting petition. Is this a troubling measure of how science has faded in the public consciousness? It is hard to think otherwise.

In his first several months in office, President Barack Obama clearly reversed some of these trends, at least at the top levels of government. His inaugural address promise to "restore science to its rightful place," his appointment of widely respected scientists to head science-related agencies, his repeal of the previous administration's restrictions on stem-cell research, and his vow to base decisions on the best scientific evidence rather than ideology have cheered scientists and rationalists. But unless and until these welcome attitudes trickle down and permeate through the general culture, which might happen only slowly at best, the problems remain.

And then there is pseudoscience, that perennial dark shadow of science, where scientific pretenders co-opt and apply science's reputation and goodwill to all manner of bizarre ideas, theories, and devices unsupported by any real scientific findings. Here scientific illiteracy again comes into play, both on the creating and consuming ends of the matter. Motivations may range all the way from (to be generous) an honest desire to break new scientific ground and make important advances (without understanding how to go about that) to achieving personal gain and wealth or political/cultural influence and power, and everything in between. It can be fairly innocuous or deeply pernicious. Pseudoscience is, to me, always disconcerting yet endlessly fascinating, in that it continually morphs into new disguises, popping up unexpectedly in new forms and new ways with new language, all seeking to confuse the public into thinking that this time, this one's the real deal.

Science Under Siege addresses a wide range of these issues. It is a timely new exploration of important public issues both within science and along its borderlands. I have selected the subjects for their significance, relevance, interest, and timeliness in today's scientific, political, social, and cultural debates. Some, I think you will find, are as timely as today's headlines. And many of them are darned good reading.

For more than thirty years the *Skeptical Inquirer*, which it has been my pleasure to edit, has steadfastly championed science and reason and been the leading voice for reliable scientific examination of all manner of questionable claims popularized by the media and mass culture. These articles originally appeared in its pages, most in the past five years, and many of them provoked widespread response and comment and controversy. A few have been slightly updated. Most stand fine on their own. All of them, it seems to me, have something important to say.

My hope is that this publication in book form will help bring these particular articles to an audience well beyond *SI*'s loyal readers: scientists and scholars in all fields, educators and teachers, students and inquirers at all levels, and citizens who long for scientifically reliable information in what has regrettably become an age of misinformation. I think they will be of interest to everyone concerned about the place of science in society, the way science works in unraveling the best evidence about nature (and eventually discarding unsupported evidence), and the damage done by those who distort this process. They have much to say about the public perception (and misperception) of science and how its findings get misunderstood and misstated to fit different ends. I strongly feel that science and its unique, time-tested methods at getting at the truth are not just a part of the noble human quest to understand, but a crucial aspect of a democratic society. Good citizenship, it seems to me, requires some understanding and support of science and its methods.

The introductory section focuses on Science and Skeptical Inquiry and our culture's apparent recent turning away from science and reason. I begin it with Paul Kurtz's expansive review of thirty years of the *Skeptical Inquirer*. I hope readers will forgive any slight self-indulgence in this regard. The point here is not what *SI* has done but what the so-called skeptical movement has investigated and accomplished over those three decades. Valuable historical perspective comes from learning of, or recalling, the battles and investigations that have preceded most of those published in this anthology. Back in the mid-1970s the modern organized skeptical movement began, and our pages became the focus for publication of evidence-based skeptical investigation into, first, paranormal, fringe-scientific, and pseudoscientific claims, then, more broadly, any and all controversial or questionable claims having some scientific component. It is all rather extraordinary. Carl Sagan's final question-and-answer session on science and scientific inquiry, published here for the first time in book form, likewise discusses a wide range of such issues in the easy, clear-thinking style that made the late astronomer and Pulitzer Prize-winning author so popular. Ann Druyan adds her own poignant and stirring postscript, "The Great Turning Away." She then continues on with her beautifully expressed "Science, Religion, Wonder, and Awe," an ode to science and discovery, including a brief look at religion in the context of science. Chris Mooney portrays some aspects of the recent Bush administration's war on science. (We are nonpartisan. All governments try to some degree to bend science to their own ends, but it is well documented that the past administration's efforts were particularly egregious; we will be interested in applying similar analyses to the new administration, and future ones.) And questions such as "can the sciences help us make wise ethical decisions?" and "do extraordinary claims require extraordinary evidence?" are explored. And a fallacy often committed by well-meaning skeptics, especially physical scientists, is enunciated—attempting to explain some proclaimed unusual phenomenon in natural terms before one knows there is a real phenomenon to explain.

The second section is on Critical Inquiry and Public Controversies. Here controversies abound—controversies about almost everything imaginable. Almost all still reverberate through the daily news and other currents of culture. Most in some way involve our best current understandings of science, opposed by intense—often very intense—advocates of other, contrary specific interests. The continuing assaults on evolution are the subject of the first five chapters. Creationists and intelligent design advocates attack this basic unifying principle of science from every quarter possible—but especially by well-funded propagandistic distortion tactics and getting supporters onto local school boards as well as local and regional governments. This, despite a landmark, decisive, and marvelously well-written 2005 federal court ruling by US District Judge John E. Jones III (included here) that ID is not science and that including it in the science classroom is unconstitutional. (Previous court decisions have said the same thing of creationism itself.) Philosopher Barbara Forrest recounts her experience testifying in that trial and exposing the IDers' now notorious "wedge" and "vise" strategies. Charles Sullivan and Cameron McPherson Smith describe four common myths about evolution (among ten in their subsequently published book) that creationists (and others merely misinformed) constantly spread. It seems clear that evolution has become the most willfully misunderstood and abused core explanatory scientific principle in all of modern science. In "Only a Theory?" David Morrison examines one of those myths of syntactical misperception, one that evolution's opponents gleefully exploit to frame the issue to their advantage.

Another controversy explored is global climate change and the denial of global warming. This article, although it was merely attempting to summarize what climate scientists had been saying for some time, surprisingly provoked an intense critical reaction from libertarian-leaning readers, prompting author and NASA scientist Stuart Jordan to add a thoughtful follow-up essay (also included here), "The Global Warming Debate: Science and Scientists in a Democracy." Other chapters explore AIDS denialism, the anti-vaccination movement, terrorism against animal researchers, and the memory wars, in which claims of repressed memories unleashed under suggestive hypnosis have resulted in false accusations of wrongdoing. The latter is by Martin Gardner, a national treasure. Gardner has been writing about science and pseudoscience for nearly sixty years, beginning with his classic (and still relevant and recommended) 1952 book, *Fads and Fallacies in the Name of Science,* and continuing through his years as a *Scientific American* and then *Skeptical Inquirer* columnist and now occasional *SI* contributor.

An especially provocative chapter by planetary scientists Clark Chapman and Alan Harris argues that Americans overreacted to the terrorist attacks of 9/11 and that this misjudgment is rooted in our innate difficulty in rationally assessing risks. This article first appeared one year after 9/11 when nerves were still raw and fears of more attacks still strong. Many respondents felt the authors underplayed the real continuing risks and otherwise failed to give justice to understandable fears. But the authors now say they believe subsequent events only reinforce and strengthen their original conclusions. Even the noted MIT cognitive scientist Steven Pinker, who originally criticized the authors in our pages, here in a brief postscript for this volume, moderates his criticisms in the perspective of events over the ensuing seven years. Still another chapter, by investigator and *SI* managing editor Ben Radford, presents evidence that fears of sexual predators abducting our children are likewise highly exaggerated. And another, by energy recycling expert Thomas R. Casten shows—contrary to our intuition—why decentralized generation of electricity is far more efficient than our current centralized methods and could also save vast amounts of energy.

A third section—Understanding Pseudoscience, Investigating Claims—begins with a thoughtful exploration by philosopher of science and physicist Mario Bunge contrasting the philosophies behind science and pseudoscience. It may come as a surprise that pseudoscience has a philosophy. It does, Bunge notes; it is just "orthogonal" to that of science. I have led off this section with Bunge's chapter because I think it is powerfully explanatory. It provides a comprehensive, intellectual, philosophical, explanatory framework for virtually every issue we address in this volume and in the pages of the *Skeptical Inquirer.* Bunge notes that every intellectual endeavor, "whether authentic or bogus," has an underlying philosophy. Science has at least six philosophical underpinnings and four other distinguishing features. Pseudoscience has its own set. The philosophies of the two just happen to be perpendicular to each other. For example, the demanding ethics of science do not tolerate the self-deceptions and frauds that plague pseudoscience. Bunge also has some pungent takes on specific fields: psychoanalysis, computationist psychology, quantum theory, string theory, and "pockets of pseudoscience ensconced in the sciences." (These provoked much subsequent comment and controversy.)

Section three then goes on to present a series of scientific examinations of specific cases of bad science, fringe-science, or pseudoscience. The first of these is a highly questionable study, unfortunately published in a medical journal and widely reported in the news media, claiming to find effects of prayer on pregnancy. Physician/investigator Bruce Flamm became suspicious and honed in on this pray-for-pregnancy "miracle" study. He found and publicly revealed so many flaws and questions that the lead author, a department chairman at Columbia University, eventually renounced his connection to the study. Another author, the man who allegedly designed and conducted the study, was eventually convicted of fraud (on other matters) and went to prison. The most troubling aspect, apart from the fact that the journal that published the study never retracted it (and so it still stands in the scientific literature, open to citation in support of its supposed findings), was that the authors never provided Flamm (or anyone else, so far as is known) any of their original data for examination. This is a no-no in science, where data are supposed to be open to examination and testing by outside researchers. In fact, the authors never provided evidence to allay concerns that their much-ballyhooed international study was even actually carried out. It seems relevant to note that the author who pleaded guilty to fraud had earlier published a series of much-publicized studies making pro-paranormal claims about "therapeutic touch," human energy fields, miraculous cures, and so on. But the pray-for-pregnancy study was the first to have the imprimatur of establishment science, if only temporarily. Now all those other studies are likewise suspect.

Also included here is physician Harriet Hall's amused/bemused examination of psychology professor Gary Schwartz's claims of "energy healing." (This is just the latest of outlandish claims the good professor has advanced, never seeming to revise his claims in response to strong critiques they inevitably draw from fellow scientists.) We also include two up-to-date examinations of "anomalous cognition," the latest euphemism for psi powers. One is by a young cognitive neuroscientist, Amir Raz, the second by veteran constructive critic of parapsychology, psychologist Ray Hyman. They complement each other nicely.

Other claims examined in these pages involve magnet therapy; oxygentherapy pseudoscience; the claims of the now-defunct Princeton Engineering Anomalies Research group alleging that human intent can remotely affect electronic devices; tests of a Russian girl who claims to have X-ray vision (this article, and a companion one not included here, won the Committee for Skeptical Inquiry's first Robert P. Balles Prize in Critical Thinking); some outlandish claims regarding vacuum energy; lessons of the "fake moon flight" myths; an undercover investigation into a notorious spiritualist enclave in Indiana; and a constructive investigation into a so-called haunted house that actually showed the worried occupants what was going on. We conclude with some fascinating insights into how the noted magician/investigator James Randi improvised under newly imposed conditions during challenging live psi-like demonstrations (very relevant to understanding how some "psychics" pull off their tricks and, more generally, how easy we can all be misled), and a recurring but still patently false patent myth. (No, no patent commissioner ever said, "Everything that can be invented has been invented" and resigned!) An industrial librarian tracks down the myth's origin.

Forty distinguished scientists, psychologists, philosophers, physicians, critics, writers, and investigators contribute these provocative and insightful essays. This book is dedicated to them, and to others like them worldwide who continue to cast the light of science into the shadows of misunderstanding, deception, and ignorance (willful or otherwise). They, to borrow a phrase enshrined by our late esteemed colleague Carl Sagan—as the subtitle of his last great work of scientific skepticism—truly embrace "science as a candle in the dark."

🏵 🏵 🏵

I thank Paul Kurtz, our founding chairman, the philosopher who started this all, who continues to this day to inspire us all with his phenomenal energy, vision, dedication, intelligence, wisdom, and diplomatic grace, and who remains one of the world's leading champions of scientific skepticism; the other founding leaders of the modern skeptical movement: Martin Gardner, James Randi, Ray Hyman, James Alcock, and the late Philip J. Klass; fallen colleagues like Isaac Asimov, Carl Sagan, and Stephen Jay Gould, all of whom eloquently championed science and reason to the masses and were always so helpful to me; the current fellows and scientific consultants of the Committee for Skeptical Inquiry (CSI, formerly CSICOP, publisher of the *Skeptical Inquirer*), whose continuing contributions to both science and scientific skepticism make this a better, more interesting world; all the tireless contributors to the *Skeptical Inquirer*, including all those whose work could not be represented here if we were to have a manageably-sized volume; the new and upcoming next generation of scientists and skeptical inquirers, our hope for the future; and other scholars and scientists worldwide who share our interest not only in advancing and defending good science, but in bringing to the public the best available scientific knowledge in the hope that it nurtures and inspires. Nature is awe-inspiring, and to experience some of the excitement of discovery and insight into its secrets is a privilege—and to use one's skills to share that with others is a high calling. And finally, I thank my longtime colleagues at CSI and the *Skeptical Inquirer*, Barry Karr and Ben Radford, whom it has been my pleasure to work with for many years, and likewise Lisa Hutter, Donna Budniewski, Julia Lavarnway, Andrea Szalanski, and Paul Loynes, who contributed mightily to the original publication of all these chapters; Ronald Lindsay and all our colleagues at the Center for Inquiry, which, like CSI and *SI*, promotes and defends reason, science, and freedom of inquiry; and finally to Steven Mitchell, Jonathan Kurtz, and their colleagues at Prometheus Books for encouraging and bringing this volume to fruition.

—KENDRICK FRAZIER

Part I

SCIENCE AND SKEPTICAL INQUIRY

SCIENCE AND THE PUBLIC
Summing Up Thirty Years of the *Skeptical Inquirer*

Paul Kurtz

W e marked the thirtieth year of publication of the *Skeptical Inquirer*, the official magazine of the Committee for the Scientific Investigation of Claims of the Paranormal, with its September/October 2006 issue. CSICOP had been founded in May 1976, six months before the first issue was published in Fall/Winter 1976 as *The Zetetic* (meaning "skeptical seeker"), under the editorship of Marcello Truzzi. The name was changed to the *Skeptical Inquirer* the following year, and Kendrick Frazier was appointed the new editor, a position he has served with brilliant virtuosity and distinction ever since. Ken had been the editor of *Science News*, and during his tenure at the *Skeptical Inquirer* he also worked full time at Sandia National Laboratories for 23 years until his retirement from there in April 2006. He has kept abreast of the many breakthroughs on the frontiers of the sciences and is eminently qualified to interpret the sciences for the general public; hence he continues to be a perfect fit for the *Skeptical Inquirer*.

In preparation for this overview, I reviewed the entire corpus published in the past thirty years, which is now available on CD-ROM. What impressed me greatly was the wide range of topics and the distinguished authors that Ken has attracted to its pages. I can highlight only some of these in this article. I wish to use this occasion to focus on what I believe we have accomplished in the past three decades and to speculate as to what directions our magazine might take in future decades. Today, many threats to science come from disparate quarters—as Ken points out in his editorial "In Defense of the Higher Values" in the July/August 2006 issue of the *Skeptical Inquirer* (the next chapter in this volume). These include efforts to undermine the integrity of science and freedom of research, and we are continually confronted by irrational antiscientific forces rooted in fundamentalist religion and ideology. Given these challenges, no doubt skeptical inquiry will continue to be necessary in the future.

The original name of CSICOP was the Committee for the Scientific Investigation of Claims of the Paranormal *and Other Phenomena*, but this mouthful was deemed too long—and the acronym would have been *CSI-COPOP*—so we shortened it! (In September 2006 we shortened it again, to the Committee for Skeptical Inquiry, but I'll retain the CSICOP designation here.) It is clear that the *Skeptical Inquirer* was never intended to confine itself solely to paranormal issues; and the topics it has dealt with have been truly wide-ranging. The subtitle that was eventually developed and now appears on every issue is "The Magazine for Science and Reason," which states succinctly what it is all about. It has encouraged "the critical investigation of paranormal and fringe-science claims," but "it also promotes science and scientific inquiry, critical thinking, science education, and the use of reason."

I.

The enduring contribution of the *Skeptical Inquirer* in its first three decades, I submit, has been its persistent efforts to raise the level of the public understanding of science. No nation or region can cope with the challenges of the global marketplace and compete effectively unless it provides a steady stream of highly educated scientific practitioners. This is true of the developing world, which wishes to catch up with the advanced industrial and informational economies; but it is true of those latter nations as well. Today, China and India have embarked upon massive efforts to increase the number of scientists in their countries—China graduates anywhere from 350,000 to 600,000 engineers annually, compared to 70,000 to 120,000 in the United States, of which some 30,000 are foreign born. Alas, we still have a tremendous task, for US students rank only twenty-fourth in scientific knowledge out of the twenty-nine industrialized countries. Only 40 percent of twelfth graders tested had any comprehension of the basic concepts and methods of science. Presumably, even fewer political figures in Washington have the requisite comprehension!

The long-standing policy of CSICOP has been four-fold: (1) to criticize claims of the paranormal and pseudoscience, (2) to explicate the methods of scientific inquiry and the nature of the scientific outlook, (3) to seek a balanced view of science in the mass media, and (4) to teach critical thinking in the schools. Unfortunately, the constant attacks on science, the rejection of evolution by creationists and intelligent design advocates (some thirty-seven states have proposed programs to teach ID and creationism in the schools), the limiting of stem-cell research by the federal government, and the refusal to accept scientific findings about global warming vividly demonstrate the uphill battle that the United States faces unless it improves the public appreciation of scientific research.

Clearly, the major focus of the *Skeptical Inquirer*, especially in its first two decades, was on the paranormal; for there was tremendous public fascination with this area of human interest, which was heavily promoted and sensationalized by an often irresponsible media. Our interest was not simply in the paranormal curiosity shop but to increase an understanding among the general public of how science works.

The term *paranormal* referred to phenomena that allegedly went "beyond normal science." Many topics were lumped under this rubric. And many credulous people believed that there was a paranormal-spiritual dimension that leaked into our universe and caused strange entities and events. Included in this mysterious realm was a wide range of phenomena, which the *Skeptical Inquirer* examined within its pages over the years: psychic claims and predictions; parapsychology (psi, ESP, clairvoyance, telepathy, precognition, psychokinesis); UFO visitations and abductions by extraterrestrials (Roswell, cattle mutilations, crop circles); monsters of the deep (the Loch Ness monster) and of the forests and mountains (Sasquatch, or Bigfoot); mysteries of the oceans (the Bermuda Triangle, Atlantis); cryptozoology (the search for unknown species); ghosts, apparitions, and haunted houses (*The Amityville Horror*); astrology and horoscopes (Jeanne Dixon, the "Mars effect," the "Jupiter effect"); spoon bending (Uri Geller); remote viewing (Targ and Puthoff); cult anthropology; von Däniken and the Nazca plains of Peru; biorhythms; spontaneous human combustion; psychic surgery and faith healing; the full moon and moon madness; firewalking; psychic detectives; Ganzfeld experiments; poltergeists; near-death and out-of-body experiences; reincarnation; Immanuel Velikovsky and catastrophes in the past; doomsday forecasts; and much, much more!

The term *paranormal* was first used by parapsychologists, but it was stretched uncritically by advocates of the New Age, the Age of Aquarius, and harmonic convergence to include bizarre phenomena largely unexamined by mainline science. Murray Gell-Mann, Nobel Laureate and Fellow of

CSICOP, at our conference at the University of Colorado in 1986—I can remember it vividly—observed that we skeptics do not really believe in the "paranormal," because it deals with things beyond science, and as skeptical inquirers, he reiterated that we were dealing with investigations amenable to scientific methods of explanation. We would refuse to stop at any point and attribute phenomena to occult or hidden causes; we would keep looking for causal explanations and never declare that they were beyond the realm of natural causation by invoking the paranormal; and if we found new explanations, we would extend science to incorporate them. Incidentally, he also denied the feeling of some New Agers "that quantum mechanics is so weird, that anything goes" (*Skeptical Inquirer*, Fall 1986).

Sociologist Marcello Truzzi, who studied satanic cults, pointed out in our very first issue that we intended to examine esoteric anomalous claims, the "damned facts," as Charles Fort called them (a rainfall of frogs, etc.), to see what we could make of them. The public was intrigued by such mysteries, and we tried to encourage scientific investigators to explain them and to find out if they ever even existed or occurred.

Almost every issue of the *Skeptical Inquirer* attempted to fathom what was really happening in one or another alleged paranormal area. Thus, Ray Hyman described the technique of "cold reading" to show how guesswork and cues were used by psychics to deceive people who thought that they were having a *bona fide* paranormal reading. Philip Klass, head of CSICOP's UFO subcommittee, tried to unravel unusual cases of alleged UFO visitations and abductions in answer to astronomer J. Allen Hynek or Bruce Maccabee or other UFO buffs, and offered alternative prosaic explanations to account for apparent misperceptions. Conjuror James Randi and *Scientific American* columnist Martin Gardner looked for fraud or deceit. This was graphically illustrated in the case of a young psychic named Suzie Cottrell, who had bamboozled Johnny Carson by card reading. Put to the test under controlled conditions, Gardner said that she used Matt Schulein's forced-card trick, and Randi caught her red-handed peeking at the bottom card (see the Spring 1979 issue).

The *Skeptical Inquirer* published what appeared to be solutions to previously unexplained mysteries. We became exasperated with the media—such as NBC's *Unsolved Mysteries*, because they would present persons as having "real" paranormal abilities in spite of the fact that those persons were fraudulent—as in the case of Tina Resch, the Columbus, Ohio, youngster. Poltergeists supposedly manifested themselves when she came on the scene, lamps shattered, lights or faucets turned off and on. She was exposed by a TV camera that the crew left on while she thought that she was alone in a room: she was seen knocking down a lamp herself and screaming "poltergeist!"

I must say that these early years were exciting and exhilarating. We loved working with James Randi, Penn and Teller, Jamy Ian Swiss, Henry Gordon, Bob Steiner, and other magicians, who could usually duplicate a supposedly paranormal feat by sleight of hand or other forms of chicanery.

Deception is unfortunately widespread in human history, and it is revealing to point it out when it is encountered, especially where loose protocol makes it easy to hoodwink a gullible experimenter. Harry Houdini performed yeoman's service earlier in the twentieth century by exposing the blatant fraud perpetuated by Marjery Crandon and other spiritualists and mediums. I surely do not wish to suggest that conscious deception is the primary explanation for all or even most paranormal beliefs. Rather, it is *self-deception* that accounts for so much credulity. There is a powerful willingness in all too many people to believe in the unbelievable in spite of a lack of evidence or even evidence to the contrary. This propensity was due in part to what I have called the *transcendental temptation*, the tendency to resort to magical thinking, the attribution of occult causes for natural phenomena. The best antidote for this, I submit, is critical thinking and the search for natural causes of such phenomena.

Some paranormalists complained that we were poking fun at them and that ridicule is no substitute for objective inquiry. Martin Gardner observed that one joke might be worth a thousand syllogisms, if it dethrones a phony or nincompoop. Editor Kendrick Frazier, in my judgment, has always attempted to be fair-minded; and if an article criticized a proponent of a paranormal claim, he would invariably give that person an opportunity to respond. We attempted to make it clear that we were interested in fair and impartial inquiry, that we were not dogmatic or closed-minded, and that skepticism did not imply *a priori* rejection of any reasonable claim. Indeed, I insisted that our skepticism was not totalistic or nihilistic about paranormal claims, but that we proposed to examine a claim by means of scientific inquiry. I called this "the new skepticism" (see the *Skeptical Inquirer,* Winter 1994), to distance it from classical Greco-Roman skepticism, which rejected virtually anything and everything; for no kind of knowledge was considered reliable. But this was before the emergence of modern science, in which hypotheses and theories are based upon rigorous methods of empirical investigation, experimental confirmation, and replication, and also by whether a paranormal claim contradicts the body of tested theories or is consistent with them. One must be prepared to overthrow an entire theoretical framework—and this has happened often in the history of science—but there has to be strong contravailing evidence that requires it. It is clear that skeptical doubt is an integral part of the method of science, and scientists should be prepared to question received scientific doctrines and reject them in the light of new evidence.

II.

Looking back to the early years of the *Skeptical Inquirer* and CSICOP, it is evident that the salient achievement was that we called for new investigations and researchers in our network of collaborators responded by engaging in them.

(1) A good illustration of this is the determined efforts by skeptics to evaluate astrology experimentally. Although not paranormal in a strict sense—it was surely on the fringe of science—nevertheless, the claim that there were astro-biological influences present at the moment of birth could be tested. The "Mars effect" was a good illustration of this. French psychologists Michel and Francois Gauquelin maintained that the positions of planets at the time and place of birth—in this case Mars (in the first and fourth sector of the heavens)—was correlated with whether or not a person would become a sports champion. Egged on by Truzzi and a British psychologist, Hans Eysenck, we attempted several tests of this claim, and scientists tested the birth dates of sports champions born in the United States and France (and similar tests were made for other planets and professions). The results were negative, but it took twenty years of patient investigation to ascertain that. The most likely explanation for the "Mars effect" is biased data selection by the Gauquelins. CSICOP encouraged other researchers (such as Shawn Carlson and Geoffrey Dean) to test classical astrological claims. The results, published in the pages of the *Skeptical Inquirer*, again were invariably negative. Astrology provided no coherent theory or mechanism for the influence of planetary bodies at the time and place of a person's birth.

(2) Similar efforts were applied to parapsychology. Ray Hyman, James Alcock, Barry Beyerstein, and others were able by serious meta-analyses to evaluate the results of experimental research. Working with Charles Honorton, Robert Morris, and other parapsychologists, they questioned the findings of parapsychological investigations, and they found badly designed protocols, data leakage, experimenter biases, and insufficient replication by independent researchers.

The significant achievement of the *Skeptical Inquirer* was that it helped crystallize an appreciation by the scientific community of the need to investigate such claims. After the establishment of CSICOP, many scientific researchers were willing to devote the time to carefully examine the data. These results were published in the *Skeptical Inquirer*, so there was an independent record of explanation. And anyone who was puzzled by the phenomena could consult this new literature to deflate the paranormal balloon. This applied to a wide range of other phenomena.

(3) Near-death experiences provided insufficient evidence for the conjecture that a conscious self or soul left the body and viewed it from afar—this is better explained by reference to physiological and psychological causes, as Susan Blackmore pointed out in the *Skeptical Inquirer* (Fall 1991).

(4) The ability of fire-walking gurus to walk over hot coals was not due to some mind-over-matter spiritual power but rather because hot embers are poor conductors of heat, and it was possible for anyone to attempt it without injury.

(5) Another area of importance was the critical evaluation of the use of hypnosis by UFO investigators, who believed they were uncovering repressed memories that depicted alleged abductions. John Mack, a professor of psychiatry at Harvard, used hypnosis to probe the unconscious minds of certain troubled people who thought they had been abducted aboard UFOs by extraterrestrials. There was a similar pattern in such cases, he said, which was repeated time and time again by his patients: a sense of lost time, flashing lights, out-of-body experiences, etc. Mack thought this provided strong evidence for the claim; skeptics maintained that these evidences were not corroborated by independent testimony. At one point, Carl Sagan wrote to us, urging CSICOP to undertake an investigation of these claims, which by then were proliferating everywhere. We invited John Mack to a CSICOP conference in Seattle in June 1994 to hear what he had to say. There was a colorful confrontation between John Mack the believer and Phil Klass the skeptic—who insisted that hypnosis was unreliable as a source of truth. The influence of urban abduction legends popularized by the mass media predisposed many fantasy-prone persons to imbibe this tale, and the suggestibility of hypnotists reinforced the reality of their subjective experiences. Some critics asked Mack whether he accepted the fantasies of his psychotic patients as true—which gave him some pause.

"The Amazing" Kreskin, who used hypnosis in his act, appeared at one of CSICOP's conferences expressing doubt that hypnosis was a genuine "trance state" or a source of truth—it seemed to work in suggestible patients because they followed the bidding of the hypnotist. (Incidentally, many skeptics were highly critical of Kreskin for suggesting that he possessed ESP.)

(6) Hypnosis was also used in so-called past-life regressions to provide supposed evidence of previous lives. The *Skeptical Inquirer* carried many articles criticizing this technique. Past life therapists maintained that the hypnotic state provided empirical support for the doctrine of reincarnation, maintaining that the memory of a previous life was lodged deep

within the unconscious. More parsimonious explanations of these experiences are available: creative imagination, suggestions implanted by the hypnotherapist, and cryptomnesia (information stored in the unconscious memory without knowledge of the true source). Again, there was no independent factual corroboration, and these methods seem to rely more on *a priori* belief in reincarnation than reliable empirical evidence.

(7) Many research issues in psychology were critically examined in the *Skeptical Inquirer*. The work of Elizabeth Loftus is especially noteworthy here. In the decade of the 1990s, the mass media focused on charges that young children had been molested by relatives, friends, and teachers. Many reputations were destroyed after lurid accounts of sexual improprieties were made public. The popularity of such confessions spread like wildfire, and thousands of people claimed that they had been likewise molested. This was dramatized by the McMartin trial in California, where teachers in a day-care center were accused of sexual assaults of young children. This was based on testimony extracted from children and extrapolated by overzealous prosecutors. It had been pointed out that there is a "false-memory syndrome," which is fed by suggestion, and that testimony based on this is highly questionable. The *Skeptical Inquirer* was among the first publications to point out the fragmentary nature of the evidence and the unreliability of such testimony. This helped to turn the tide against such accusations, many of which had been exaggerated.

It would be useful at this point to sum up the pitfalls that skeptical inquirers encountered in studying paranormal and fringe-science claims and of guidelines that emerged as a consequence:

- Eyewitness subjective testimony uncritically accepted without corroboration is a potential source of deception (in accounts of molestation, reports of apparitions, past-life regression, UFO visitations, etc.).
- Extraordinary claims demand extraordinary evidence.
- The burden of proof rests with the claimant, not the investigators.
- Paranormal reports are like unsinkable rubber ducks: no matter how many times they are submerged, they tend to surface again.
- There is widespread gullibility and will to believe expressed by certain segments of the population, fascinated by mystery and magical thinking and willingness to accept tales of the occult or supernatural.
- In some cases, but surely not all, blatant fraud and chicanery may be observed, even in young children.
- In evaluating evidence, watch out for hidden bias and self-deception pro and con (including your own) to determine if something is a pseudoscience or not.

- There is no easily drawn demarcation line between science and pseudo-science, for one may be dealing with a proto-science. In my view, we need to descend to the concrete data and we cannot always judge *a priori* on purely philosophical grounds whether something is a pseudoscience or not (although I agree in general with Mario Bunge's views about the characteristics of a pseudoscience; see *Skeptical Inquirer*, Fall 1984 and July/August 2006; the latter appears in this volume).

III.

In recent years, popular interest in the paranormal has declined markedly, at least in comparison with its heyday. I do not deny that belief in paranormal phenomena is widespread; however, there are fewer manifestations of it in the mass media, and apparently less scientific interest. In previous decades, there were huge best sellers whose sales figures numbered in the millions: Raymond Moody's *Life After Life*, Charles Berlitz's *The Bermuda Triangle*, Erich von Däniken's *Chariot of the Gods*, etc.

Today, very few such books make *The New York Times* best-sellers list; and a top-selling paranormal book is likely to sell only 200,000 to 300,000 copies. (Sylvia Browne is the current best-selling guru, but there are few others besides her.) And there are very few major television programs devoted to the paranormal, though there are smaller-market cable shows.

Attention has turned to other areas. First, alternative medicine has grown by leaps and bounds. Prior to 1996, very few medical schools taught courses or offered programs in alternative medicine—and the medical profession was highly skeptical of the therapeutic value of remedies such as homeopathy, acupuncture, Therapeutic Touch, herbal medicines, iridology, and chiropractic. This magazine published many articles critical of these areas. It may be that such therapies are useful—the criterion we suggested was to conduct random, double-blind tests of their efficacy. Until there is sufficient data to support a therapy, the public should be cautious of its use. The medical profession needs to be open-minded yet suspicious of therapies until they are demonstrated to work—notwithstanding the evidential value of placebos.

Interestingly, the skeptical movement in Europe has concentrated on alternative medicines, though this is not strictly paranormal but is on the borderline of fringe medicine. I must confess that we are dismayed by the rapid growth of alternative and complementary medicine, which has had enormous acceptance virtually overnight. This is helped no doubt by the fact that it is a highly profitable source of income for both practitioners and the companies that market herbal remedies. Homeopathy is very strong in Europe and is now making inroads in the United States, though its remedies have never been adequately tested. Therapeutic Touch is so widespread in the nursing profession that it has gained great acceptance, though the basis of its curative powers has not been adequately demonstrated. The role of intercessory prayer as a healing method has provoked considerable controversy. Some advocates of prayer have claimed positive results; however, skeptics have seriously questioned the methodology of these tests. The most systematic tests were recently conducted by a team of scientists led by Herbert Benson (see the July/August 2006 *Skeptical Inquirer*). Using fairly rigorous protocol, these tests produced negative results.

Many skeptics have likewise been very critical of schools of psychotherapy, notably psychoanalysis, for lacking clinical data about the efficacy of their methods. In this regard, the Center for Inquiry has taken over the journal *The Scientific Review of Mental Health Practice* edited by Scott Lilienfeld, which evaluates the scientific validity of mental health treatment modalities. Some people say that the change from evidence-based medicine to other forms of medicine spells the emergence of a new paradigm; Marcia Angel has observed that this shift is toward a kind of spiritual medicine, influenced by the growth of religiosity in the culture.

Over the years, the *Skeptical Inquirer* has dealt with many other areas that needed critical scrutiny, including the efficacy of dowsing, graphology, facilitated communication, SETI, animal speech, the Atkins Diet, obesity, the Rorschach test, holistic medicine, and veterinary medicine. In addition, there were many articles on the philosophy of science, the nature of consciousness, and the evidence for evolution.

IV.

Numerous distinguished scientists have contributed to *Skeptical Inquirer*, including Richard Feynman, Glenn Seaborg, Leon Lederman, Gerald Holton, Steve Weinberg, Carl Sagan, Richard Dawkins, Jill Tarter, Steven Pinker, Carol Tavris, Neil deGrasse Tyson, and Victor Stenger. Among the topics examined have been quantum mechanics, the brain and consciousness, and cold fusion. Thus, the scope of the *Skeptical Inquirer* under Kendrick Frazier's editorship has been impressively comprehensive. And I should add that his fine editorials in every issue have pinpointed central questions of concern to science.

In a very real sense, the most important controversy in the past decade has been the relationship between religion and the paranormal and whether and to what extent CSICOP and the *Skeptical Inquirer* should deal with religious claims. As a matter of fact, evangelical and fundamentalist religion have grown to such proportions that religion and the paranormal overlap and one cannot easily deal with one without the other. The *Skeptical Inquirer* has dealt with religious claims from the earliest. First, it was in the vanguard of responding to the attacks on the theory of evolution coming from the creationists. Eugenie Scott, who served on the CSICOP Executive Council for a period of time, has done great service in critically analyzing "creation science," and the *Skeptical Inquirer* was among the first magazines to do so, demonstrating that creationism is not a science, for it does not provide a testable theory. The young-earth view maintains that Earth and the species on it are of recent origin, a view so preposterous that it is difficult to take it seriously. Most recently, intelligent design theory (which rejects the young-earth theory) claims that the complexity of biological systems is evidence for design. Numerous articles in the *Skeptical Inquirer* have pointed out that evolution is supported by overwhelming evidence drawn from many sciences. The existence of vestigial organs in many species, including the human species, is hardly evidence for design; for they have no discernible function. And the extinction of millions of species on the planet is perhaps evidence for *un*intelligent design.

Second, the *Skeptical Inquirer* was always willing to deal with religious questions, insofar as there are empirical claims that are amenable to scientific treatment. Thus, the Shroud of Turin has been readily investigated in the *Skeptical Inquirer* (see, for example, November/December 1999), presenting evidence (such as carbon-14 dating) that indicated that it was a thirteenth-century cloth on which an image had been contrived. Joe Nickell (CSICOP's Senior Research Fellow for the past decade) has said for years that the shroud is a forgery—as did the bishop of the area of France where it first turned up. Moreover, Nickell has shown how such a shroud could easily have been concocted. Similarly, the so-called "Bible code" was easily refuted by Dave Thomas (see the *Skeptical Inquirer*, November/December 1997 and March/April 1998).

In recent years, reports of miracles have proliferated, much to the surprise of rationalists, who deplore the apparent reversion of society to the thinking of the Middle Ages. David Hume offered powerful arguments questioning miracles, which he said were due to ignorance of the causes of such phenomena. There is abundant evidence, said Hume, to infer that nature exhibits regularities; hence, we should reject any exception to the laws of nature. In the late eighteenth century, showers of meteorites were interpreted by religious believers as signs of God's wrath. A special commission of scientists in France was appointed to investigate whether such reports of objects falling from the sky were authentic, and if so, if they were caused by natural events.

The *Skeptical Inquirer* has dealt with miracles in its pages, given the great public interest in them. The so-called miracle at Medjugorje, Yugoslavia, at a shrine where the Virgin Mary appeared before young children was critically discussed (November/December 2002). The conclusion was that the children's testimony had not been corroborated by independent testimony and was hence suspect. But as a result of the attention the children received, they became media celebrities. Oddly, the Virgin never warned about the terrible war that was about to engulf Bosnia and Kosovo. The cases were similar for the numerous other sightings of Mary and those of Jesus, which have attracted great public fascination. The investigations of Joe Nickell are models to follow; Nickell refuses to declare a priori that any miracle claim is false, but instead, he attempts where he can to conduct an on-site inquest into the facts surrounding the case. If, after investigation, he can show that the alleged miracle was due to misperception or deception, his analysis is far more effective.

The one area of interest in the paranormal that has also had a resurgence in recent years is "communicating with the dead." The form it has taken is reminiscent of the spiritualism of the nineteenth century, which had been thoroughly discredited because of fraud and deceit. The new wave of interest is fed by appearances on radio and television by such people as Sylvia Browne, James Van Praagh, and John Edwards. The techniques that the most popular psychics use are the crudest form of cold reading—which they seem to get away with easily. In some cases, they have resorted to doing hot readings (using information surreptitiously gleaned beforehand). This latter-day revival of spiritualism is no doubt fueled by the resurgence of religiosity in the United States, but it also shows a decline of respect for the rigorous standards of evidence used in the sciences.

The question of the relationship between science and religion intrigues many people today. It is especially encouraged by grants bestowed by the Templeton Foundation. Indeed, three special issues of the *Skeptical Inquirer*, beginning with July/August 1999, were devoted to explorations of the relationship, or lack of, between these two perennial areas of human interest.

These issues of the *Skeptical Inquirer* proved to be the most popular that we have ever published. Most skeptics have taken a rather strong view that science and religion are two separate domains and that science needs latitude for freedom of research, without ecclesiastical or moral censorship. This is one of the most burning issues today. Stephen Jay Gould defended a dissenting viewpoint of two magisteria: religion, which included ethics within its domain, and science. The *Skeptical Inquirer* has consistently brought philosophers to its pages to discuss a range of philosophical questions on the borderlines of science, religion, and morality. Susan Haack, Mario Bunge, and myself, Paul Kurtz, among others, have argued that the scientific approach is relevant to ethics and therefore ethics should not be left to the exclusive domain of religion (see the September/ October 2004 *Skeptical Inquirer,* also published as a chapter in this volume).

V.

Skeptics have often felt isolated in a popular culture that is often impervious to or fails to fully appreciate the great discoveries of science on the frontiers of research. They have done arduous work attempting to convince producers, directors, and publishers to present the scientific outlook fairly. When pro-para normal views are blithely expressed as true, we have urged that scientific cri- tiques also be presented to provide some balance. Our goal is to inform the public about the scientific outlook. We believe that we still have a long way to go to achieve some measure of fairness in the media. Almost the first official act of CSICOP was to challenge NBC for its program *Exploring the Unknown,* narrated by Burt Lancaster, which presented pro-paranormal propaganda on topics such as psychic surgery and astrology, without any scientific dissent at all. Our suit against NBC citing the Fairness Doctrine was turned down by a federal judge, and our subsequent appeal to the First District Court in Washington was also rejected (see the Fall/Winter 1977 *Skeptical Inquirer*). Conversely, the *Skeptical Inquirer* has been the victim of many legal suits or threats of suits over the years. The most notorious was Uri Geller's protracted legal suits against James Randi, CSICOP, and Prometheus Books. The most recent suit has named Elizabeth Loftus and CSICOP for an article that she authored (with Melvin J. Guyer) on the case of alleged sexual abuse of "Jane Doe" in the *Skeptical Inquirer* (May/June 2002). So the struggle that we have waged still continues.

On a more positive note, it is a source of great satisfaction that the *Skep- tical Inquirer* is read throughout the world and that CSICOP has helped gen- erate new skeptics groups, magazines, and newsletters almost everywhere— from Australia and China to Argentina, Peru, Mexico, and Nigeria; from India, Eastern Europe, and Russia to Germany, France, Spain, Italy, and the United Kingdom, so that the Center for Inquiry/Transnational (including CSICOP) has become truly planetary in scope. Especially gratifying is the fact that CSICOP has convened meetings in places all over the world, including China, England, France, Russia, Australia, India, Germany, Africa, etc.

Looking ahead, I submit that the *Skeptical Inquirer* and CSICOP should investigate other kinds of intellectually challenging and controversial claims. It is difficult to know beforehand where the greatest needs will emerge. In my view, we cannot limit our agenda to the issues that were dominant thirty, twenty, or even ten years ago, interesting as they have been. I think that we should of course continue to investigate paranormal claims, given our skilled expertise in that area. But we need to widen our net by entering into new arenas we've never touched on before, and we should be ever-willing to apply the skeptical eye wherever it is needed. Actually, Editor Kendrick Frazier has already embarked in new directions, for recent issues of the *Skeptical Inquirer* have dealt with topics such as cyberterrorism, "A Skeptical Look at September 11," "The Luck Factor," and critical thinking about power plants and the waste of energy in our current distribution systems. But there are many other issues that we have not dealt with that would benefit from skeptical scrutiny—and these include issues in biogenetic engineering, religion, economics, ethics, and politics.

Perhaps we have already become the Committee for Scientific Investigation or the Committee for Skeptical Inquiry (CSI), to denote that we are moving in new directions. This fulfills our general commitment to science and reason that's stated in the masthead of the *Skeptical Inquirer*. But one may say, there are so many intellectually controversial issues at large in society, which do we select? May I suggest the following criteria: we should endeavor to enter into an area, first, if there is considerable public interest and controversy; second, where there has not been adequate scientific research nor rigorous peer review; third, where some kind of interdisciplinary cooperative efforts would be useful; and fourth, where we can enlist the help of specialists in a variety of fields who can apply their skills to help resolve the issues.

We originally criticized pseudoscientific, paranormal claims because we thought that they trivialized and distorted the meaning of genuine science. Many of the attacks on the integrity and independence of science today come from powerful political-theological-moral doctrines. For example, one of the key objections to stem-cell research is that researchers allegedly destroy innocent human life—even when they deal with the earliest stage of fetal development or when a cell begins to divide in a petri dish. First is the claim that the "soul" is implanted at the moment of conception and that human life begins at the first division of a cell, and second, that it is "immoral" for biogeneticists working in the laboratory to intervene. The first claim is surely an *occult* notion if there ever was one, and it urgently needs to be carefully evaluated by people working in the fields of biology, genetics, and medical ethics; a similar response can be made to the second claim that it is immoral to intervene. There are many other challenges that have emerged in the rapidly expanding field of biogenetic research that might benefit from careful scrutiny: among these are the ethics of organ transplants, the use of mind-enhancing drugs, life-extension technologies, etc. The "new singularity," says Ray Kurzweil, portends great opportunities for humankind but also perplexing moral issues that need examination.

VI.

In closing, permit me to touch on another practical problem that looms larger every day for the *Skeptical Inquirer* and other serious magazines like it. I am here referring to a double whammy: the growth of the Internet on the one hand and the steady decline of reading of magazines on the other. No doubt, the Internet provides an unparalleled resource for everyone, but at the same time, it has eroded the financial base of the *Skeptical Inquirer*, and we do not see any easy solution to the deficit gap that increasingly imperils our survival.

Recognizing these dangers, we have extended our public outreach, first by offering for the first time an academic program, "science and the public," at the graduate level. Second, we have just opened an Office of Public Policy at our new Center for Inquiry in Washington, D.C., the purpose of which is to defend the integrity of science in the nation's capital and to try to convince our political leaders of the vital importance of supporting science education and the public understanding of science.

Finally, the most gratifying factor in all of this has been the unfailing support of the readers of the *Skeptical Inquirer*, who have helped to sustain us throughout our first three decades. Your support has been especially encouraging to Kendrick Frazier, who has worked so diligently in editing and publishing an outstanding magazine. It has been a rare privilege—an honor—for me to have worked with him so closely over these valiant years. We look forward to continuing the great legacy of the *Skeptical Inquirer* into its fourth decade—and beyond.

IN DEFENSE OF THE HIGHER VALUES

Kendrick Frazier

W hen Paul Kurtz brought us to the SUNY–Buffalo campus thirty years ago to found CSICOP, the nation was awash in what he called "The New Irrationalisms."

Velikovskyism saw ancient world history through a bizarre prism of civilization-affecting planetary pinballs.

Von Dänikenism attributed major achievements of ancient history, especially in the New World, not to the ingenuity of indigenous peoples but to ancient astronauts visiting Earth and stimulating creation of its archaeological wonders.

Gellerism promoted a showman conjuror as a real psychic with an ability to bend spoons with his mind and to cause several prominent but credulous physicists to lose their grips on reality.

Astrology had gained such a foothold on thought that astronomer Bart Bok and Paul Kurtz provoked worldwide controversy over a simple "Objections to Astrology" statement signed by 192 prominent scientists saying that astrology was bunk and had nothing to do with astronomy.

Paranormalism seemed everywhere, and New Age mystical thought that arose as part of the counterculture revolution of the late '60s influenced and entwined broad segments of society.

And reports of UFOs, despite the critical Condon report only seven years earlier, flew in regularly and gained credulous publicity in the press.

In the intervening three decades the specific claims that we might broadly label paranormal or pseudoscientific have changed dramatically. Most of the specific manifestations of the enthusiasms I just mentioned have waned. Some have even disappeared.

The situation has changed so much that Paul sometimes argues that no one is interested in the paranormal anymore. (I almost detect a certain longing!) We have some interesting internal debates about that, but to the degree it is true I have argued, and still do argue, that one key reason has been the remarkable work of CSICOP and the *Skeptical Inquirer* . . . and the network of scientists, scholars, and investigators worldwide it invigorated in forcefully addressing these and similar claims, providing detailed scientific analyses that showed their empirical poverty, and—in the end—debunking them. Solidly. Convincingly. Comprehensively. I think it has been a remarkable achievement.

Of course, as many of us note, although the cultural climate has shifted a lot, and in some respects for the better, the modern communications revolution has multiplied almost exponentially the number and types of outlets now available for the rapid promulgation of all new information and ideas, good and bad, reliable and unreliable—and that goes for pseudoscientific and paranormal nonsense and all its popular manifestations.

I won't even begin to detail all this here. We've dealt with all these matters in the *Skeptical Inquirer* now for years—and of course will continue to do so. Also, we have never limited ourselves to just the paranormal and pseudoscience. We deal with all topics at the intersection of science, public perception, and public misperception, with emphasis on those that attract large notice or that raise important public issues.

What I want to do here is sketch out some new and—I think—disturbing aspects of the cultural climate we find ourselves in, and emphasize the higher values that CSICOP and the *Skeptical Inquirer* exemplify and promote—values that seem essential to a modern, progressive, humane society; values that are under assault from broad quarters of society, here and abroad. No matter the specific, topical subjects we analyze and critique, the defense of these values is what we are really all about.

If thirty years ago Paul Kurtz and others were worried that we were in danger of descending toward a new dark age of superstition, paranormalism, mysticism, and pseudoscience, we in fact seem now to be in danger of descending toward a new dark age of a slightly different—and perhaps even more dangerous—sort. The first was more one of credulous, wide-eyed acceptance of wondrous, incredible claims. In retrospect, it all has a certain air of innocence. These claims all had their counterparts, after all, in science, and could be seen or interpreted as just misguided but understandable fascinations, uninformed by real science. Show people the real science and they might easily—at least in principle—shift their support to it. UFOs → the Search for Extraterrestrial Intelligence. Astrology → astronomy. Van Daniken pseudoarchaeology → real archaeology. Psychic powers and parapsychology → experimental psychology and modern studies of neuroscience and cognition. And so on.

But the new areas I am most concerned about now aren't like that. Not at all. They arise from deep-seated ideologies. They arise from a dangerous capturing of mainstream, liberal, open-minded, religious viewpoints by those with far more extreme, narrow, rigid, authoritarian, judgmental religious viewpoints. They arise from a devout determination to impose those viewpoints on everyone else. We've seen this abroad, but it is happening here in America too.

Their attacks are on many things, but among those that concern me are:

- The open-minded tolerance of others slightly different from oneself that marks a progressive society.
- The love of learning and the quest for new knowledge that mark a progressive society.
- The willingness to entertain and examine new ideas that marks a progressive society.
- A free and open society's distrust of authoritarian dogma, whatever the source—biblical or otherwise.
- Freedom of expression and the clear separation of church and state on which this nation was founded.
- Reliance on science-based evidence over unexamined belief and prejudice.
- The basic rights of women to make their own choices, to be educated, and to shape their own futures.
- A deep appreciation for education as a progressive force and a nurturing of environments for creativity and achieving novel solutions to problems.
- A related deep appreciation for not just the useful achievements of science but for the methods of science in determining and advancing provisional new truths, small and large, about the natural world.
- An acceptance that those methods of science often result in reliable judgments about what is real and what is not.
- A realization that we humans—while unique in our humanity—are nevertheless part of the natural world, and derived from and influenced by a long co-evolutionary history with the other life forms, large to microscopic, of the natural world.

In short, these attacks are on many key aspects of the modern world first shaped by the Enlightenment and the beginnings of science—when we began to develop the first abilities to actually deeply understand nature and, to some degree, exert some fledgling, limited controls of it for the well-being of people. They are attacks on curiosity and learning and on the scientific outlook itself. They are attacks on intellectual inquiry and thought—the open-minded, no-holds-barred examination of competing ideas and claims that is essential to an open, democratic society.

In many respects—although their proponents in America would no doubt dispute being so characterized—these are attacks on democracy itself. For these fundamentalist partisans would—if allowed—willingly impose their own, very specific ideological views on those they oppose.

We have to fight these trends.

We *will* fight these trends.

Our efforts at CSICOP and in the *Skeptical Inquirer* can't and don't deal with these issues in the abstract. Instead, we examine, critique, review, and report on specific, concrete topics, within the broad context of science and reason.

But it is important to keep in mind the higher values we are nevertheless defending:

- Reason and rationality—among the highest, most hard-won attributes of thinking, independent people.
- The scientific outlook—with its rich tradition of creative, open minded, empirical inquiry and evidence-based probing of nature's secrets.
- The skeptical attitude—a key component of the scientific outlook, with its obligation to put all new assertions to tests of empirical evidence.
- The traditions of learning—real learning, deep and broad—and the value of education not only in achieving a better life for each person but in creating reflective minds and in improving the lot of society.
- The deepest traditions of democracy—valuing human dignity and rights, drawing on the free and open interplay of ideas, and depending upon an educated, informed citizenry for making wise decisions.

No small matters. No small challenges.

Are we up to it? We have to be. There is no choice.

ANN DRUYAN TALKS ABOUT SCIENCE, RELIGION, WONDER, AND AWE
...and Carl Sagan

Ann Druyan

I've been thinking about the distorted view of science that prevails in our culture. I've been wondering about this, because our civilization is completely dependent on science and high technology, yet most of us are alienated from science. We are estranged from its methods, its values, and its language. Who is the scientist in our culture? He is Dr. Faustus, Dr. Frankenstein, Dr. Strangelove. He's the maker of the Faustian bargain that is bound to end badly. Where does that come from? We've had a long period of unprecedented success in scientific discovery. We can do things that even our recent ancestors would consider magic, and yet our self-esteem as a species seems low. We hate and fear science. We fear science and we fear the scientist. A common theme of popular movies is some crazed scientist somewhere setting about ruining what is most precious to all of us.

I think the roots of this antagonism to science run very deep. They're ancient. We see them in Genesis, this first story, this founding myth of ours, in which the first humans are doomed and cursed eternally for asking a question, for partaking of the fruit of the Tree of Knowledge.

It's puzzling that Eden is synonymous with paradise when, if you think about it at all, it's more like a maximum-security prison with twenty-four hour surveillance. It's a horrible place. Adam and Eve have no childhood. They awaken full-grown. What is a human being without a childhood? Our long childhood is a critical feature of our species. It differentiates us, to a degree, from most other species. We take a longer time to mature. We depend upon these formative years and the social fabric to learn many of the things we need to know.

So here are Adam and Eve, who have awakened full grown, without the tenderness and memory of childhood. They have no mother, nor did they ever have one. The idea of a mammal without a mother is, by definition, tragic. It's the deepest kind of wound for our species; antithetical to our flourishing, to who we are.

Their father is a terrifying, disembodied voice who is furious with them from the moment they first awaken. He doesn't say, "Welcome to the planet Earth, my beautiful children! Welcome to this paradise. Billions of years of evolution have shaped you to be happier here than anywhere else in the vast universe. This is your paradise." No, instead God places Adam and Eve in a place where there can be no love; only fear and fear-based behavior, obedience. God threatens to kill Adam and Eve if they disobey his wishes. God tells them that the worst crime, a capital offense, is to ask a question; to partake of the fruit of the Tree of Knowledge. What kind of father is this? As Diderot observed, the God of Genesis "loved his apples more than he did his children."

This imperative not to be curious is probably the most self-hating aspect of all, because what is our selective advantage as a species? We're not the fastest. We're not the strongest. We're not the biggest. However, we do have one selective advantage that has enabled us to survive and prosper and endure: A fairly large brain relative to our body size. This has made it possible for us to ask questions and to recognize patterns. And slowly over the generations we've turned this aptitude into an ability to reconstruct our distant past, to question the very origins of the universe and life itself. It's our only advantage, and yet this is the one thing that God does not want us to have: consciousness, self-awareness.

Perhaps Genesis should be read as an ironic story. Here's a god who does not give us the knowledge of good and evil. He knows we don't know right from wrong. Yet he tells us not to do something anyway. How can someone who doesn't know right from wrong be expected to do the right thing? By disobeying god, we escape from his totalitarian prison where you cannot ask any questions, where you must never question authority. We become our human selves.

Our nation was founded on a heroic act of disobedience to a king who was presumed to rule by divine right. We created social and legal mechanisms to institutionalize the questioning of authority and the participation of every person in the decision-making process. It's the most original thing about us, our greatest contribution to global civilization. Today, our not-exactly-elected officials try to make it seem as if questioning this ancient story is wrong. . . . That the teaching of our evolving understanding of nature, which is a product of what we have been able to discover over generations, is somehow un-American or disrespectful of strongly held beliefs. As if we should not teach our children what we've learned about our origins, but rather we should continue to teach them this story which demonizes the best qualities of our founding fathers.

This makes no sense and it leads me to a question: Why do we separate the scientific, which is just a way of searching for truth, from what we hold sacred, which are those truths that inspire love and awe? Science is nothing more than a neverending search for truth. What could be more profoundly sacred than that? I'm sure most of what we all hold dearest and cherish most, believing at this very moment, will be revealed at some future time to be merely a product of our age and our history and our understanding of reality. So here's this process, this way, this mechanism for finding bits of reality. No single bit is sacred. But the search is.

And so we pursue knowledge by using the scientific method to constantly ferret out all the mistakes that human beings chronically make, all of the lies we tell ourselves to combat our fears, all of the lies we tell each other. Here's science, just working like a tireless machine. It's a phenomenally successful one, but its work will never be finished.

In four hundred years, we evolved from a planet of people who are absolutely convinced that the universe revolves around us. No inkling that the Sun doesn't revolve around us, let alone that we are but a minuscule part of a galaxy that contains roughly a hundred billion stars. If scientists are correct, if recent findings of planets that revolve around other stars are correct, there are perhaps five hundred billion worlds in this galaxy, in a universe of perhaps another hundred billion galaxies. And it is conceivable, even possible, that this universe might one day be revealed to be nothing more than an electron in a much greater universe. And here's a civilization that was absolutely clueless four or five hundred years ago about its own tiny world and the impossibly greater vastness surrounding it. We were like a little bunch of fruit flies going around a grape, and thinking this grape is the center of everything that is. To our ancestors the universe was created for one particular gender of one particular species of one particular group among all the stunning variety of life to be found on this tiny little world.

There was only one problem. These very special beings for whom the universe was created had a holiday called Easter and they wanted to be able to celebrate it on the same day at the same time. But in this geocentric universe that they blissfully inhabited, there was no way to create a workable calendar that was coherent. At this time, there was a phrase to describe what science was. It is suffused with disarming candor and not a bit of self-consciousness at all. It was called *saving the appearances.* That was the task of science: To save the appearances. Figure out a way to take the reported appearances of the stars and the planets in the sky and predict with some reliability where they would be in the future. It's almost as if they knew they were living a cosmic lie. To call it saving the appearances is wonderful.

So the Lateran Council of 1514 was convened, and one of its main goals was to figure out a calendar that everybody could use so that they won't be celebrating Easter on different days. A man named Nicolas Copernicus, who was a very religious guy, whose lifelong career was in the church, had already figured out what the problem was. He was invited to present this information at the Council, but he declined because he knew how dangerous it would be to puncture this cosmological illusion. Even though the pope at that moment was not actually terribly exercised about this idea, Copernicus's fears were not baseless. Even sixty years later, a man named Giordano Bruno was burned alive for one reason: he would not utter the phrase, "There are no other worlds."

I've thought about this a lot. How could you have the guts to be willing to be burned alive? Bruno had no community of peers to egg him on. He wasn't even a scientist, he didn't really have any scientific evidence, but he chose this horrible death because he refused to say this phrase: "There are no other worlds." It's a magnificent thing, it's a wondrous mystery to me, and I don't think I completely understand how it was possible.

Copernicus did find the courage to publish his idea when he was comfortably near a natural death. When in 1543, *On The Revolutions of Celestial Spheres* was published, something unprecedented happened: a trauma from which we have never recovered. Up until that time, the sacred and the scientific had been one. Priests and scientists had been one in the same. It is true that two millennia before Copernicus there had been the pre-Socratic philosophers, who really were the inventors of science and the democratic values of our society. These ancient Greeks could imagine a universe and a world without God. But they were very much the exception, flourishing too briefly before being almost completely extirpated philosophically by the Platonists. Many of their books were destroyed. Plato loathed their materialism and egalitarian ideals. So there really wasn't a vibrant school of thought with a continuous tradition that survived down through the ages, daring to explain the wonder of nature without resorting to the God hypothesis.

It was actually initiated by a group of uncommonly religious men like Copernicus, Newton, Kepler, and (much later) even Darwin, who catalyzed that separation between our knowledge of nature and what we held in our hearts. All four of them either had religious careers or were contemplating such a profession. They were brilliant questioners, and they used the sharpest tools they had to search for what was holy. They had enough confidence in the reality of the sacred to be willing to look at it as deeply as humanly possible. This unflinching search led to our greatest spiritual awakening—the modern scientific revolution. It was a spiritual breakthrough, and I think that it is our failure to recognize it as such that explains so much of the loneliness and madness in our civilization, so much of the conflict and self-hatred. At that time, the public and their religious institutions, of course, rejected out of hand their most profound insights into nature. It was several hundred years before the public really thought about this, and took seriously what Copernicus was saying. The last four centuries of disconnect between what our elders told us and what we knew was true has been costly for our civilization.

I think we still have an acute case of post-Copernican-stress syndrome. We have not resolved the trauma of losing our infantile sense of centrality in the universe. And so as a society we lie to our children. We tell them a palliative story, almost to ensure that they will be infantile for all of their lives. Why? Is the notion that we die so unacceptable? Is the notion that we are tiny and the universe is vast too much of a blow to our shaky self-esteem?

It has only been through science that we have been able to pierce this infantile, dysfunctional need to be the center of the universe, the only love object of its creator. Science has made it possible to reconstruct our distant past without the need to idealize it, like some adult unable to deal with the abuse of childhood. We've been able to view our tiny little home as it is. Our conception of our surroundings need not remain the disproportionate view of the still-small child. Science has brought us to the threshold of acceptance of the vastness. It has carried us to the gateway of the universe. However, we are spiritually and culturally paralyzed and unable to move forward; to embrace the vastness, to embrace our lack of centrality and find our actual place in the fabric of nature. That we even *do* science is hopeful evidence for our mental health. It's a breakthrough. However, it's not enough to allow these insights; we must take them to heart.

What happened four or five hundred years ago? During this period there was a great bifurcation. We made a kind of settlement with ourselves. We said, okay, so much of what we believed and what our parents and our ancestors taught us has been rendered untenable. The Bible says that the Earth is flat. The Bible says that we were created separately from the rest of life. If you look at it honestly, you have to give up these basic ideas, you have to admit that the Bible is not infallible, it's not the gospel truth of the creator of the universe. So what did we do? We made a corrupt treaty that resulted in a troubled peace: We built a wall inside ourselves.

It made us sick. In our souls we cherished a myth that was rootless in nature. What we actually knew of nature we compartmentalized into a place that could not touch our souls. The churches agreed to stop torturing and murdering scientists. The scientists pretended that knowledge of the universe has no spiritual implications.

It's a catastrophic tragedy that science ceded the spiritual uplift of its central revelations: the vastness of the universe, the immensity of time, the relatedness of all life and it's preciousness on this tiny world.

When I say "spiritual," it's a complicated word that has some unpleasant associations. Still, there has to be a word for that soaring feeling that we experience when we contemplate 13 billion years of cosmic evolution and four and a half billion years of the story of life on this planet. Why should we give that up? Why do we not give this to our children? Why is it that in a city like Los Angeles, a city of so many churches and temples and mosques, there's only one place like this Center for Inquiry? And that it's only us here today? Fewer than a hundred people in a city of millions? Why is that? Why does the message of science not grab people in their souls and give them the kind of emotional gratification that religion has given to so many?

This is something that I think we have to come to grips with. There's a confusion generally in our society. There is a great wall that separates what we *know* from what we *feel*.

Medicine has had an oath that goes back to Hippocrates. Hippocrates is an amazing figure, both a father of scientific ethics and first articulator of the insight that frees humankind to discover the universe. He's one of those pre-Socratic philosophers I was talking about earlier, and he said something that resonated for me at a moment in my life when I realized what my path would be. His words inspired me to try as hard as I could in my own life to make it matter what is true. Hippocrates was writing in an essay called *Sacred Disease* 2,500 years ago. He was writing about the sacred disease that is now called epilepsy, and very matter-of-factly he said something that struck me like a lightning bolt. I'll paraphrase: "People believe that this disease is sacred simply because they don't know what causes it? But some day I believe they will, and the moment they figure out why people have epilepsy, it will cease to be considered divine." Why don't we have schools everywhere that teach children about Hippocrates, about the power of asking questions, rather than cautionary tales about the punishment for doing so. Our kids are not taught in school about Hippocrates, not taught about this multigenerational process of divesting ourselves of superstitions, false pattern recognition, and all the things that go with it, racism, sexism, xenophobia, all that constellation of baggage that we carry with us. We live in a society now where our leadership is all about promoting superstition, promoting xenophobia.

It seems to me that the biggest challenge we face is to evolve a language that couples the cold-eyed skepticism and rigor of science with a sense of community, a sense of belonging that religion provides. We have to make it matter, what is true. If instead we say that what really matters is to have faith, what really matters is to believe, we'll never get there. It's not enough to have forty minutes of science in the daily school program, because science shouldn't be compartmentalized that way. Science is a way of looking at absolutely everything.

What I find disappointing about most religious beliefs is that they are a kind of statement of contempt for nature and reality. It's absurdly hubristic. It holds the myths of a few thousand years above nature's many billion-yeared journey. It says reality is inferior and less satisfying than the stories we make up.

We need to create a community of skepticism for people of all ages. We desperately need some good music. We don't have to cut any corners on our ethos of skepticism. We do have to learn how to instill a sense of community, a rational experience of communion with nature and each other.

I would love to see, actually, not so much building more Centers for Inquiry, which would be great, but why don't we take over the planetaria of the country, of which there are hundreds, and turn them into places of worship. Not worship of the science that we know of this moment. Always give the message, over and over again, that our understanding could be wrong, this is what we think at this moment. The wonder of science is that we may find out that all of this is untrue. Why don't we take over these places and have services in the planetaria? We can connect. We can find inspiration in the revelations of science. We can have skepticism and wonder, both.

To me, faith is antithetical to the values of science. Not hope, which is very different from faith. I have a lot of hope. Faith is saying that you can know the outcome of things based on what you hope is true. And science is saying in the absence of evidence, we must withhold judgment. It's so hard to do. It's so tempting to believe in the lie detector or in heaven or that you know who you are based on the day of the month that you were born. It's a sort of unearned self-esteem. It's an identity that you can slip right into, and it's tremendously reassuring. So, I don't have any faith, but I have a lot of hope, and I have a lot of dreams of what we could do with our intelligence if we had the will and the leadership and the understanding of how we could take all of our intelligence and our resources and create a world for our kids that is hopeful.

I had a wonderful experience writing for the relatively new Rose Center at the Hayden Planetarium in New York. It's the greatest virtual reality theatre on Earth; completely immersive in the experience of traveling through the universe. I was honored to cowrite, with our *Cosmos* cowriter Steve Soter, the first two shows that inaugurated the planetarium center. And this is what got me thinking about how we might offer something that would be at least as compelling as whatever anyone else in the religion business is offering. We get to take you through the universe, and through the history of not only the Milky Way Galaxy but also the larger universe, and to tell something—the second one's called *The Search for Life, Are We Alone?*—something about the nature of life. It's a very uncompromising message about evolution and I think very directly promotes the kind of values and ideas that I think we share. Every kid who goes to a city public school gets taken to these shows. It was eye-opening to me, first of all, how far you could go in this direction, and what you could do with music and a fantastic technical capability that lets you tour that part of the universe we have come to know something about. You really hold on to your chair. You feel like you're traveling through the galaxies. It's uplifting. I constantly get mail about this and everyone is saying the same thing: you made me feel a part of something. You made me feel, even though I'm really small, that I'm a part of this greater fabric of life, which is so beautiful. And that's the kind of stuff that Cosmos Studios is working on, all of our projects. If they don't combine rigorous science with that soaring, uplifting feeling, then they don't qualify as a project for us. So I would say that that there's a lot in the entertainment world that we could be doing that I think has the power to really reach people.

Since we founded Cosmos Studios in the spring of 2000, we have accomplished the following: We are launching Cosmos 1, the first solar sailing spacecraft later this year. Our partners are The Planetary Society and the Babakin Space Research Center of Russia. We are actually launching the spacecraft from an intercontinental ballistic missile based on a Russian submarine. We have taken this weapon of mass destruction and converted it to a means of advancing the dream of space exploration. Solar sailing is an idea that has been around in science since the 1920s, but it's never been tried before. If we succeed, we will have demonstrated a practical means of literally riding light all the way to the stars. We liken our solar sail to what the Wright brothers did at Kitty Hawk, because although they were aloft for only twelve seconds and went 165 feet, they demonstrated that powered flight in a heavier than air vehicle was possible. What we're trying to demonstrate is that solar sailing is possible, and solar sailing is the only physically sound way of which we know to travel so quickly that it begins to be feasible to do interstellar flight on human time scales—two thousand years to the nearest star instead of twenty thousand years.

Cosmos Studios has funded research that has resulted in two papers published in the journal *Science*. We have produced a spiffed-up version of the thirteen-hour *Cosmos* TV series on DVD. We have produced three full-length documentaries. Perhaps our most promising project is an ambitious new way of teaching science from pre-kindergarten through high school. This involves a whole new approach to curricula. We hope to engage people from early childhood in science as a way of thinking.

I'm also at work on a book dealing with the themes I've tried to cover here.

[In answer to a question about Carl Sagan's role in garnering support for the legitimate scientific search for extraterrestrial intelligence (SETI) and taking on the creationists]:

Congress cut off federal funding for SETI years ago. I was with Carl when he went into Senator William Proxmire's office after Proxmire had given the Golden Fleece Award to the SETI program. Carl sat down with him. I didn't say a word. I was just a witness. And I just watched Carl. I was inspired by him, by not only the breadth of his knowledge, but his patience, his lack of arrogance, his willingness to hear the other person out. Senator Proxmire did a complete turnabout as a result of that meeting.

And there were other instances of Carl's remarkable persuasiveness. One was a great story of a so-called "creation scientist" who watched Carl testify at a hearing about creationism in schools. Carl testified for about four hours. It was somewhere in the South, I can't remember where. And six months later a letter came from the "creation scientist" expert who had also testified that day, saying that he had given up his daytime job and realized the error of what he was doing. It was only because Carl was so patient and so willing to hear the other person out. He did it with such kindness and then, very gently but without compromising, laid out all of the things that were wrong with what this guy thought was true. That is a lesson that I wish that all of us in our effort to promote skepticism could learn, because I know that very often the anger I feel when confronting this kind of thinking makes me want to start cutting off the other person. But to do so is to abandon all hope of changing minds.

When my husband died, because he was so famous and known for not being a believer, many people would come up to me—it still sometimes happens—and ask me if Carl changed at the end and converted to a belief in an afterlife. They also frequently ask me if I think I will see him again. Carl faced his death with unflagging courage and never sought refuge in illusions. The tragedy was that we knew we would never see each other again. I don't ever expect to be reunited with Carl. But, the great thing is that when we were together, for nearly twenty years, we lived with a vivid appreciation of how brief and precious life is. We never trivialized the meaning of death by pretending it was anything other than a final parting. Every single moment that we were alive and we were together was miraculous/not miraculous in the sense of inexplicable or supernatural. We knew we were beneficiaries of chance. . . . That pure chance could be so generous and so kind. . . . That we could find each other, as Carl wrote so beautifully in *Cosmos,* you know, in the vastness of space and the immensity of time. . . . That we could be together for twenty years. That is something which sustains me and it's much more meaningful. . . . The way he treated me and the way I treated him, the way we took care of each other and our family, while he lived. That is so much more important than the idea I will see him someday. I don't think I'll ever see Carl again. But I saw him. We saw each other. We found each other in the cosmos, and that was wonderful.

CARL SAGAN'S LAST Q&A ON SCIENCE AND SKEPTICAL INQUIRY

When Carl Sagan delivered his keynote address "Wonder and Skepticism" before a large audience at the CSICOP Conference in Seattle, Washington, June 23–26, 1994, a lively question-and-answer session followed. The Skeptical Inquirer *published Sagan's adaption of his talk as the cover article in its first bimonthly, magazine-format issue, January/February 1995. (It was also republished after Sagan's death in December 1996 as the lead chapter in the fourth general SI anthology,* Encounters with the Paranormal: Science, Knowledge, and Belief, *Prometheus 1998, with a two-page epilogue.) The Q/A session had been transcribed at the time along with the talk but put away and never published. In 2005 it was relocated, and Carl's wife and collaborator, Ann Druyan, readily agreed that it should be published. It appears here, with omission of only a few nonsubstantive exchanges. If some of the specifics discussed seem dated, others are as topical as today's news. And the general themes remain current. We then publish following it a passionately felt postscript, "The Great Turning Away," written specially for us by Ann Druyan.*

—KENDRICK FRAZIER, Editor

QUESTION: Dr. Sagan, you mentioned in your talk that one of the most important functions in science is to reward those who disprove our most closely held beliefs. Sir, if you were to look ahead two or three or four generations, which of our most closely held beliefs today do you think are the most likely candidates for disproof?

SAGAN: Maybe the belief that challenges our most closely held beliefs. Prophecy is a lost art. I have no way of doing that. If I could do that, think of the enormous effort we could save. The question flies in the very face of what I was just saying about how the most obvious points, the things we're absolutely sure of, may turn out to be wrong. So I am not immune to that fallibility and frailty. Let me give another example. It's the middle of the nineteenth century. The leading futurologist—although the word didn't exist then; it's a terrible word—was Jules Verne. He was asked to project a century ahead. What would be the means of transportation, the most exotic means of transportation, in the middle of the twentieth century? He then did whatever he did, looked into his crystal ball metaphorically speaking, and then gave the following conclusion: by 1950 there would be Victorian living rooms with lots of red velvet plush, I imagine, in the gondolas of great airships (they were called, but essentially dirigibles) which would cross the Atlantic Ocean in no more than a week. And people said, "Whew! That Jules Verne, he sure is far-seeing. Who could have thought of that?" And he was grossly off. Why was he off? Was he stupid? Was he not a good futurologist? No. He didn't foresee heavier-than-air aircraft, nor did anybody else. The view in the middle of the nineteenth century was that it was impossible. And in just the same way, whatever I would tell you, where we would be in space or something like that, is bound to be, whatever I would tell you, where we would be in space or something like that, is bound to be, unless we destroy ourselves, overtaken by scientific ideas and technological developments that I haven't a ghost of a chance of foreseeing. So forgive me for not being able to answer your question.

QUESTION: Thank you for your talk. I just wanted to challenge the idea that the reward system in science is essentially different from any other system. A person who successfully challenges the emperor gets the greatest rewards. An entrepreneur who successfully displaces some other technology or some other entrepreneur gets the greatest rewards. And a scientist who fails to successfully challenge the head of his discipline can see his head rolling, professionally, just as quickly, I think, as the unsuccessful coup d'état. Thank you.

SAGAN: Thank you. I think you raise a good point; permit me to disagree. There are certainly similarities along the lines that you say and, for example, maybe you remember the novel and television series *Shogun*, in which the English sailor, washed ashore in Japan, is brought to meet

Tokanaga, the future Shogun, who is very autocratic and authoritarian, hierarchical, as of course all military leaders are. And when he discovers that the Dutch were revolting from their Spanish overlords, he immediately identifies with the Spanish. He never met a Spaniard in his life, but they were in charge, and anybody challenging them must be doing something wrong. The hero then says, "The only mitigating condition is that the upstarts win." And Tokanaga says "Yes, yes, very true," and then they are friends. That's the point you just made. But that doesn't mean that there's a reward structure that encouraged the Dutch to revolt against the Spanish. It just means that if they succeed, then they succeed. It's a tautology. Whereas in science, there is a reward structure from the beginning. It doesn't mean that if somebody challenges Newton he is immediately rewarded. Einstein had some difficulties with special relativity. His Nobel prize was not even for relativity, it was for the photoelectric effect, because relativity was considered to be worrisome. Nevertheless, there were many scientists who recognized the value of what Einstein said. He was not challenging Isaac Newton; Isaac Newton was dead. The value of what Einstein said was there plain for anyone to see; nobody had thought of it before. As soon as people had worked through the arguments on the idea that simultaneity was a nonsensical idea, many were converted on the spot. I don't say that everybody was; I don't say that there weren't some problems with it, but there is a reward structure built in. And Einstein, just a few years after his 1905 relativity paper, was Full Professor and at the top of his profession.

QUESTION: Did you really say billions and billions of — (Laughter)

SAGAN: Never. Johnny Carson said it. I once saw him put on a wig and a corduroy jacket and pretend to be me, but I no more said it than Sherlock Holmes, in any of the writings of Arthur Conan Doyle, said "Elementary, my dear Watson."

QUESTION: I find it a little surprising that you use the words "science" and "truth" together in the same sentence. You said that science doesn't seek absolute truths, but asymptomatically tries to approach truth. I find truth is something that is very anthropocentric, relative to human being at a given time and a given place. I usually think of science more as seeking asymptomatically a better understanding of reality, not of truth.

SAGAN: I won't quibble on words. There are as many people who argue about the existence of reality as about the existence of truth. I encourage them to debate each other. (Laughter)

QUESTION: If I understand the theory of relativity, the space/time viewpoint, a causal violation should not be able to create a paradox. Do you think we may ever have as much control over space/time geometry as we do over electricity?

SAGAN: That's an awfully good question and I don't know the answer. But yes, a topic that is being hotly debated these days in the gravitational physics community is whether producing a paradox is a contradiction, or whether a paradox of the sort you referred to is just something we are going to have to live with. Can effects precede causes, for example. We have a tendency just to throw up our hands in amazement and despair: "What are you talking about? It's nonsense!" But there are certain sciences that seem to be in a funny way internally consistent with what else we know about physics and which may say effect can precede causes. I don't guarantee it's true, but if it is true it's just another one of those cases where our common sense doesn't apply everywhere.

QUESTION: Richard Hoagland has recently got hold of some pictures, Hasselblad pictures from NASA, which were taken some twenty years ago of the moon, and he has been describing those in great detail. He gave a talk at Ohio State University a couple of weeks ago and he had video cameras on and they were supposed to have videos available. I wonder if you've heard about this and had previous knowledge of . . . .

SAGAN: You forgot to mention what is on those videos.

QUESTION: Structures on the moon.

SAGAN: Richard Hoagland is a fabulist. By the way, it's not difficult getting hold of the hand-held Hasselblad camera pictures; NASA freely releases them to everybody. These are in the public domain, they're available to anybody. You don't have to do something remarkable to get the pictures. The aspect of this story I know best has to do with the so-called Face on Mars. There is a place on Mars called Sidonia, which was photographed in a mission I was deeply involved in, the Viking mission to Mars in 1976.

And there is one picture in which along a range of hulking mesas and hillocks, there is what looks very much like a face, about three kilometers across at the base and a kilometer high. It's flat on the ground, looking up. It has a helmet or a hair-do, depending on how you look at it, it has a nose, a forehead, one eye—the other half is in shadow—pretty eerie looking. You could almost imagine it was done by Praxiteles on a monumental scale. And this gentleman deduces from this that there was a race of ancient Martians. He has dated them, he purports to have deduced when they were around, and it was 500,000 years ago or something like that, when our ancestors were certainly not able to do space flights, and then all sorts of wonderful conclusions are deduced and "we came from Mars" or "guys from other star systems came here and left a statue on Mars and left some of them on Earth." By the way, all of which fails to explain how it is that humans share 99.6 percent of their active genes with chimpanzees. If we were just dropped here, how come we're so closely related to them? What is the basis of the argument? How good is it? My standard way of approaching this is to point out that there is an eggplant that looks exactly like former President Richard Nixon. The eggplant has this ski nose and, "that's Richard Nixon, I'd know him anywhere."

What shall we deduce from this eggplant phenomenon? Extraterrestrials messing with our eggplants? A miracle? God is talking to us through the eggplant? Or, that there have been tens, hundreds of thousands, millions of eggplants in history, and they all have funny little knobs, and every now and then there is going to be one that by accident looks like a human face. Humans are very good at recognizing human faces. I think clearly the latter. Now let's go to Mars. Thousands of low, hilly mesas have all sorts of features. Here's one that looks a little like a human face. When you bring out the contrast in the shadowed area it doesn't look as good. Now, we're very good at picking out human faces. We have so many of these blocky mesas. Is it really a compelling sign of extraterrestrial intelligence that there's one that looks a little like a human face? I think not. But I don't blame people who are going into the NASA archives and trying to find things there; that is in the scientific spirit. I don't blame people who are trying to find signs of extraterrestrial intelligence—I think it's a good idea, in fact. But I do object to people who consider shoddy and insufficient evidence as compelling.

QUESTION: May we hear your opinion on the canceling of the Superconducting Supercollider in Texas?

SAGAN: Yes. There are many physicists who think that that latter was a great tragedy. My own view is that it was not nearly explained well enough. We're talking about eight, ten, twelve, fourteen billion dollars to do very arcane experiments — and I don't think physicists did a good job at all in explaining to Congress why at a time of many pressures on the discretionary federal budget so much money should go to this. It doesn't build weapons, it doesn't cure diseases, it isn't generally known or understood. What is it about and why should we spend money on it? I think the physicists have, not altogether but to a significant degree, themselves to blame.

QUESTION: A question concerning the search for extraterrestrial intelligence. It seems that we as skeptics, there's an argument that seems very disappointing and maybe a bit persuasive in the Fermi paradox, the idea that if civilizations were to arise at any significant level, that even given a very extremely slow rate of expansion in the galaxy, that there's been more than enough time for them to have populated the galaxy several times over. What's your view on the Fermi paradox?

SAGAN: The Fermi paradox essentially says, as you said, that if there's extraterrestrial high technology intelligence anywhere they should have been here because if they travel at the speed of light, the galaxy is 100,000 light-years across, it takes you 100,000 years to cross the galaxy. The galaxy is 10 billion years old, they should be here. And if you say you can't travel at the speed of light, take a tenth of the speed of light, a hundredth of the speed of light, still much less than the age of the galaxy. William Newman and I published a paper on this very point, in which we point out: Imagine there is a civilization that has capable interstellar spacecraft and now they start exploring. What are we talking about? That they're sending out 400 billion spacecraft, all at once, simultaneously, to every star in the galaxy? Not at all. Interstellar space flight is going to be hard, you're going to go slow, you're going to go to the nearest star systems first, you're going to explore those stars. It is not a straight line but a diffusion question And when you do the diffusion physics with the appropriate diffusivity, that is, the time to random walk, there are many cases in which the time for an advanced civilization to fully explore the galaxy in the sense of visiting every star system is considerably longer than the age of the galaxy. It's just a bad model, we claim, the straight line, dedicated exploration of every star in the galaxy.

QUESTION: Dr. Sagan, you've spoken about the need to, as you say, be defenders of science, or to spread the wonders of science and the value of science among those who are perhaps less well educated or have less of an appreciation of it. It seems to be quite a challenge, and I was wondering, in particular, there are many people, of course, plus the people in this room, perhaps a fairly large portion have some background in science. Amongst people who have what is called a liberal education, who may be in the arts or in the humanities, science has among many of them something of a bad name. I wonder if you have any thought on what path might be taken to remedy that situation.

SAGAN: I think one, perhaps, is to present science as it is, as something dazzling, as something tremendously exciting, as something eliciting feelings of reverence and awe, as something that our lives depend on. If it isn't presented that way, if it's presented in very dull textbook fashion, then of course people will be turned off. If the chemistry teacher is the basketball coach, if the school boards are unable to get support for the new school bond issue, if teachers' salaries, especially in science, are very low, if very little is demanded of our students in terms of homework and original class time, if virtually every newspaper in the country has a daily astrology column and hardly any of them has a weekly science column, if the Sunday morning pundit shows never discuss science, if every one of the commercial television networks has somebody designated as a science reporter but he (it's always he) never presents any science, it's all technology and medicine, if an intelligent remark on science has never been uttered in living memory by a President of the United States, if in all of television there are no action-adventure series in which the hero or heroine is someone devoted to finding out how the universe works, if spiffy jackets attractive to the opposite sex are given to students who do well in football, basketball, and baseball but none in chemistry, physics, and mathematics, if we do all of that, then it is not surprising that a lot of people come out of the American educational system turned off, or having never experienced, science. That was a very long sentence.

QUESTION: Good evening, Dr. Sagan. Just one point first. Both the Canadian Broadcasting Corporation and the CTV private network have female science reporters.

SAGAN: Excellent. I was only talking about the U.S. and I recognize that you are within range of Canadian broadcasting here in Seattle.

SAME QUESTIONER: I'm a Canadian myself.

SAGAN: I'm very glad to hear it. David Suzuki has done for many years an excellent job on Canadian television.

SAME QUESTIONER: Absolutely. With the debacle of cold fusion, which may be said to be the ultimate proof of the scientific method with its peer review and its replicability or lack of same, if you were a person who is interested in the question of developing energy sources that would be both safer than the ones we use now and less expensive, would you continue to work in the area of fusion, and if not where would you work?

SAGAN: Cold fusion or hot fusion?

SAME QUESTIONER: I understand that hot fusion takes up a lot more energy than it ultimately produces.

SAGAN: But the margin is shrinking. If it were up to me, there's nothing in the way of compelling evidence for cold fusion, but if there were such a thing as cold fusion—you know, desktop conversion into enormous energy—we need that. So I can understand why there are companies, especially abroad, that are devoting small resources to it. I don't think that's cause for apoplexy. It'll probably come to nothing, but if there are scientists who want to spend their time on that, let them do it. Maybe they'll find something else that's interesting. On hot fusion, the margin, as I said, is shrinking, but the predicted, even optimistic estimates when commercial, large-scale, worldwide hot fusion would be available is too far into the twenty-first century to solve the energy problems we have today.

The energy problem I'm talking about is in particular global warming, the burning of fossil fuels. So what I would encourage is first of all, much greater emphasis on efficient use of fossil fuels—fluorescent rather than incandescent bulbs, you save a factor of several, or to put it another way, with the same amount of photons you put three or four or five times fewer carbon dioxide molecules into the atmosphere from the coal-burning power plant that provides the electricity. And I would put the money into forcing the automobile companies to produce cars that get 75, 85, 95 miles a gallon. Why are we satisfied with 25 miles a gallon when it is commercially perfectly possible to have safe, quick acceleration, spunky-looking cars that are efficient in their burning of petroleum? And then the other area where I would put emphasis is in non-nuclear alternatives to fossil fuel, of which I would stress biomass conversion, solar-electric power, and wind turbines, all of which are technologies that are coming along very swiftly despite, until recently, real hostility in the US government. Let me give you just one political story. There was once a president of the United States recently in the news named Jimmy Carter. He thought that there was an energy problem and he gave, in effect, talks to the nation in his cardigan sweater saying about how you should save electricity. He put into the roof of the White House a solar thermal converter which circulated cold water to the roof, and on sunny days in Washington sunlight heated this water and in repeated passes it made it very hot, and when it was time for a Presidential shower, here was hot water that did not rely on a power plant. He was succeeded by a President named Ronald Reagan. One of the first acts in office of President Reagan was to rip out the solar thermal converter from the roof of the White House at considerable cost—after all, it was in there and working—because he was ideologically opposed to alternatives to fossil fuels. We lost twelve years in research into these alternatives during the Reagan-Bush administration.

QUESTION: The dapper gentleman there, Bill Nye, his work on television bodes well for science education; he's to be applauded. I also want to thank you for answering all my questions about Richard Nixon; it explains a lot. You expressed some encouragement about the age mixture represented here in this audience. I wonder if you would comment on the conspicuous lack of racial diversity and the implications for science education in general.

SAGAN: Thank you. We also might ask how it is that of the first ten or twelve questioners only one was a woman in an audience in which women are much more strongly represented. These are wide-ranging, difficult questions. I don't claim to have the answers except to say that I know of no evidence that women and what in the United States are called racial minorities are not as competent as anybody else in doing science. It has to do, I think, entirely, or almost entirely, with the built-in biases and prejudices of the educational system and the way the society trains people. Nothing more than that. Women, for example, who are told that they're too stupid for science, that science isn't for them, that science is a male thing, are turned off. And women who despite that try to go into science and then find hostility from the high school math teacher—"What are you doing in my class?"—find hostility from the 95 percent male science classes, with the kind of raucous male culture in which they find themselves excluded, those are powerful social pressures to leave science. I wrote a novel once, *Contact*, in which I tried to describe what women dedicated to science have to face, that men don't, in order to make a career in science.

QUESTION: I would like to challenge you to answer the questions without ridicule. . . .

SAGAN: Fire away.

SAME QUESTIONER: . . . whether they be about crop circles, Richard Hoagland, or the abductees.

SAGAN: I didn't think I had any ridicule there.

SAME QUESTIONER: I think you had quite a lot. I was quite offended.

SAGAN: Which one particularly?

SAME QUESTIONER: The crop circles, the jokes you started with, the answer about Richard Hoagland offended me.

SAGAN: Okay, let's take one. Let's take Richard Hoagland —

SAME QUESTIONER: I would like to ask you in general to watch the ridicule. There are so many people here that think such ideas are worthy of ridicule. You have spoken about the need for compassion. I would like to see you model that here.

SAGAN: I appreciate that remark, and if I had not done what I preached I apologize. However, you must recognize —

SAME QUESTIONER: I accept your apology.

SAGAN: There was an "if" in front of the apology: However, you must recognize that vigorous debate is an essential aspect of getting to the truth, and the fact that Mr. Hoagland, for example, is not here—unless he is somewhere—I had nothing to do with it. Someone asked me a question about Richard Hoagland; I said what I thought. I happen to know that when Mr. Hoagland is asked questions and I'm not present, he says things about me, that I sometimes wish I had a chance to —

SAME QUESTIONER: (interrupting): Are you capable of modeling him?

SAGAN: I don't understand the question. What do you mean "modeling?"

SAME QUESTIONER: Modeling. Modeling compassion.

SAGAN: I've known Richard Hoagland for many, many years. I think I have just the right measure of compassion. (Laughter)

QUESTION: After the lady's question I don't know if mine is appropriate. I was going to ask you: We have Scott Peck, psychiatrist, Dr. [Brian] Weiss, he wrote *Many Lives, Many Masters*, Dr. [John] Mack [the Harvard psychiatrist who contended patients who say they were abducted by aliens are describing real events, and who spoke at the conference], we saw him yesterday, Dr. [Raymond] Moody—they're all mighty good thinkers. How do you think they went wrong?

SAGAN: I'm being asked to speculate offering psychiatric matters—hard to do. Some of those people I know very well, some I have never met. I don't think it would be right for me to guess why it is that they don't agree with me. I think that's all I want to say about it. I tried to stress before that it doesn't matter what the character of the debater is, it doesn't matter what reservations we have about him or her, what matters is the quality of the argument presented. For each of these people, I think the issue is: is there evidence? Yes, Dr. Moody has an M.D. but he uses, as I said, my own memories of my parents speaking to me as evidence of life after death. I know that's not a good argument. I know better than he what those voices are about, and so, by extrapolation, I think maybe the rest of his argument isn't so good. To the extent that I have some way of hooking onto the arguments I try to use what I know and see if there's a good case or not. I want to stress that there are some claims in the areas of parascience or pseudoscience that may well turn out to be right. And I don't think that is a reason for us not to demand the highest standards of argument about it.

For example, one thing I didn't mention to the last questioner: when Mars Observer was on its way to Mars with the high-resolution camera that might photograph things about the size of this, I thought that among the many other targets, it ought to take a look at the so-called face and settle the issue. If it's just some odd aspect of eolian abrasion on Mars, let's find that out. If it's something else, let's find that out. The fact that I think I understand, via the Richard Nixon eggplant, what the face on Mars is, doesn't mean that I don't want anyone to check it out. I could be wrong. If we have the tool to, with a few pictures, find out what the answer is, for heaven's sake let's do it. So each of these cases. In Johnny Mack's case I would say: "Never mind anecdotes, let's ask about physical evidence."

The claim is that many abductees have probes inserted up their nostrils, into their sinuses, which are, who knows, monitoring devices telling about where they've been and what's happening to their bodies. I say — and I've said to Mack a number of times—you give me one of those and we'll give it a really close scrutiny, and let's see if can we find evidence of alien manufacture. Are there principles or physical laws we don't understand? Are there isotopic ratios or the immiscibility of metals that we don't know about on Earth? Are there elements from the so-called island of stability, heavy elements, transuranic elements that are thought to be stable but we don't have any of them on Earth? There are many possibilities and you've certainly guessed that in some way an object of manufacture by aliens of extremely advanced capability —they travel from interstellar space, they effortlessly slither through walls, those guys really have powerful technology. Let's look at this. Never has there been one made available. There's always one about to be made available, there's always one that is going to be given to a laboratory, but it never happens.

What is that standard story that I get from Mack and others about the implants? It's that the abductees, going about his or her everyday life, and in many cases like this it is alleged the implant dropped out, clunk. The abductee picks it up, looks at it incuriously, and throws it in the garbage. Never once—and as a rule, this may prove my case—does he give it to some chemist or physicist, a chemist or physicist who could demonstrate the existence of alien technology. They'd give their eye teeth for that. They would be crawling all over each other to be able to examine the artifact. How come we've never had one case like that which really works out? I think that is a telling counterargument to all the anecdotal claims.

QUESTION: Dr. Sagan, we are fourteen years into the AIDS epidemic, HIV epidemic right now in this country, and apparently scientifically we are not coming any closer to finding a cure, creating a vaccine, even though there's lots of money being expended. And apparently also now there are new superbugs or new strains of bacteria that are becoming resistant to many of the antibiotics. You spoke about a concern for your children and your grandchildren in terms of what's happening. I'm just curious; what implications you see with these newfound illnesses, viruses, that are all of a sudden coming to us who believe that we were conquering everything in this day and age.

SAGAN: This is natural selection in action. If we overdose ourselves with antibiotics, we wipe out all the microorganisms not resistant to antibiotics and preferentially amplify the ones that are resistant. Eventually we arrange things, very cleverly, so that the entire population of microorganisms inside our bodies—including the disease-causing ones—are resistant to antibiotics. So overdosing antibiotics, which physicians have done routinely for reasons that are not hard to understand, is a mistake. Part of the answer is of course not to overdose anymore, and also to develop new strains of antibiotics. There ought to be major efforts to do that. On AIDS, my impression is that while there is nothing like a vaccine or a cure, there are substantial advances in the molecular biology of the HIV virus, and I take that to be a sign of significant hope, but of course not on the time scale of someone who is dying of AIDS. It's very slow in that respect. I don't think this is a money-driven situation. I don't think there just isn't enough money. I think there is enough money and some things maybe are not supported well enough, but in general there is, and it's a matter of not enough wisdom, not the right experiments, not having progressed far enough, not having done it swiftly enough. There's nothing magical about the HIV virus. It will succumb eventually to the ministrations of molecular biology. And I hope, for the reasons you mentioned, that that time will come soon.

QUESTION: Dr. Sagan, my question is in regard to the future of the skeptical movement.

SAGAN: Thank you. That's a good thing to end on.

SAME QUESTIONER: The responsibility that we have now, I feel, is as great as ever. The skeptical movement has been around for a number of years, perhaps thousands. First of all, I'd like to applaud the leadership of the skeptical movement we have here with you and the panel of speakers we have this weekend. But also important is the grassroots movements, the consensus of opinion of those that do adhere to the tenet, logic, reason, skeptical inquiry. We're at a point in time now that it's very important, and after having read the *Skeptical Inquirer* over a number of years and other articles and books and so forth, I find it somewhat amusing to see some of the investigations on a number of subjects like Paul Kurtz mentioned, we have hundreds of them—crystal power, pyramid power, a Loch Ness monster, whatever, I could go on and on. To be most effective in the long run I would think that would be something we would need to look at.

SAGAN: What would be?

SAME QUESTIONER: To be effective in fostering the logic and reason in skeptical thinking. I have found these various subject matters to be interesting and yet probably a greater area we could look at is the investigation into the major religions of the world.

SAGAN: Now we're to it. Okay.

SAME QUESTIONER: There are billions and billions of people that adhere to the tenets of these religions and I would imagine that we could spend more time in the skeptical movement—question is: Should the skeptical movement devote some of its attention to religion?

SAME QUESTIONER: Well said.

SAGAN: This is a really good question, and I know that Richard Dawkins talked about this a year or so ago, and drew the conclusion that many religious beliefs were not noticeably different from any of the parasciences or pseudosciences beliefs, and why one of them is the object of our attention and the other is off-limits, and he urged that we be, if I may use the expression, more ecumenical in our hostility. I will answer in the following way: first, that there is no human culture without religion. That being the case, that immediately says that religion provides some essential meat, and if that's the case shouldn't we be a little careful about condemning something that is desperately needed? For example, if I am with someone who has just lost a loved one, I do not think it is appropriate for me to say, "You know, there's no scientific evidence for life after death." If that person is gaining some degree of support, stability, from the thought that the loved one has gone to heaven and that they will be joined after the person I'm talking to, himself or herself, dies. That would be uncompassionate and foolish. Science provides a great deal, but there are some things that it doesn't provide. Religion is an attempt to provide, whether truly or falsely, some solutions to those problems. Human mortality is one of those where there isn't a smidgeon of help from science. Yes, it's a grand and glorious universe, yes it's amazing to be part of it, yes we weren't alive before we were born (not much before we were born) so we hope we're alive after we're dead. We won't know about it. It's a big deal. But that's not too reassuring, at least to many people.

Take the issue of the Bible. The Bible is in my view a magnificent work of poetry, has some good history in it, has some good ethical and moral scriptures—but by no means everywhere, the book of Joshua is a horror, for example— and on those grounds is well worth our respect. But on the other hand, the Book of Genesis was written in the sixth century B.C. during the Babylonian captivity of the Jews. The Babylonians were the chief scientists of the time. The Jews picked up the best science available and put it in the book of Genesis, but we have learned something in the intervening two and a half millennia, and to believe in the literal truth of the attempted science in the Bible, is to believe too much. I know there are Biblical literalists who believe that every jot and tittle in the Bible is the direct word of God, given to a scrupulous and flawless stenographer, and with no attempt to use the understanding of the time, or metaphor or allegory, but just straight-out truth. I know there are people who think that. That seems to me highly unlikely. I think the way to approach the Bible is with some critical wits about us, but not dismissing it out of hand. There's a lot of good stuff in the Bible. Case-by-case basis is what I'm saying. Where religion does not pretend to do science, I think we should be open within the boundaries of good sense. I think that you cannot extract an "ought" from an "is," and therefore science per se does not tell us how we should behave, although it can certainly shed considerable light on the consequences of alternative kinds of behavior. From that we can decide how to arrange our legal codes and what to do. So the idea of an all-out attack on religion I think on many grounds would be foolish, but the idea of treating Biblical literalism, for example, with some skeptical scrutiny is an excellent idea. But it is being done, has been done for the last century by Biblical scholars themselves. I don't think there's any particular expertise in this movement for a critical examination of the Bible. There are other people who are doing it just fine.

I hope that sort of middle ground is not too different from what you were asking about, but I certainly don't think that religion should be off-limits. I don't think anything should be off limits. We should feel free to discuss and debate everything. That's what the Bill of Rights is about. And in that sense, and many other senses, the Constitution of the United States, particularly the Bill of Rights, particularly the First Amendment, and the scientific method are very mutually supportive approaches to knowledge. Both of them recognize the extreme dangers of having to pay attention to and do whatever the authority says.

THE GREAT TURNING AWAY

Ann Druyan

I vividly recall the Seattle CSICOP conference. It was my first visit to that beautiful city. Humanism seemed on the ascendant then. Science and reason had a remarkably effective champion in Carl Sagan, who was known and respected in every country on earth. At that conference we had no idea that Carl was ill and that Seattle was about to become our home for the next two years. Carl endured three bone marrow transplants at a cancer research hospital there before dying in that city in 1996.

How I miss that voice with its mesmerizing blend of passion, brilliance, warmth, humor, and honesty. Carl spoke and wrote with equal measures of skepticism and wonder; never one at the expense of the other. He managed to maintain an exquisite balance between these two competing values. His life's work to awaken us to the *wonders* of the universe revealed by rigorous, *skeptical* science was a joyful labor of love.

Here we are, a mere ten years later in a radically different cultural atmosphere. Now, we seem to be engaged in a great turning away from reality. The engine of science continues to churn out discoveries at an astonishing rate, and yet our culture seems to have lost the ability to translate these insights into a grander perspective. It's as if our first forays into the immensity of the universe have sent us flying into a panic. The dawning reality of our tiny portion of space and time has been too much for us to bear. We turn away, seeking refuge in the discredited myths of our centrality.

Evidence of this failure of nerve is plentiful: Public school science educators shrink from teaching the fundamentals of biology. Science museums in the South feel the need to shield their visitors from Imax films that give the true age of volcanoes. Our president declares himself an instrument of God and with impunity makes illegal and baseless unprovoked war. An unctuous piety suffuses every public utterance. Radical religious fundamentalists have seized the national conversation. The rest of us are left to re-fight battles and arguments won decades ago.

Most worrisome is the steady erosion of the Bill of Rights made acceptable by the fearful attacks of September 11, 2001. Our precious constitution is in danger of being tainted with religious-based homophobia, the so-called "defense of marriage" amendment. More and more frequently we are told that we live in a Christian country. In the frightening logic of this moment, the separation of church and state—which was the founding source of our national genius—is seen as nothing more than an obstacle to a proper commitment to God.

It's enough to make you long for the days when skeptics could afford the luxury of sparring with mild-mannered Harvard psychiatrists who claimed feebly that their patients had been abducted by aliens. And for countless reasons, I do. However, my overall sense about our future is cautiously optimistic.

My hunch is that we are living during the twilight of the magical thinking phase of human history. Lest you think this is mere faith, I offer some evidence: Consider all the futures depicted in science fiction that you have ever seen or read; whether of life on this world or any other. How many of them imagine a future in which the dominant religious traditions and beliefs of the present survive? Remember: This is the output of countless independent imaginations of every conceivable point of view. Yet, when we imagine the future, the gods of our childhood are long gone.

As the man said, "Prophecy is a lost art." And yet, I'm happy to venture one. I believe that new Carl Sagans will emerge to uplift us once again with the beauty and truth of natural reality. We will overcome our fears and turn back to look unflinchingly at the vastness. We will find our home in the cosmos. It is only a matter of time.

CAN THE SCIENCES HELP US TO MAKE WISE ETHICAL JUDGMENTS?

Paul Kurtz

I.

C an science and reason be used to develop ethical judgments? Many theists claim that without religious foundations, "anything goes," and social chaos will ensue. Scientific naturalists believe that secular societies already have developed responsible ethical norms and that science and reason have helped us to solve moral dilemmas. How and in what sense this occurs are vital issues that need to be discussed in contemporary society, for this may very well be the hottest issue of the twenty-first century.

Dramatic breakthroughs on the frontiers of science provide new powers to humans, but they also pose perplexing moral quandaries. Should we use or limit these scientific discoveries, such as the cloning of humans? Much of this research is banned in the United States and restricted in Canada. Should scientists be permitted to reproduce humans by cloning (as we now do with animals), or is this too dangerous? Should we be allowed to make "designer babies?" Many theologians and politicians are horrified by this; many scientists and philosophers believe that it is not only inevitable but justifiable under certain conditions. There were loud cries against in vitro fertilization, or artificial insemination, only two generations ago, but the procedure proved to be a great

boon to childless couples. Many religious conservatives are opposed to therapeutic stem-cell research on fetal tissues, because they think that "ensoulment" occurs with the first division of cells. Scientists are appalled by this censorship of scientific research, since the research has the potential to cure many
illnesses; they believe those who oppose it have ignored the welfare of countless numbers of human beings. There are other equally controversial issues on
the frontiers of science: Organ transplants—who should get them and why? Is
the use of animal organs to supply parts for human bodies wrong? Is transhu-
manism reforming what it means to be human? How shall we control AIDS—
is it wicked to use condoms, as some religious conservatives think, or should
this be a high priority in Africa and elsewhere? Does global warming mean we
need a radical transformation of industry in affluent countries? Is homosexu-
ality genetic, and if so, is the denial of same-sex marriage morally wrong?
How can we decide such questions? What criteria may we draw upon?

II.

Many adhere today to the view that ethical choices are merely relativistic and
subjective, expressing tastes; and you cannot dispute tastes (*de gustibus non dis-
putandum est*). If they are emotive at root, no set of values is better than any
other. If there is a conflict, then the best option is to persuade others to accept
our moral attitudes, to convert them to our moral feelings, or, if this fails, to
resort to force.

Classical skeptics denied the validity of all knowledge, including ethical
knowledge. The logical positivists earlier in the twentieth century made a distinction between fact, the appropriate realm of science, and value, the realm
of expressive discourse and imperatives, claiming that though we can resolve
descriptive and theoretical questions by using the methods of science, we
cannot use science to adjudicate moral disputes. Most recently, postmod-
ernists, following the German philosopher Heidegger and his French followers, have gone further in their skepticism, denying that there is any special
validity to humanistic ethics or indeed to science itself. They say that science
is merely one mythological construct among others. They insist that there are
no objective epistemological standards; that gender, race, class, or cultural
biases likewise infect our ethical programs and any narratives of social eman-
cipation that we may propose. Who is to say that one normative viewpoint is
any better than any other, they demand. Thus have many disciples of multi-
cultural relativism and subjectivism often given up in despair, becoming
nihilists or cynics. Interestingly, most of these well-intentioned folk hold pas-

sionate moral and political convictions, but when pushed to the wall, will they concede that their own epistemological and moral recommendations likewise express only their own personal preferences?

The problem with this position is apparent, for it is impaled on one horn of a dilemma, and the consequence of this option is difficult to accept. If it is the case that there are no ethical standards, then who can say that the Nazi Holocaust and the Rwandan, Cambodian, or Armenian genocides are evil? Is it only a question of taste that divides sadists and masochists on one side from all the rest on the other? Are slavery, the repression of women, the degradation of the environment by profit-hungry corporations, or the killing of handicapped people morally impermissible, if there are no reliable normative standards? If we accept cultural relativity as our guide, then we have no grounds to object to Muslim law (*sharia*), which condones the stoning to death of adulteresses.

III.

What is the position of those who wish to draw upon science and reason to formulate ethical judgment? Is it possible to bridge the gap and recognize that values are relative to human interests yet allow that they are open to some objective criticisms? I submit that it is, and that upon reflection, most educated people would accept them. I choose to call this third position "objective relativism" or "objective contextualism;" namely, values are related to human interests, needs, desires, and passions—whether individual or socio-cultural—but they are nonetheless open to scientific evaluation. By this, I mean a form of reflective intelligence that applies to questions of principles and values and that is open to modification of them in the light of criticism. In other words, there is a Tree of Knowledge of Good and Evil, which bears fruit, and which, if eaten and digested, can impart to us moral knowledge and wisdom.

In what sense can scientific inquiry help us to make moral choices? My answer to that is it does so all the time. This is especially the case with the applied sciences: medicine, dentistry, nursing, pharmacology, psychiatry, and social psychology; and also in the policy sciences: economics, education, political science; and such interdisciplinary fields as criminology, gerontology, etc. Modern society could not function without the advice drawn from these fields of knowledge, which make evaluative judgments and recommend prescriptions. They advise what we ought to do on a contextual basis.

Nonetheless, there are the skeptical critics of this position, who deny that science per se can help us or that naturalistic ethics is possible. I think that those critics are likewise mistaken and that naturalism is directly relevant to ethics. My thesis is that an increase in knowledge can help us to make wiser decisions. By knowledge, I do not refer simply to philosophical analysis, but scientific evidence. It would answer both the religionist, who insists that you cannot be moral unless you are religious, and the subjectivist, who denies there is any such thing as ethical knowledge or wisdom.

Before I outline this position, let me concede that the skeptical philosophical objections to deriving ethics from science have some merit. Basically, what are they? The critics assert that we cannot *deduce* ethics from science, i.e., what *ought to be* the case from what *is* the case. A whole series of philosophers from David Hume to the emotivists have pointed out this fallacy. G. E. Moore, at the beginning of the twentieth century, characterized this as "the naturalistic fallacy"[1] (mistakenly, I think).

But they are essentially correct. The fact that science discovers that something is the case factually does not make it *ipso facto* good or right. To illustrate: (a) Charles Darwin noted the role of natural selection and the struggle for survival as key ingredients in the evolution of species. Should we conclude, therefore, as Herbert Spencer did, that *laissez-faire* doctrines ought to apply, that we ought to allow nature to take its course and not help the handicapped or the poorer classes? (b) Eugenicists concluded earlier in the century that some people are brighter and more talented than others. Does this justify an elitist hierarchical society in which only the best rule or eugenic methods of reproduction be followed? This was widely held by many liberals until the fascists began applying it in Germany with dire consequences.

There have been abundant illustrations of pseudoscientific theories—monocausal theories of human behavior that were hailed as "scientific"—that have been applied with disastrous results. Examples: (a) The racial theories of Chamberlain and Gobineau alleging Aryan superiority led to genocide by the Nazis. (b) Many racists today point to IQ to justify a menial role for blacks in society and their opposition to affirmative action. (c) The dialectical interpretation of history was taken as "scientific" by Marxists and used to justify class warfare. (d) Environmentalists decried genetics as "racist" and thought that changes in species should only be induced by modifications of the environment. Thus, one has to be cautious about applying the latest scientific fad to social policy.

We ought not to consider scientific specialists to be especially gifted or possessed with ethical knowledge nor empower them to apply this knowledge to society—as B. F. Skinner in *Walden II* and other utopianists have attempted to do. Neither scientist-kings nor philosopher-kings should be entrusted to design a better world. We have learned the risks and dangers of abandoning democracy to those wishing to create a *Brave New World*. Alas, all humans—

including scientists—are fallible, and excessive power may corrupt human judgment. Given these caveats, I nevertheless hold that *scientific knowledge has a vital, if limited, role to play in shaping our moral values and helping us to frame wiser judgments of practice*—surely more, I would add, than our current reliance on theologians, politicians, military pundits, corporate CEOs, and celebrities!

IV.

How and in what sense can scientific inquiry help us?

I wish to present a modified form of naturalistic ethics. By this, I mean that ethical values are *natural*; they grow out of and fulfill human purposes, interests, desires, and needs. They are forms of preferential behavior evinced in human life. "Good," "bad," "right," and "wrong" relate to sentient beings, whether human or otherwise. These values do not reside in a far-off heaven, nor are they deeply embedded in the hidden recesses of reality; they are empirical phenomena.

The principle of naturalism is based on a key methodological criterion: We ought to consider our moral principles and values, like other beliefs, *open to examination in the light of evidence and reason and hence amenable to modification.*

We are all born into a sociocultural context; and we imbibe the values passed on to us, inculcated by our peers, parents, teachers, leaders, and colleagues in the community.

I submit that ethical values should be amenable to inquiry. We need to ask, are they reliable? How do they stack up comparatively? Have they been tested in practice? Are they consistent? Many people seek to protect them as inviolable truths, immune to inquiry. This is particularly true of transcendental values based on religious faith and supported by custom and tradition. In this sense, ethical inquiry is similar to other forms of scientific inquiry. We should not presuppose that what we have inherited is true and beyond question. But where do we begin our inquiry? My response is, in the *midst* of life itself, focused on the practical problems, the concrete dilemmas, and contextual quandaries that we confront.

Let me illustrate by referring to three dilemmas. I do so not in order to solve them but to point out *a method of inquiry in ethics*. First, should we exact the death penalty for people convicted of murder? The United States is the only major democracy that still demands capital punishment. What is the argument for the death penalty? It rests on two basic premises: (a) A factual question is at issue: capital punishment is effective in deterring crime, especially murder; and (b) the principle of justice that applies is retributive. As the Old Testament adage reads, "Whatever hurt is done, you shall give life for life, eye for eye, tooth for tooth. . . ."[2]

The first factual premise can be resolved by sociological studies, by comparing the incidence of murder in those states and nations that have the death penalty in force and those that do not and by states and nations before and after the enactment or abrogation of the death penalty. We ask, has there been an increase or a decrease in murder? If, as a matter of fact, the death penalty does *not* restrain or inhibit murder, would a person still hold his view that the death penalty ought to be retained? The evidence suggests that the death penalty does not to any significant extent reduce the murder rate, especially since most acts of murder are not deliberate but due to passion or are an unexpected result of another crime, such as robbery. Thus, *if* one bases his or her belief in capital punishment primarily on the deterrence factor, and it does not deter, would one change one's belief? The same consideration should apply to those who are opposed to the death penalty: Would they change their belief if they thought it would deter excessive murder rates? These are empirical questions at issue. And the test of a policy are its consequences in the real world. Does it achieve what it sets out to do?

There are, of course, other factual considerations, such as: Are many innocent people convicted of crimes they did not commit (as was recently concluded in the State of Illinois)? Is capital punishment unfairly applied primarily to minorities? This points to the fact that belief in capital punishment is, to some extent, *a function of scientific knowledge concerning the facts of the case.* This often means that such measures should not be left to politicians or jurors alone to decide; the scientific facts of the case are directly relevant.

The second moral principle of retributive justice is far more difficult to deal with, for this may be rooted in religious conviction or in a deep-seated tribal sense of retaliation. If you injure my kin, it is said, I can injure yours; and this is not purely a factual issue. There are other principles of justice that are immediately thrown into consideration. Those opposed to the death penalty say that society "should set a humane tone and not itself resort to killing." Or again, the purpose of justice should be to protect the community from future crimes, and alternative forms of punishment, perhaps lifetime imprisonment without the right of parole, might suffice to deter crime. Still another principle of justice is relevant: Should we attempt, where possible, to rehabilitate the offender? All of the above principles are open to debate. The point is, we should not block inquiry; we should not say that some moral principles are beyond any kind of re-evaluation or modification. Here, a process of deliberation enters in, and a kind of moral knowledge emerges about what is comparatively the best policy to adopt.

Another example of the methods of resolving moral disputes is the argument for assisted suicide in terminal cases, in which people are suffering intolerable pain. This has become a central issue in the field of medical ethics, where medical science is able to keep people alive who might normally die. I first saw the emergence of this field thirty years ago, when I sponsored a conference in biomedical ethics at my university and could find very few, if any, scholars or scientists who had thought about the questions or were qualified experts. Today, it is an essential area in medicine. The doctor is no longer taken as a patriarchal figure. His or her judgments need to be critically examined, and others within the community, especially patients, need to be consulted. There are here, of course, many factual questions at issue: Is the illness genuinely terminal? Is there great suffering? Is the patient competent in expressing his or her long-standing convictions regarding his or her right to die with dignity? Are there medical and legal safeguards to protect this system against abuse?

Our decision depends on several further ethical principles: (a) the informed consent of patients in deciding whether they wish treatment to continue; (b) the right of privacy, including the right of individuals to have control of their own bodies and health; and (c) the criterion of the quality of life.

One problem we encounter in this area is the role, again, of transcendental principles. Some people insist, "God alone should decide life-and-death questions, not humans." This principle, when invoked, is beyond examination, and for many people it is final. Passive euthanasia means that we will not use extraordinary methods to keep a person alive, where there is a long-standing intent expressed in a living will not to do so. Active euthanasia will, under certain conditions, allow the patient, in consultation with his physician, to hasten the dying process (as practiced in Oregon and the Netherlands). The point is, there is an interweaving of factual considerations with ethical principles, and these may be modified in the light of inquiry, by comparing alternatives and examining consequences in each concrete case.

I wish to illustrate this process again by referring to another issue that is hotly debated today: Should all cloning research be banned? The Canadian legislature, in March 2004, passed legislation that will put severe restrictions on such scientific research. The bill is called "An Act Respecting Assisted Human Reproduction" (known as C-56), and it makes it a criminal offense to engage in therapeutic cloning, to maintain an embryo outside a woman's body for more than fourteen days, to genetically manipulate embryos, to choose the gender of offspring, to sell human eggs and sperm, or to engage in commercial surrogacy. It also requires that in vitro embryos be created only for the purpose of creating human beings or for improving assisted human-reproductive procedures. Similar legislation was passed by the US House of

Representatives and is now before the Senate. It is still being heatedly debated. It includes the prohibition of reproductive cloning as well as therapeutic stem-cell research. Two arguments against reproductive cloning are as follows: (a) It may be unsafe (at the present stage of medical technology) and infants born may be defective. This factual objection has some merit. (b) There is also a moral objection saying that we should not seek to design children. Yet we do so all the time, with artificial insemination, in vitro fertilization, and surrogate motherhood. We already are involved in "designer-baby" technology, with amniocentesis, pre-implantation, genetic testing, and chorionic villus sampling (the avoiding of unwanted genes by aborting fetuses and implanting desirable embryos).

If it were to become safe, would reproductive cloning become permissible? I can think of situations where we might find it acceptable—for example, if couples are unable to conceive by normal methods.

It is the second area I mentioned above that is especially telling—the opposition to any forms of embryonic stem-cell research. Proponents maintain that this line of research may lead to enormous benefits by curing a wide range of diseases such as Parkinson's disease, Alzheimer's, or juvenile diabetes. Adult stem cells are now being used, but embryonic stem cells may provide important new materials. The criterion here is consequential: that positive outcomes may result. Opponents maintain that this type of research is "immoral" because it is tampering with human persons possessed of souls. Under this interpretation, "ensoulment" occurs at the moment of conception. This is said to apply to embryos, many of which, however, are products of miscarriages or abortions. Does it also apply once the division of stem cells occurs? Surely a small collection of cells, which is called a *blastocyst*, is not a person, a sentient being, or a moral agent prior to implantation. Leon Kass, chair of President Bush's Council on Bioethics, believes that human life cannot be treated as a commodity and it is evil to manufacture life. He maintains that all human life, including a cloned embryo, has the same moral status and dignity as a person from the moment of conception.

This controversy pits two opposing moral claims: (a) the view that stem-cell research may be beneficent because of its possible contributions to human health (i.e., it might eliminate debilitating diseases) versus (b) an ethic of revulsion against tampering with natural reproductive processes. At issue here are the questions of whether ensoulment makes any sense in biology and whether personhood can be said to have begun at such an early stage, basically a transcendental claim that naturalists object to on empirical grounds. These arguments are familiar in the abortion debate; it would be unfortunate if they could be used to censor scientific stem-cell research.

This issue is especially relevant today, for transhumanists say that we are discovering new powers every day that modify human nature, enhance human capacities, and extend life spans. We may be able to extend memory and increase human perception and intelligence dramatically by silicon implants. Traditionalists recoil in horror, saying that post-humanists would have us transgress human nature. We would become cyborgs.

But we already *are*, to some extent: we wear false teeth, eyeglasses, and hearing aids; we have hair grafts, pacemakers, organ transplants, artificial limbs, and sex-change/sexual reassignment operations and injections; we use Viagra to enhance sexual potency or mega-vitamins and hormone therapy. Why not go further? Each advance raises ethical issues: Do we have the reproductive freedom and responsibility to design our children by knowing possible genetic disorders and correcting them before reproduction or birth?

V.

This leads to an important distinction between two kinds of values within human experience. Let me suggest two possible sources: (a) values rooted in unexamined feelings, faith, custom, or authority, held as deep-seated convictions beyond question, and (b) values that are influenced by cognition and informed by rational inquiry.

Naturalists say that scientific inquiry enables us to revise our values, if need be, and to develop, where appropriate, new ones. We already possess a body of prescriptive judgments that have been tested in practice in the applied sciences of medicine, psychiatry, engineering, educational counseling, and other fields. Similarly, I submit that there is a body of prescriptive ethical judgments that has been tested in practice and that constitutes normative knowledge; and new normative prescriptions are introduced all the time as the sciences progress.

The question is thus raised, what criteria should we use to make ethical choices? This issue is especially pertinent today for those living in pluralistic societies such as ours, where there is diversity of values and principles.

In formulating ethical judgments, we need to refer to what I have called a "valuational base."[3] Packed into this referent are the pre-existing *de facto* values and principles that we are committed to; but we also need to consider empirical data, means-ends relationships, causal knowledge, and the consequences of various courses of action. It is *inquiry* that is the instrument by which we decide what we ought to do and that we should develop in the young. We need to focus on moral education for children; we wish to structure positive traits of character and also the capacity for making reflective decisions. There are no easy recipes or simple formulae that we can appeal to, telling us what we ought to do in every case. There are, however, what W. D. Ross called *prima facie* general principles of right conduct, the common moral decencies, a list of virtues, precepts, and prescriptions, ethical excellences, obligations, and responsibilities, which are intrinsic to our social roles. But how they work out in practice depends on the context at hand, *and the most reliable guide for mature persons is cognitive inquiry and deliberation.*

Conservative theists have often objected to this approach to morality as dangerous, given to "debauchery" and "immorality." Here, there is a contrast between two different senses of morality: (a) the obedience/authoritarian model, in which humans are expected to follow moral absolutes derived from ancient creeds, and (b) the encouragement of moral growth, implying that there are within the human species potential moral tendencies and cognitive capacities that can help us to frame judgments.

For a naturalistic approach, in the last analysis, ethics is a product of a long evolutionary process. Evolutionary psychology has pointed out that moral rules have enabled human communities to adapt to threats to their survival. This Darwinian interpretation implies a biological basis for reciprocal behavior—epigenetic rules—according to E. O. Wilson (1998).[4] The social groups that possessed these rules transmitted them to their offspring. Such moral behavior provides a selective advantage. There is accordingly an inward propensity for moral behavior, moral sentiments, empathy, and altruism within the species.

This does not deny that there are at the same time impulses for selfish and aggressive tendencies. It is a mistake, however, to read in a doctrine of "original sin" and to say that human beings are by nature sinful and corrupt. I grant that there are individuals who lack moral empathy; they are morally handicapped. Some may even be sociopaths. The salient point is that there are genetic potentialities for good and evil; but how they work out and whether beneficent behavior prevails is dependent on cultural conditions. Both our genes (genetics) and memes (social patterns of enculturation) are factors that determine how and why we behave the way we do. We cannot simply deduce from the evolutionary process what we ought to do. What we do depends in part upon the choices we make. Thus, we still have some capacity for free choice. Though we are conditioned by environmental and biogenetic determinants, we are still capable of cognitive processes of selection, and rationality and intentionality play a causative role. (Note: There is a considerable amount of scientific literature that supports this evolutionary view. See Daniel Dennett, *Freedom Evolves* (New York: Viking, 2003) and *Darwin's Dangerous Idea* (New York: Simon and Schuster, 1995); Brian Skyrm, *Evolution of the Social Con-*

tract (Cambridge: Cambridge University Press, 1996); Robert Wright, *The Moral Animal* (New York: Pantheon Books, 1994) and *Nonzero* (New York: Vintage Books, 2001); Matt Ridley, *The Origins of Virtue* (New York: Viking, 1996); and Elliott Sober and David Sloan Wilson, *Unto Others* (Cambridge, MA: Harvard University Press, 1998).)

Ethical precepts need *not* be based upon transcendental grounds or dependent upon religious faith. Undoubtedly, the belief that they are sacred may strengthen moral duties for many persons, but it is not necessary for everyone.

I submit that it is time for scientists to recognize that they have an opportunity to contribute to naturalistic ethics. We stand at an interesting time in human history. We have great power to ameliorate the human condition. Biogenetic engineering, nanotechnology, and space research open new opportunities for humankind to create a better world.

Yet there are those today who wish to abandon human reason and freedom and return to mythological legends of our premodern existence, including their impulses of aggression and self-righteous vengeance. I submit that the Enlightenment is a beacon whose promise has not been fulfilled and that humankind needs to accept the responsibility for its own future.

CONCLUSION

A caveat is in order. In the last analysis, some degree of skepticism is a necessary antidote to all forms of moral dogmatism. We are continually surrounded by self-righteous moralists who claim that they have the Absolute Truth, Moral Virtue, or Piety or know the secret path to salvation and wish to impose their convictions on all others. They are puffed up with an inflated sense of their own rectitude as they rail against unbenighted immoral sinners who lack their moral faith. These moral zealots are willing to repress or even sacrifice anyone who stands in their way. They have in the past unleashed conquering armies in the name of God, the Dialectic, Racial Superiority, Posterity, or Imperial Design. Skepticism needs to be applied not only to religious and paranormal fantasies but to other forms of moral and political illusions. These dogmas become especially dangerous when they are appealed to in order to legislate morality and are used by powerful social institutions, such as a state or church or corporation, to enforce a particular brand of moral virtue. Hell hath no fury like the self-righteous moral fanatic scorned.

The best antidote for this is some skepticism and a willingness to engage in ethical inquiry, not only about *others'* moral zeal, but about *our own*, especially if we are tempted to translate the results of our own ethical inquiries into commandments. The epistemological theory that I propose is based upon methodological principles of skeptical scientific inquiry, and it has important moral implications. For in recognizing our own fallibility, we thereby can learn to *tolerate* other human beings and to appreciate their diversity and the plurality of lifestyles. If we are prepared to engage in cooperative ethical inquiry, then perhaps we are better prepared to allow other individuals and groups some measure of liberty to pursue their own preferred lifestyles. If we are able to live and let live, then this can best be achieved in a free and open democratic society. Where we differ, we should try to negotiate our divergent views and perhaps reach common ground; and if this is impractical, we should at least attempt to compromise for the sake of our common interests. The method of ethical inquiry requires some intelligent and informed examination of our own values as well as the values of others. Here we can attempt to modify attitudes by an appeal to cognitive beliefs and to reconstruct them by an examination of the relevant scientific evidence. Such a give-and-take of constructive criticism is essential for a harmonious society. In learning to appreciate different conceptions of the good life, we are able to expand our own dimensions of moral awareness; and this is more apt to lead to a peaceful world.

By this, I surely do not mean to imply that anything and everything can or should be tolerated or that one thing is as good as the next. We should be prepared to criticize moral nonsense parading as virtue. We should not tolerate the intolerable. We have a right to strongly object, if need be, to those values or practices that we think are based on miscalculation, misconception, or that are patently false or harmful. Nonetheless, we might live in a better world if *inquiry* were to replace faith; *deliberation*, passionate commitment; and *education and persuasion*, force and war. We should be aware of the powers of intelligent behavior, but also of the limitations of the human animal and of the need to mitigate the cold, indifferent intellect with the compassionate and empathic heart. Thus, I conclude that within the ethical life, we are capable of developing a body of melioristic principles and values and a method of coping with problems intelligently. When our ethical judgments are based on rational and scientific inquiry, they are more apt to express the highest reaches of excellence and nobility and of civilized human conduct. We are in sore need of that today.

NOTES

1. G. E. Moore, *Principia Ethica* (Cambridge: Cambridge University Press, 1903). http://www.csicop.org/si/2004-09/scientific-ethics.html#notes.

2.   See Exodus 21.

3.   Paul Kurtz (ed.), *The New Skepticism: Inquiry and Reliable Knowledge* (Amherst, New York: Prometheus, 1992), chapter 9.

4.   E. O. Wilson, *Consilience* (New York: Alfred Knopf, 1998).

SCIENCE WARS
Science and the Bush Administration

Chris Mooney

For policy wonks and issue advocates, a new area of specialization has recently arrived on the scene: "scientific integrity." Bills on the subject have been introduced in Congress. Interest groups, such as the Union of Concerned Scientists (UCS) and Public Employees for Environmental Responsibility (PEER), now specialize in tracking political interference with science. Foundations are dedicating energy and funding to the area; journalists, commentators, pundits, and bloggers have also climbed on board. One (yours truly) even has a book published on the subject. There's room, it almost seems, for a career here.

All of this activity has been triggered by repeated charges that the Bush administration has reached a new low in its willingness to twist and undermine scientific information to suit desired policy objectives. Such accusations have a four-year history, stretching from early concerns over whether the administration would even name a science adviser, through 2001 debates over stem cells and global warming, past reports complied by members of Congress denouncing the administration's meddling with science going on at federal agencies and the composition of scientific advisory committees, and up to a landmark moment: a February 2004 statement by the Union of Concerned Scientists (and assorted scientific community superstars) that denounced the Bush administration for unprecedented and systematic abuses and misuses of science.

However, the story doesn't end there. If anything, it has gathered momentum since the pivotal UCS statement, as new anecdotes and examples have repeatedly popped up, suggesting that the Bush administration hasn't learned the error of its ways. Whistleblowers from branches of government ranging from the Climate Change Science Program to the Bureau of Land Management have come forward with stories of cynical informational meddling that have made the front pages of papers including *The New York Times* and *The Los Angeles Times*. Meanwhile, the UCS and PEER have begun to survey scientists within federal agencies—so far, they've tackled the Fish and Wildlife Service and the National Marine Fisheries Service—to determine whether they think political players are meddling with scientific information. Scores of surveys have now come back with answers in the affirmative.

Perhaps most important of all in focusing attention on the issue of "scientific integrity" have been the climate-change fiascoes in the run up to the G8 summit. Shortly before President Bush departed for Gleneagles, Scotland, whistleblower Rick Piltz dropped a bomb with his revelations, reported in *The New York Times*, that a political appointee at the White House Council on Environmental Quality (who had formerly worked at the American Petroleum Institute) had taken a metaphorical red pen to government climate-science reports and inserted language that had the effect of magnifying uncertainty about various conclusions. A media frenzy began as this same individual—Philip Cooney—then resigned and promptly went to work at ExxonMobil, now perhaps the leading corporation encouraging skepticism about the ongoing climate crisis.

All of these events have had a cumulative effect, making it virtually impossible to take seriously the ongoing denials from the White House that any undue influence is being brought to bear on science. Those denials do, of course, persist; with each new revelation comes a dutiful response: "There's nothing out of the ordinary here"; "This is a typical interagency review process"; "The debate here is really over policy, not science"; and so forth. But such replies don't hold up very well when you consider that the critics of the administration are themselves current or former government-agency scientists who know very well what an "interagency review process" is and nevertheless insist that such processes have been corrupted in this administration. Even if we concede that some of these whistleblowers may have axes to grind, we're nevertheless left with a huge horde of disgruntled government scientists who can't possibly all be wrong.

Where does that leave us? Assuming—as I think we must, given all of the evidence—that something alarming is happening here at the interface between science and politics, it's worth asking why exactly that might be so. My conclusion is that what we're seeing is the result of a certain type of constituency-driven politics, in which federal agencies get staffed with Republican political appointees who know very well who their friends are and are willing to listen to them on matters of science. So business interests get their "scientific" arguments privileged at agencies that are supposed to be protecting endangered species and the environment, even as religious conservative interests get their "science" humored at agencies dedicated to public health and even, to some extent, medical research.

We don't have to postulate a nefarious conspiracy, then, to explain the war on science that has manifested itself during the second Bush administration. We need only point to an army of political appointees in government agencies who are going about their jobs the only way they know how—i.e., talking a lot to their industry or Religious Right allies and frequently rewarding their lobbying attempts in scientific areas. In short, it's a politico-scientific spoils system. And as this particular spoils system proceeds to allocate rewards, it simultaneously undermines, cheapens, and compromises federal agencies as reliable, public-oriented sources of scientific analysis and information.

But if we're looking at a government-wide problem based on staffing and a culture that has developed within federal agencies, that suggests it won't be easily solved. In fact, the damage done could long outlast the Bush administration, because the integrity of the federal government will have been compromised and because taxpayer-funded agencies may not recover quickly (or at all) from the traumas they've been put through. Here's where the political abuse of science becomes a core issue for the nation's future: The crisis promises to leave Americans with a less reliable, less effective, less professional, and ultimately less respectable government. The consequences will be felt in a wide range of areas, ranging from public health to the environment.

In conclusion, then, "scientific integrity" emerged virtually out of nowhere as a central issue under the Bush administration, and has since transmogrified into a broad-scale concern about good governance and the effectiveness and integrity of agencies funded by the public purse. The standard way to address concerns about good government is to initiate reform, and momentum has now begun to build in support of precisely that outcome, at least among Democrats in Congress. (Though there are prominent exceptions, most GOP representatives remain unwilling to seriously investigate or criticize the Bush administration.) In the meantime, however, the abuse of science by politics shows no sign of going away. And already, the wounds it has inflicted may take a very long time to heal.

DO EXTRAORDINARY CLAIMS REALLY REQUIRE EXTRAORDINARY EVIDENCE?

Massimo Pigliucci

arl Sagan had a rare gift for making clear rather abstruse ideas. To the readers of *Skeptical Inquirer*, perhaps no example is more familiar than the succinct rendition of David Hume's treatment of miracles that Sagan popularized: "extraordinary claims require extraordinary evidence."[1] Hume wrote a highly controversial essay on miracles in 1748, and philosophers and theologians are still talking about it. The essay is important not only to skeptics but to scientists at large, because it lays out some important observations about the nature of evidence and how it relates to the acceptance of hypotheses.

A recent caustic critique of Hume's work has been published by John Earman (in a collection entitled *Bayes's Theorem*, edited by Christian theologian Richard Swinburne). In essence, Earman claims that Hume is vague about what he says, and that his arguments can be interpreted in a fashion that ranges from the trivial (one ought to be careful about accepting eyewitness testimony in the case of miracles) to the absurd (no testimony will ever be sufficient to establish a miracle). Earman couches his critique in terms of Bayes's theorem on conditional probability (see my Thinking about Science column in the May/June 2004 *Skeptical Inquirer*), claiming that Bayes's theorem can be interpreted as a devastating blow to Hume's "pompous" opinion on the matter.

I think Hume can be read precisely as proposing a rather sophisticated Bayesian account of miracles and the nature of evidence, and that there is nothing vague, trivial, or absurd in the Scottish philosopher's essay. Let us proceed first by quoting (as does Earman) Hume himself on the definition of miracles: "A miracle is a violation of the laws of nature," i.e., miracles are in Hume's view a wholly different matter from unusual or exceptional natural phenomena. The skeptical philosopher then goes on to say that "no testimony is sufficient to establish a miracle, unless the testimony be of such a kind, that its falsehood would be more miraculous than the fact which it endeavors to establish." This should make it clear that Hume is *not* saying that testimony could *never* establish the truth of a miracle, only that the standard of acceptance should be exceedingly high; as Sagan put it, given that we are trying to establish the truth of a very extraordinary claim, it stands to reason that we need an extraordinary degree of evidence to back up such a claim.

But what of highly unusual, naturalistic phenomena, such as some quantum events? Earman charges that when accepting Hume's dictum about miracles, one is then forced to either throw away a significant number of scientific findings or to adopt a double standard as far as evidence is concerned. I suggest that Hume's writing is consistent with the second option, and that, contrary to how it may appear at first glance, this is a very reasonable position indeed.

Earman himself points to the following excerpt from Hume's essay: "suppose all authors, in all languages, agree, that, from the first of January 1600, there was a total darkness over the whole earth for eight days. Suppose that the tradition of this extraordinary event is still strong and lively among the people: that all travelers, who return from foreign countries bring us accounts of the same tradition without the least variation or contradiction: it is evident, that our present philosophers, instead of doubting the fact, ought to receive it as certain. . . ." That is, testimony *can* be sufficient to establish the truth of a highly unlikely, even unique, event; no need to throw away quantum mechanics just because some of it deals with unusual occurrences.

What, then, is the difference between the case of a highly unusual, even never before observed, natural phenomenon and a miracle? Why do we need so much more testimony to accept the latter than the former? Aren't we just ideologically biased against miracles? Hume's distinction can make perfect sense within the framework proposed by his contemporary, Thomas Bayes. Bayes's theorem essentially says that the probability we attach to a hypothesis, given the available evidence—let's call it $P(h \mid e)$—depends on the relationship between three quantities: (A) the probability of observing that evidence, if the hypothesis is in fact true; (B) the *a priori* probability of the hypothesis, based on prior knowledge we have of the world; and (C) the probability of observing the evidence given our prior knowledge of the world. In fact, Bayes showed that $P(h|e)=(A \cdot B)/C$.

Hume's argument about the proportionality between the extraordinariness of claims and the evidence necessary to back them up can be interpreted (although Hume did not, of course, use a Bayesian framework) as saying that the *a priori* probability we attach to miracles (quantity B above) is much lower than the *a priori* probability we grant to any natural phenomenon, because we have daily experience of phenomena that can be explained naturally, and rarely, if ever, do we even need to consider supernatural causes. Furthermore, quantity C, the probability of observing the evidence given our prior knowledge, is also much lower in the case of alleged miracles, simply because nobody has ever, in fact, made a convincing case for the occurrence of a miracle, so far. Hence, the only way to increase our *a posteriori* probability that the hypothesis (the miracle) is true, given the evidence, is to increase term A in the Bayesian equation. This means to attach a high probability to the evidence, given the hypothesis in question. The only way to do that is if, as Hume put it, the chances of the testimony being false would be more "miraculous" (i.e., improbable) than the miracle itself.

Hume's argument against miracles, therefore, is neither trivial nor absurd, and it can be couched in terms of modern conditional probability. Furthermore, it illuminates important points about the relationship between evidence and hypothesis in both science and pseudoscience. Sagan (and other scholars before and after him) was right: extraordinary claims do require extraordinary evidence.

NOTE

1. Before Sagan formulated that statement in *Cosmos* and later works, it was made by the late sociologist Marcello Truzzi in his opening editorial in the first issue of the *Skeptical Inquirer,* then called *The Zetetic,* vol. 1, no. 1 (Fall/Winter 1976).

THE FALLACY OF MISPLACED RATIONALISM

Robert Sheaffer

There seem to be so many fallacious arguments making the rounds these days that it would be surprising if skeptics were totally free from them. But I'm not talking about supposed skeptics' fallacies that the paranormal proponents keep talking about, such as "skeptics are so closed-minded, they ignore all the evidence that doesn't fit their beliefs." Nope, sorry, we ignore so-called "evidence" that isn't really evidence at all—the anecdotal and the unverified. That's not a fallacy; it's just good science.

The fallacy that I'm talking about is the tendency among many skeptics to pursue elaborate, rational explanations for reported phenomena *when it is by no means certain that we even have such phenomena.* Skeptics sometimes talk of what James Alcock has termed Ray Hyman's Categorical Directive (with apologies to Kant), usually stated as "before we try to explain something, we should be sure it actually happened." Unfortunately, this directive is violated all the time, even by skeptics, and the fruit of its violation is what I term the Fallacy of Misplaced Rationalism: the generation of elaborate, unlikely-but-still-possible explanations for "facts" that may not really be facts.

For example, science-minded folks have spent countless hours trying to answer the question, "What was the Star of Bethlehem?" Perhaps comet so-and-so was visible that year? Perhaps it was a rare conjunction of bright planets, which can be shown to have occurred at a certain time? Or possibly a supernova?

But the proper question to ask is: *"Was there really a Star of Bethlehem?"* From a strictly historical perspective, the answer appears to be no. Most New Testament scholars accept that there is little if any historical content in the birth narratives in Matthew and in Luke. For one thing, when read carefully, the two accounts contradict each other on most major points, such as the city of Joseph's residence, the reason for being in Bethlehem, what happened when Jesus was born, and what his family did afterward. Professional historians of the ancient world generally agree that Herod's Massacre of the Innocents (in Matthew) and the Decree of Augustus (in Luke) are both fictional events. Unbiased scholarship leads inevitably to the conclusion that the Birth Narratives are largely fictional accounts composed long after Jesus' birth to answer perceived doctrinal problems for the new religion (for example, if Jesus was born in Nazareth, it would pose a problem for the Messianic claim on his behalf, whereas if the story has him born in Bethlehem, that problem vanishes), as well as to create the belief that cosmically important things happened when Jesus was born. This is much more interesting than what almost certainly was the case: Jesus of Nazareth was born in Nazareth, and nothing out of the ordinary happened when he was born. What was the Star of Bethlehem? Most serious New Testament scholars would agree that in all probability, it never existed.

For more examples in the Biblical realm, see the article "Explaining the Plagues of Egypt" by Jeffrey A. Lee (*SI*, November/December 2004). He reviews many proposed, rational explanations by various scholars for the ten plagues of the book of Exodus. The "red tide" was said to be caused by dinoflagellates; the plague of flies was probably *Stomoxys calcitrans,* and so on. But as Lee notes, "scholars sometimes hang their conclusions on rather flimsy evidence." Indeed: we have no historical evidence whatsoever that the Israelites were ever in slavery in Egypt in the first place. Therefore, to speculate on the true nature of remarkable events said to accompany the Israelites' alleged escape from slavery is definitely a Fallacy of Misplaced Rationalism. This fallacy is very common in skeptical literature discussing purported miracles in the Bible. Before formulating any unusual-but-possible hypothesis to explain a passage in a religious text, it's wise to research the best scholarship on that text to make sure that the skeptic does not join the religious believer in proposing improbable events to explain a story that is probably fictional.

Consider the naturalistic explanations set forth to explain crop circles that were allegedly forming by themselves in fields of grain. The best known of these is the "ion plasma vortex" theory by meteorologist Terrence Meaden of England. According to Meaden, highly ionized air high in the atmosphere creates a vortex of plasma that descends to the surface, creating intense winds in a very small area. Is there any proof that an "ion plasma vortex" exists in our atmosphere? If not, we are trying to explain one unknown phenomena by invoking another. Can an "ion plasma vortex" create one circle inside another, inside yet another like a set of Russian dolls? Do we even *have* a phenomenon of self-forming crop circles, or do we need any explanation other than nocturnal hoaxers using ropes and boards to create the circles? As a general rule, when a would-be rationalist proposes a previously unknown action of some natural phenomenon to explain some claimed unusual or paranormal phenomenon, it's highly likely that we're seeing the Fallacy of Misplaced Rationalism at work.

When Israeli spoon-bender Uri Geller first achieved fame in the early 1970s by performing supposedly paranormal feats indistinguishable from magicians' tricks, there was an effort in some scientific circles to conjure up explanations of how such apparent miracles could be performed. There was widespread speculation that Geller used a concealed chemical that softened metal, enabling spoons to be easily broken or bent. These theorists apparently never tried to bend a spoon with their hands—it isn't difficult—and once started, simple metal fatigue will do the rest. Since magicians can readily perform the same feat using deception, there is no need to hypothesize anything more complicated. To explain Geller's claimed ability to see inside sealed envelopes, physicist Joseph Hanlon, writing in *New Scientist* (October 1974), noted that Geller's chief promoter, Andrija Puharich, had patented a radio receiver designed to be hidden in a tooth. The suggestion was that if one looked inside Geller's mouth, one might find such a device. One difficulty with that hypothesis is that some tests took place in rooms utilizing electromagnetic shielding. Ascertaining that it would be *possible* for such electromagnetic shielding to be breached is not the same as showing how Puharich and Geller could likely have done so without assistance. Before suggesting such a bizarre and complex yet still ultimately rational explanation, it would be more rational to inquire whether there is really any mystery in need of explaining. Did Geller succeed in any such test under sufficiently strict controls that would have precluded a professional magician from cheating? If not (and indeed Geller refused to work when controls became too strict), no additional explanation is needed.

UFOlogy has seen more than its share of misplaced rationalism at work. Perhaps the most obvious example is found in the earliest writings on the subject by the late Philip J. Klass. He began with the idea that many classic UFO photos and sightings might be caused by plasma phenomena arising from electrical lines, an idea underlying his first book on the subject, *UFOs Identified* (1966). Unfortunately, the glowing plasma balls suggested by Klass were not those occurring in laboratories under extreme, controlled conditions, but supposedly formed out in the open air from the electrical leakage of power lines. Apart from claimed UFO sightings, do we know whether such bright, glowing plasma balls form and persist in our everyday environment? If not, then we are using one unknown phenomena to explain another. Fortunately, as Klass continued studying and investigating UFO claims, he was perceptive enough to realize the enormous role being played by human error, misperception, and hoaxing. He was especially impressed by accounts in which trained flight crews reported a near-collision with an object that turned out to be a bright meteor more than one hundred miles away. Klass was also swayed by a demonstration that the "classic" UFO photos he had once considered solid proof of a plasma phenomenon—the Lucci brothers' two 1965 photos from Beaver County, Pennsylvania—could easily be duplicated using a simple camera trick. He realized that this made any "plasma hypothesis" unnecessary. By the time Klass wrote his second UFO book (in my opinion the best), *UFOs Explained* (1974), his earlier "plasma hypothesis" was largely forgotten, replaced by a sophisticated understanding of human knavery and error (two phenomena whose existence is extremely well-documented).

UFO skepticism got a black eye when the Air Force released its 1997 report, *Roswell—Case Closed*, suggesting that those witnesses who claimed to have seen little bodies of space aliens may have actually seen crash dummies dropped in the desert years later to test parachutes. But are there any credible accounts of seeing alien bodies that are not already rendered dubious by other considerations? In other words, are there any facts here requiring an explanation? There might be if the witnesses reported only seeing bodies in the desert and nothing else. But instead, they regale us with rich fables about huge and complex military retrieval operations, hastily erected morgues, draconian, secret arrangements enforced by naked threats, as well as little dead bodies. Such elaborate fables are most likely the result of innocent confusion. Indeed, the very "eyewitness sightings" cited as supporting the crash-dummy hypothesis have been shown to be suspect because of other inconsistencies and are disbelieved even by many Roswell proponents.[1] Even if we accept the Air Force suggestion that the Roswell "witnesses" saw crash dummies during the 1950s and confused that sighting with an alleged saucer crash in the 1940s, that still does not explain the accounts of large-scale, secretive military crash retrieval operations in the desert. Even if the Air Force suggestion was correct, it still would not explain other very bizarre aspects of the accounts. With stories like this, the skeptic must confront the likelihood of self-deception or mendacity (made all the more likely by the fact that the witnesses generally contradict each other on major details). So, while the Air Force investigators thought their "explanation" resolved a problem, they created a new problem in public confidence when an implausible explanation was put forth as an explanation for "observations" (eyewitness reports of *little dead bodies*) that really were not observations at all, but carefully contrived fables.

So, what insight does the skeptic take away from this? Do not neglect Hyman's Categorical Directive: one must ensure, in formulating an explanation, that one does not end up creating a preposterous scenario that is little better than the claimed phenomenon it is supposed to explain. Is the existence of the alleged phenomenon established sufficiently beyond doubt, or is it likely that the alleged phenomenon, as described, is exaggerated, misperceived, or even wholly fabricated? Do we have solid, credible evidence that cries out for an explanation? If not, trying to formulate a rational explanation is actually counterproductive. Before the skeptic expends time, effort, and credibility to come up with an explanation for some alleged phenomenon, he or she needs to be quite certain that there is indeed a phenomenon in need of explanation.

NOTE

1.   The principal Roswell "eyewitness" stories cited as possibly representing dummy-drops were those of Gerald Anderson and the late James Ragsdale. Anderson has been caught fabricating a document to substantiate his story, and even the "Flying Saucer Physicist" Stanton T. Friedman, once one of Anderson's biggest supporters, has publicly disavowed his story. Ragsdale (among other things) signed two different and mutually contradictory affidavits concerning the supposed "crash" that differ on major details, especially concerning the location of the site. The latter affidavit is suspected of being motivated by financial considerations in the promotion of a tourist attraction "landing site" and a book deal (Klass, *Skeptics UFO Newsletter* 45, May 1997).

Part II

CRITICAL INQUIRY AND PUBLIC CONTROVERSIES

"WE FIND THAT INTELLIGENT DESIGN IS NOT SCIENCE"

Excerpts from The Dover Federal Court Ruling on Intelligent Design

It was clear, strong, eloquent, and decisive. It ruled broadly on the merits of the case. And it dealt a stinging rebuke to "intelligent design" advocates and a brought a welcome legal defense of good science in the classroom.

Judge John E. Jones III's much-awaited decision in the case of parents Tammy Kitzmiller, et al. against the Dover [Pennsylvania] Area School District (outside of Harrisburg) was just what defenders of science had hoped for. Perhaps even better.

Jones, US District Court Judge for the Middle District of Pennsylvania and a Republican, ruled on December 20, 2005, that intelligent design is not science and that teaching it in the public school science classroom as such violates the US Constitution (and the Pennsylvania Constitution).

He also excoriated the board for duplicity in pretending ID isn't just warmed-over creationism (whose teaching had already been ruled unconstitutional) and for "breathtaking inanity" in adopting an unconstitutional policy that wasted taxpayers' time and money. The board had accepted a policy in November 2004 that required ninth-grade biology teachers at Dover High School to read a statement referring to evolution as a theory ("not a fact") with gaps and calling attention to ID as "an explanation for the origin of life that differs from Darwin's view."

In what seemed to be an intentional effort to deter other school boards from trying similar pro-ID tactics, Jones awarded damages, costs, and attorney fees to the plaintiffs who brought the suit against the board (whose pro-ID members were subsequently thrown out).

The full text of his 139-page decision, which came after a six-week trial at which a number of scientists and science scholars testified is on the web, and merits everyone's reading. Following are extended excerpts from the body and Conclusions section of the ruling. Numerals in brackets refer to page numbers of the official document.

—KENDRICK FRAZIER, Editor

JUDGE JOHN E. JONES III

Dramatic evidence of ID's religious nature and aspirations is found in what is referred to as the "Wedge Document." The Wedge Document, developed by the Discovery Institute's Center for Renewal of Science and Culture (hereinafter "CRSC"), represents from an institutional standpoint the IDM's [intelligent design movement's] goals and objectives, much as writings from the Institute for Creation Research did for the earlier creation-science movement, as discussed in *McLean*. . . . [28]

The CSRC expressly announces, in the Wedge Document, a program of Christian apologetics to promote ID. A careful review of the Wedge Document's goals and language throughout the document reveals cultural and religious goals, as opposed to scientific ones. [29]

[W]e find that the classroom presentation of the disclaimer, including school administrators making a special appearance in the science classrooms to deliver the statement, the complete prohibition on discussion or questioning ID, and the "opt out" feature, all convey a strong message of religious endorsement. [47]

In summary, the disclaimer singles out the theory of evolution for special treatment, misrepresents its status in the scientific community, causes students to doubt its validity without scientific justification, presents students with a religious alternative masquerading as a scientific theory, directs them to consult a creationist text as though it were a science resource, and instructs students to forego scientific inquiry in the public school classroom and instead to seek out religious instruction elsewhere. [49]

WHETHER ID IS SCIENCE

After a searching review of the record and applicable case law, we find that while ID arguments may be true—a proposition on which the Court takes no position—ID is not science. We find that ID fails on three different levels, any one of which is sufficient to preclude a determination that ID is science. They are: (1) ID violates the centuries-old ground rules of science by invoking and permitting supernatural causation; (2) the argument of irreducible complexity, central to ID, employs the same flawed and illogical contrived dualism that doomed creation science in the 1980s; and (3) ID's negative attacks on evolution have been refuted by the scientific community. As we will discuss in more detail below, it is additionally important to note that ID has failed to gain acceptance in the scientific community, it has not generated peer-reviewed publications, nor has it been the subject of testing and research. [64]

It is notable that defense experts' own mission, which mirrors that of the IDM itself, is to change the ground rules of science to allow supernatural causation of the natural world, which the Supreme Court in *Edwards* and the Court in *McLean* correctly recognized as an inherently religious concept. First, defense expert Professor Fuller agreed that ID aspires to "change the ground rules" of science, and lead defense expert Professor [Michael] Behe admitted that his broadened definition of science, which encompasses ID, would also embrace astrology. Moreover, defense expert Professor Minnich acknowledged that for ID to be considered science, the ground rules of science have to be broadened to allow consideration of supernatural forces. [67–68]

In addition to Professor Behe's admitted failure to properly address the very phenomenon that irreducible complexity purports to place at issue, natural selection, Drs. [Kenneth] Miller and [Kevin] Padian testified that Professor Behe's concept of irreducible complexity depends on ignoring ways in which evolution is known to occur. . . .

As expert testimony revealed, the qualification on what is meant by "irreducible complexity" renders it meaningless as a criticism of evolution. In fact, the theory of evolution proffers exaptation as a well-recognized, well-documented explanation for how systems with multiple parts could have evolved through natural means. Exaptation means that some precursor of the subject system had a different, selectable function before experiencing the change or addition that resulted in the subject system with its present function. For instance, Dr. Padian identified the evolution of the mammalian middle ear bones from what had been jawbones as an example of the process. By defining irreducible complexity in the way that he has, Professor Behe attempts to exclude the phenomenon by definitional fiat, ignoring as he does so abundant evidence which refutes his argument. [75]

As irreducible complexity is only a negative argument against evolution, it is refutable and accordingly testable, unlike ID, by showing that there are intermediate structures with selectable functions that could have evolved into the allegedly irreducibly complex systems. Importantly, however, the fact that the negative argument of irreducible complexity is testable does not make testable the argument for ID. Professor Behe has applied the concept of irreducible complexity to only a few select systems: (1) the bacterial flagellum; (2) the blood-clotting cascade; and (3) the immune system. Contrary to Professor Behe's assertions, with respect to these few biochemical systems among the myriad existing in nature, however, Dr. Miller presented evidence, based upon peer-reviewed studies, that they are not in fact irreducibly complex. [76]

First, with regard to the bacterial flagellum, Dr. Miller pointed to peer-reviewed studies that identified a possible precursor to the bacterial flagellum, a subsystem that was fully functional, namely the Type-III Secretory System.....

Second, with regard to the blood-clotting cascade, Dr. Miller demonstrated that the alleged irreducible complexity of the blood-clotting cascade has been disproven by peer-reviewed studies dating back to 1969, which show that dolphins' and whales' blood clots—despite missing a part of the cascade, a study that was confirmed by molecular testing in 1998. Additionally and more recently, scientists published studies showing that in puffer fish, blood clots despite the cascade missing not only one, but three parts. Accordingly, scientists in peer-reviewed publications have refuted Professor Behe's predication about the alleged irreducible complexity of the blood-clotting cascade.... [77]

The immune system is the third system to which Professor Behe has applied the definition of irreducible complexity.... However, Dr. Miller presented peer-reviewed studies refuting Professor Behe's claims that the immune system was irreducibly complex. Between 1996 and 2002, various studies confirmed each element of the evolutionary hypothesis explaining the origin of the immune system.... [78]

We therefore find that Professor Behe's claim for irreducible complexity has been refuted in peer-reviewed research papers and has been rejected by the scientific community at large. Additionally, even if irreducible complexity had not been rejected, it still does not support ID as it is merely a test for evolution, not design. [79]

Before discussing Defendants' claims about evolution, we initially note that an overwhelming number of scientists, as reflected by every scientific association that has spoken on the matter, have rejected the ID proponents' challenge to evolution. Moreover, Plaintiffs' expert in biology, Dr. Miller, a widely recognized biology professor at Brown University who has written university-level and high school biology textbooks used prominently throughout the nation, provided unrebutted testimony that evolution, including common descent and natural selection, is "overwhelmingly accepted" by the scientific community and that every major scientific association agrees. [83]

Plaintiffs' science experts, Drs. Miller and Padian, clearly explained how ID proponents generally and [the ID book] *Of Pandas and People* specifically, distort and misrepresent scientific knowledge in making their anti-evolution argument. [84]

The evidence presented in this case demonstrates that ID is not supported by any peer-reviewed research, data, or publications. Both Drs. Padian and [Barbara] Forrest testified that recent literature reviews of scientific and medical-electronic databases disclosed no studies supporting a biological concept of ID. On cross-examination, Professor Behe admitted that: "There are no peer-reviewed articles by anyone advocating for intelligent design supported by pertinent experiments or calculations which provide detailed rigorous accounts of how intelligent design of any biological system occurred." Additionally, Professor Behe conceded that there are no peer-reviewed papers supporting his claims that complex molecular systems, like the bacterial flagellum, the blood-clotting cascade, and the immune system, were intelligently designed. In that regard, there are no peer-reviewed articles supporting Professor Behe's argument that certain complex molecular structures are "irreducibly complex." In addition to failing to produce papers in peer-reviewed journals, ID also features no scientific research or testing.

After this searching and careful review of ID as espoused by its proponents, as elaborated upon in submissions to the Court, and as scrutinized over a six-week trial, we find that ID is not science and cannot be adjudged a valid, accepted scientific theory, as it has failed to publish in peer-reviewed journals, engage in research and testing, and gain acceptance in the scientific community. ID, as noted, is grounded in theology, not science. Accepting for the sake of argument its proponents', as well as Defendants' argument that to introduce ID to students will encourage critical thinking, it still has utterly no place in a science curriculum. Moreover, ID's backers have sought to avoid the scientific scrutiny which we have now determined that it cannot withstand by advocating that the *controversy*, but not ID itself, should be taught in science class. This tactic is at best disingenuous, and at worst a canard. The goal of the IDM is not to encourage critical thought, but to foment a revolution which would supplant evolutionary theory with ID.

To conclude and reiterate, we express no opinion on the ultimate veracity of ID as a supernatural explanation. However, we commend to the attention of those who are inclined to superficially consider ID to be a true "scientific" alternative to evolution without a true understanding of the concept the foregoing detailed analysis. It is our view that a reasonable, objective observer would, after reviewing both the voluminous record in this case, and our narrative, reach the inescapable conclusion that ID is an interesting theological argument, but that it is not science. [87–89]

The Board consulted no scientific materials. The Board contacted no scientists or scientific organizations. The Board failed to consider the views of the District's science teachers. The Board relied solely on legal advice from two organizations with demonstrably religious, cultural, and legal missions, the Discovery Institute and the TMLC. Moreover, Defendants' asserted secular purpose of improving science education is belied by the fact that most if not all of the Board members who voted in favor of the biology curriculum change conceded that they still do not know, nor have they ever known, precisely what ID is. To assert a secular purpose against this backdrop is ludicrous. [131]

Any asserted secular purposes by the Board are a sham and are merely secondary to a religious objective. [132]

Defendants' previously referenced flagrant and insulting falsehoods to the Court provide sufficient and compelling evidence for us to deduce that any allegedly secular purposes that have been offered in support of the ID Policy are equally insincere. Accordingly, we find that the secular purposes claimed by the Board amount to a pretext for the Board's real purpose, which was to promote religion in the public school classroom, in violation of the Establishment Clause. [132]

CONCLUSION [136–139]

The proper application of both the endorsement and Lemon tests to the facts of this case makes it abundantly clear that the Board's ID Policy violates the Establishment Clause. In making this determination, we have addressed the seminal question of whether ID is science. We have concluded that it is not, and moreover that ID cannot uncouple itself from its creationist, and thus religious, antecedents.

Both Defendants and many of the leading proponents of ID make a bedrock assumption which is utterly false. Their presupposition is that evolutionary theory is antithetical to a belief in the existence of a supreme being and to religion in general. Repeatedly in this trial, Plaintiffs' scientific experts testified that the theory of evolution represents good science, is overwhelmingly accepted by the scientific community, and that it in no way conflicts with, nor does it deny, the existence of a divine creator.

To be sure, Darwin's theory of evolution is imperfect. However, the fact that a scientific theory cannot yet render an explanation on every point should not be used as a pretext to thrust an untestable alternative hypothesis grounded in religion into the science classroom or to misrepresent well-established scientific propositions.

The citizens of the Dover area were poorly served by the members of the Board who voted for the ID Policy. It is ironic that several of these individuals, who so staunchly and proudly touted their religious convictions in public, would time and again lie to cover their tracks and disguise the real purpose behind the ID Policy.

With that said, we do not question that many of the leading advocates of ID have bona fide and deeply held beliefs which drive their scholarly endeavors. Nor do we controvert that ID should continue to be studied, debated, and discussed. As stated, our conclusion today is that it is unconstitutional to teach ID as an alternative to evolution in a public school science classroom.

Those who disagree with our holding will likely mark it as the product of an activist judge. If so, they will have erred as this is manifestly not an activist Court. Rather, this case came to us as the result of the activism of an ill-informed faction on a school board, aided by a national public interest law firm eager to find a constitutional test case on ID, who in combination drove the Board to adopt an imprudent and ultimately unconstitutional policy. The breathtaking inanity of the Board's decision is evident when considered against the factual backdrop which has now been fully revealed through this trial. The students, parents, and teachers of the Dover Area School District deserved better than to be dragged into this legal maelstrom, with its resulting utter waste of monetary and personal resources.

To preserve the separation of church and state mandated by the Establishment Clause of the First Amendment to the United States Constitution, and Art. I, § 3 of the Pennsylvania Constitution, we will enter an order permanently enjoining Defendants from maintaining the ID Policy in any school within the Dover Area School District, from requiring teachers to denigrate or disparage the scientific theory of evolution, and from requiring teachers to refer to a religious, alternative theory known as ID. We will also issue a declaratory judgment that Plaintiffs' rights under the Constitutions of the United States and the Commonwealth of Pennsylvania have been violated by Defendants' actions. Defendants' actions in violation of Plaintiffs' civil rights as guaranteed to them by the Constitution of the United States and 42 U.S.C. § 1983 subject Defendants to liability with respect to injunctive and declaratory relief, but also for nominal damages and the reasonable value of Plaintiffs' attorneys' services and costs incurred in vindicating Plaintiffs' constitutional rights.

NOW, THEREFORE, IT IS ORDERED THAT:

1. A declaratory judgment is hereby issued in favor of Plaintiffs pursuant to 28 U.S.C. §§ 2201, 2202, and 42 U.S.C. § 1983 such that Defendants' ID Policy violates the Establishment Clause of the First Amendment of the Constitution of the United States and Art. I, § 3 of the Constitution of the Commonwealth of Pennsylvania.

2. Pursuant to Fed.R.Civ.P. 65, Defendants are permanently enjoined from maintaining the ID Policy in any school within the Dover Area School District.

3. Because Plaintiffs seek nominal damages, Plaintiffs shall file with the Court and serve on Defendants, their claim for damages and a verified statement of any fees and/or costs to which they claim entitlement. Defendants shall have the right to object to any such fees and costs to the extent provided in the applicable statutes and court rules.

s/John E. Jones III
John E. Jones III
United States District Judge

THE "VISE STRATEGY" UNDONE
Kitzmiller et al. v. Dover Area School District

Barbara Forrest

In a post on his Uncommon Descent (UD) blog on May 6, 2005, intelligent-design (ID) creationist William Dembski was talking tough. He offered a lesson for "Darwinists" drawn from the then-ongoing hearings held before the Kansas Board of Education on May 5–7 to discuss the Kansas science standards. The creationist-dominated board had hoped that pro-evolution scientists and ID creationists would debate revisions proposed by the creationist minority on the board's Science Curriculum Writing Committee. These revisions included redefining science to allow the supernatural as a scientific explanation. Refusing to lend legitimacy to this "Kansas kangaroo court," scientists boycotted the hearings. The only pro-evolution participant representing pro-science groups, was attorney Pedro Irigonegaray, who cross-examined many of the twenty-three creationists who were brought in at taxpayer expense to testify. These twenty-three are supporters of Dembski and his associates, who have promoted ID for a decade from the Center for Science and Culture (CSC), the creationist arm of the Discovery Institute (DI), a conservative Seattle think tank.

Grousing that "only the evolution critics are being interrogated," Dembski was "waiting for the day when the hearings are not voluntary but involve subpoenas in which evolutionists are deposed at length." When "that happy day" comes, Dembski predicted, the Darwinists "won't come off looking well." On

May 11, Dembski portrayed "evolutionists" as too chicken to participate: ". . . evolutionists escaped critical scrutiny by not having to undergo cross-examination . . . by boycotting the hearings." He proposed a "vise strategy" for "interrogating the Darwinists to, as it were, squeeze the truth out of them," childishly illustrated with a photograph of a Darwin doll with its head compressed in a bench vise. On May 16, he outlined his strategy: "interrogating Darwinists" about "five terms: science, nature, creation, design, and evolution." Under subpoena, they would be compelled to answer, hence the "vise" metaphor.

Dembski already knew that such a day of legal reckoning was approaching. Exactly one month later, on June 6, he sat across from me when I was deposed as an expert witness for the plaintiffs in the first ID legal case, *Kitzmiller et al. v. Dover Area School District*. He attended my deposition as the adviser to the lead defense attorney, Richard Thompson of the Thomas More Law Center, and was scheduled to be deposed himself on June 13 as a defense witness. Besides being on opposite sides, there was another big difference between us: I showed up for my deposition. Dembski "escaped critical scrutiny by not having to undergo cross-examination" when he withdrew from the case on June 10.

Not only did I show up for my deposition, but I also testified at the trial despite being delayed by Hurricanes Katrina and Rita. Moreover, I had the distinction of being the only witness whom the defense tried to exclude from the case. When they failed, the Discovery Institute tried to discredit me with ridicule.

A BRIEF HISTORY OF THE KITZMILLER CASE

In October 2004, the Dover [Pennsylvania] Area School District Board of Directors decided that "Students will be made aware of gaps/problems in Darwin's theory and of other theories of evolution including, but not limited to, intelligent design." In November, they announced that Dover High School's ninth-grade biology teachers would read a statement informing students that "Darwin's Theory . . . is not a fact" and that "intelligent design is an explanation of the origin of life that differs from Darwin's view." The statement referred students to the creationist textbook, *Of Pandas and People*, to learn "what intelligent design actually involves." On December 14, eleven parents filed suit in the Middle District of Pennsylvania, represented by attorneys from the ACLU, Americans United for Separation of Church and State, and Pepper Hamilton, a Philadelphia law firm. The Thomas More Law Center (TMLC), a Michigan Religious Right legal organization, represented the board. United States District Judge John E. Jones III was assigned to hear the case.

Dover's problems actually started in 2002. Bertha Spahr, chair of Dover High

School's science department, began to encounter animosity from Dover residents toward the teaching of evolution. In January 2002, board member Alan Bonsell began pressing for the teaching of creationism. In August, a mural depicting human evolution, painted by a 1998 graduating senior and donated to the science department, disappeared from a science classroom. The four-by-sixteen-foot painting had been propped on a chalkboard tray because custodians refused to mount it on the wall. Spahr learned that the building and grounds supervisor had ordered it burned. In June 2004, board member William Buckingham, Bonsell's co-instigator of the ID policy, told Spahr that he "gleefully watched it burn" because he disliked its portrayal of evolution. He also blocked purchase of a new science textbook that included evolution, forcing teachers to accept *Pandas* as a reference book in exchange for new textbooks. In January 2005, science teachers refused to read the ID statement; administrators read it themselves. The situation worsened. When the next school year began in September 2005, the board's policy and ID itself were on trial in Harrisburg, Pennsylvania.

TROUBLE BEHIND THE SCENES

Besides myself, the other expert witnesses for the plaintiffs were scientists Kevin Padian and Kenneth Miller, theologian John Haught, science-education expert Brian Alters, and philosopher of science Robert Pennock. Our combined work addressed all relevant aspects of ID creationism. The National Center for Science Education (NCSE) staff were consultants for the plaintiffs' team. Everyone on the plaintiffs' team, including the attorneys, served pro bono. (None of the defense's expert witnesses participated pro bono. All listed their fees—typically $100 per hour, but $200 per hour in Dembski's case—in their expert witness reports.) Our pre-trial preparations went smoothly, but things weren't going as well for the defense.

Dembski's CSC associates, Stephen C. Meyer, John Angus Campbell, Scott Minnich, and Michael Behe, were also to be witnesses for the defense, along with ID supporters Warren Nord, Steve William Fuller, and Dick M. Carpenter II. When TMLC rejected Meyer, Dembski, and Campbell's demand for legal representation independent of TMLC, the three withdrew from the case by refusing to continue without their own attorneys.[1] In the case of Campbell, Pepper Hamilton attorney Thomas Schmidt had flown to Memphis with a legal assistant and had hired a court reporter to take Campbell's long-scheduled deposition. Everything was proceeding on schedule until only minutes before the deposition was to begin, when defense attorney Patrick Gillen announced that TMLC would "no longer retain" Campbell as a witness because Campbell had "retained counsel through Discovery Institute" and had "discussed matters [with DI] to which I am not privy." Gillen learned of "these developments" only the night before. Behe and Minnich, already deposed, remained as witnesses along with Nord, Fuller, and Carpenter.

Why would three of the most important ID experts withdraw? They had already submitted expert-witness reports and scheduled depositions. Moreover, DI had hoped for years to precipitate a test case and had even prepared legal arguments. The problem, however, was that DI did not want this case, because the Dover board, urged on by TMLC, had explicitly crafted its policy to promote "intelligent design." Having come to view that term as a legal liability after encountering opposition in Ohio, Kansas, and elsewhere, DI tried unsuccessfully to persuade the board to either restate the ID policy in sanitized language or withdraw it. (DI now strategically sanitizes its terminology, using euphemisms and code words in appeals for the teaching of ID.) They were scared to death of a case they had not initiated and could not control.

The exodus of Dembski, Meyer, and Campbell is explained by their fear of cross-examination. The public shredding that Irigonegaray had given ID creationists in Kansas one month earlier was still fresh. As Pat Hayes stated it in a report on his blog, dated May 9: "The knockout punch came when . . . Pedro Irigonegaray compelled the intelligent design witnesses to confess, during a series of withering cross-examinations, that they hadn't bothered to read the science standards draft . . . before coming to Kansas at taxpayer expense." Moreover, Dembski, Meyer, and Campbell knew what the plaintiffs' expert witnesses would say in court, because they had our reports. DI must have known that our case would be devastating to the defense—and thus to ID—if it was argued before a judge who respected the truth and the Constitution.

MY ROLE IN THE KITZMILLER TRIAL

I was called as a witness because of my coauthorship, with Paul R. Gross, of *Creationism's Trojan Horse: The Wedge of Intelligent Design* (Oxford University Press, 2004) and other publications about ID. In our book, we analyze a document titled "The Wedge Strategy," CSC's tactical plan, showing how CSC creationists are executing every phase except producing scientific data to support ID. We show that ID is creationism, thus a religious belief, using the best evidence available: the words of ID leaders such as Phillip E. Johnson, William Dembski, and their ID colleagues. We also show ID's continuity with earlier creationism. My job was to present this evidence to the judge.

It probably wasn't difficult for DI and TMLC to figure out that, armed with my work and that of the other witnesses for the plaintiffs, halfway decent attorneys would make legal mincemeat of them. And the plaintiffs' attorneys were more than decent—they were superb. Pepper Hamilton's Eric Rothschild, one of the lead attorneys, prepared me well.[2] Dembski had seen a preview of my testimony at my deposition. DI also knew I was inspecting documents related to *Of Pandas and People* that our attorneys had subpoenaed from the Foundation for Thought and Ethics (FTE), which holds the *Pandas* copyright. Consequently, although DI and TMLC were squabbling behind the scenes, they shared a common goal: to either forestall or discredit my testimony.

On September 6, 2005, the defense filed a motion requesting my exclusion from the case. DI and TMLC had apparently overcome their differences long enough to collaborate on the accompanying brief, because it contained clear evidence of DI's input. Although I was not called as a scientific expert, the defense argued that I should be excluded because I had no scientific expertise and because I am, in their words, "little more than a conspiracy theorist and a web-surfing, 'cyber-stalker' of the Discovery Institute . . . and its supporters and allies." Five years earlier, Dembski had accused two of his critics, Wesley Elsberry and Richard Wein, of being "Internet stalkers who seem to monitor my every move." DI had responded to *Creationism's Trojan Horse* by labeling me as a conspiracy theorist.[3] Judge Jones denied the motion on September 22.

Scheduled to testify the following week but delayed by Hurricane Rita, I used the extra time to prepare for my testimony and to stay current on ID activities by visiting DI's Web site. On September 29, I noticed that DI had posted a transcript of an interview I had done—except that I hadn't done it. The transcript was fake. Apparently meant (though not marked) as a parody, the organization whose self-described goal is "to support high quality scholarship . . . relevant to the question of evidence for intelligent design in nature" ridiculed me by, among other things, having fictitious radio host "Marvin Waldburger" refer to me as "Dr. Barking Forrest Ph.D." If DI thought this would unsettle me, they were ignoring the fact that I had just been through two killer hurricanes. I could only shake my head at their doing something so jaw-droppingly stupid. If they were hoping Judge Jones would see and be influenced by this silliness, it was just another sign of the disrespect for his intelligence and integrity that began before the trial and continues today (more on this shortly).

ON THE WITNESS STAND

When I was sworn in on October 5, the defense spent the entire morning presenting arguments as to why I should not be qualified as an expert witness.

Judge Jones again denied the motion, meaning that he—and the whole world—would hear what both DI and TMLC had hoped they could bar from the record: the truth about ID.

I had two tasks: to demonstrate to Judge Jones (1) that ID is creationism, thus a religious belief, and (2) that *Of Pandas and People* is a creationist textbook. As part of the evidence for my first task, I included the words of two leading ID proponents, Phillip E. Johnson and William Dembski. Under direct examination by Eric Rothschild, I related Johnson's definition of ID as "theistic realism" or "mere creation," by which he means that "we affirm that God is objectively real as Creator, and that the reality of God is tangibly recorded in evidence accessible to science, particularly in biology." To that, I added Dembski's definition: "Intelligent design is just the Logos theology of John's Gospel restated in the idiom of information theory." If the judge had heard nothing except these two quotes, he would have had all the evidence he needed that ID's own leaders regard it as not only creationism but also as a *sectarian Christian* belief. But I had much more, such as CSC fellow Mark Hartwig's 1995 *Moody Magazine* article, in which he referred to a 1992 ID conference at Southern Methodist University as a meeting of "creationists and evolutionists," calling Dembski and Stephen Meyer "evangelical scholars." During those early years, when they needed money and supporters, ID proponents openly advertised both their religiosity and their creationism.

However, none of the evidence for ID's religious, creationist identity was more important than "The Wedge Strategy," probably written in 1996, when the CSC was established, but revised in 1998. Known informally as the "Wedge Document," it was leaked from a Seattle office and posted on the Internet in early 1999. DI did not acknowledge ownership of it until 2002, after I independently authenticated it and wrote about it in 2001. The technical team hired by Pepper Hamilton to create computer "demonstratives" projected the Wedge Document onto a screen in court, and I walked Judge Jones through it, explaining the most important parts. My first slide made its significance clear: ". . . could I have the first slide, please? This is the first page of the Wedge Strategy, and this is the opening paragraph of it. Quote, 'The proposition that human beings are created in the image of God is one of the bedrock principles on which Western civilization was built.' This . . . states very well the foundational belief behind the intelligent design movement and the reason that they have rejected the theory of evolution." As I continued, the judge heard the strategy's explicitly Christian goals: "Design theory promises to reverse the stifling dominance of the materialistic worldview and to replace it with a science consonant with Christian and theistic convictions." The folks at DI probably never imagined that an "obscure philosopher in Louisiana," as they once called me, would be using their strategy document in a trial—or that it would be so effective in their legal undoing. (Though actually, they probably did suspect its legal significance, which explains their taking three years to acknowledge it.)

To counter the defense's predictable denials that ID is creationism, I also explained, using an account by ID proponent and CSC fellow Paul Nelson, how Phillip Johnson had masterminded creationism's transformation into "intelligent design" after the US Supreme Court outlawed creationism in public schools in its 1987 *Edwards v. Aguillard* ruling. According to Nelson, creationists believed that *Edwards* meant the death of the "two-model approach to origins," in which creationists recognize only two alternatives, either evolution or creation, hoping to win by default after undermining evolution. But Nelson explained that "a revolution from an unexpected quarter . . . was about to occur." The revolutionary was Johnson, who decided that, for creationism to survive *Edwards*, creationists had to redefine science: "Definitions of science, [Johnson] argued, could be contrived to exclude any conclusion we dislike or to include any we favor." Not only was Johnson's deliberate but nominal transformation of creationism into ID important for demonstrating ID's true identity, but it also provided important support for my testimony about *Pandas:* to survive after *Edwards, Pandas* would require a similar transformation. (When the book was first published in 1989, Johnson was already allied with chemist Charles Thaxton, author of the creationist book *The Mystery of Life's Origin* and "academic editor" of *Pandas.*) The subpoenaed FTE documents, which contained several earlier *Pandas* drafts, revealed that precisely such a transformation had been effected.

A *Pandas* coauthor, CSC fellow Dean H. Kenyon, had been a creationist witness in the *Edwards* case and had submitted a sworn affidavit testifying that "creation-science is as scientific as evolution." I discovered a letter Kenyon wrote to FTE president Jon Buell, showing that he was working on the 1986 draft of *Pandas*, then called *Biology and Creation*, while also assisting in the *Edwards* case! All pre-*Edwards* drafts of *Pandas* (there were at least five) were written using creationist terminology. The earliest drafts had overtly creationist titles. In 1987, the title was changed to *Of Pandas and People*, and there were two drafts in that year. One was written in creationist language; in the other, creationist terminology had been replaced by "intelligent design" and other design-related terms, suggesting that the *Edwards* decision prompted this change. The clincher was a new footnote in the latter draft explicitly referencing *Edwards*, indicating that this draft was produced *after* the June 19, 1987, decision in an effort to evade the ruling. I also found a letter from Buell to a prospective publisher in which Buell made profit projections for *Pandas* contingent upon the Court's decision: "The enclosed projection showing revenues of over 6.5 million in five years are based upon modest expectations for the market, provided the US Supreme Court does not uphold the Louisiana Balanced Treatment Act. If by chance it should uphold it, then you can throw out these projections. The nationwide market would be explosive."

These excerpts reflect the general content of the evidence I produced during my day and a half of testimony. When Richard Thompson cross-examined me, he did what we anticipated: avoiding the substance of my testimony and my published work, he attacked my credibility. He apparently hoped the judge would consider my association with civil-liberties and humanist organizations unsavory enough to discredit me, asking questions such as, "When did you become a card-carrying member of the ACLU?"

THE DEFENSE

The combined testimony of the plaintiffs' expert witnesses presented a formidable obstacle for the defense—as did the testimony of defense witnesses themselves. Nord and Carpenter were withdrawn without testifying, leaving only Behe, Fuller, and Minnich.[4] Behe and TMLC attorney Robert Muise escorted the judge through a long explanation of irreducible complexity using Behe's stock example, the bacterial flagellum. During Rothschild's cross-examination, however, Behe admitted that under his own definition of a scientific theory (which he has conveniently loosened in order to classify ID as science), astrology also qualifies. Most unhelpfully, Fuller had affirmed in his deposition—under oath—that ID is creationism. Presented by ACLU attorney Vic Walczak with the relevant statements, he had no choice but to admit this:

> Walczak: "And then your answer beginning on Line 24, 'It [ID] is a kind of creationism, it is a kind of creationism.' I didn't read the same passage twice. It's actually twice on there. Did I read that accurately?"

> Fuller: "Well, it looks like that is what the sentences say."

Fuller also described his role in the trial as that of an advocate for "disadvantaged theories" needing an "affirmative action strategy." By the time Minnich, the last witness, was asked to offer still more testimony about bacterial flagella, he understood fully the position in which he found himself: "I kind of feel like Zsa Zsa's fifth husband, you know? . . . I know what to do but I just can't make it exciting. I'll try."

THE VERDICT ON ID

On December 20, 2005, Judge Jones delivered a powerful opinion—a marvel of clarity and forthrightness—giving no quarter to either the school board or ID. He was not fooled by ID proponents' denials that they are creationists: "ID cannot uncouple itself from its creationist, and thus religious, antecedents." He was especially displeased that board members Buckingham and Bonsell had lied under oath during their depositions: ". . . the inescapable truth is that both Bonsell and Buckingham lied at their January 3, 2005 depositions about their knowledge of the source of the donation for *Pandas*. . . . This mendacity was a clear and deliberate attempt to hide the source of the donations . . . to further ensure that Dover students received a creationist alternative to Darwin's theory of evolution." Presented with the truth about the board's policy and the ID creationism it promoted, Jones ruled accordingly: "A declaratory judgment is hereby issued in favor of Plaintiffs . . . such that Defendants' ID Policy violates the Establishment Clause of the First Amendment of the Constitution of the United States and . . . the Constitution of the Commonwealth of Pennsylvania."

JUDGING THE JUDGE

Reflecting the ID movement's status as an integral part of the Religious Right, their mean-spirited attacks on Judge Jones have amplified right-wing screeching about federal judges, revealing their contempt for the judicial system they have long wished to exploit on their own behalf. As Kevin Padian and Nick Matzke wrote in a commentary for NCSE, "DI immediately tried to 'swift-boat' the judge." (ID sympathizers were similarly enraged: harsh e-mails sent to Jones's office led federal marshals to place him and his family under guard for a short period.) Given the warmed-over creationism that was the only thing Dembski, Meyer, and Campbell could have offered in court, they would have needed either a political partisan or a fool on the bench. Antics such as the "Barking Forrest" interview suggest that DI regarded Judge Jones as the latter. One of Dembski's regular *Uncommon Descent* bloggers apparently regarded him as both.

Dembski declined to comment on his blog when the opinion was filed: "I have little to add to what I wrote in September, so I'll just leave it there." On September 30, he had calculated probabilities for various outcomes: (1) 20 percent that Jones would uphold the ID policy; (2) 70 percent that he would overturn it but leave the "scientific status of ID" unchallenged; and (3) less than 10 percent that he would both overturn the policy and rule ID unscientific. He hedged his bets by appealing for intelligent assistance: "I trust that Providence will bring about the outcome that will best foster ID's ultimate success." Blogger "DaveScot" urged optimism: "Have more faith, Bill! This is all about Judge Jones. If it were about the merits of the case we know we'd win. It's about politics. . . . Judge John E. Jones . . . is a good old boy brought up through the conservative ranks . . . appointed by GW hisself. . . . Unless Judge Jones wants to cut his career off at the knees he isn't going to rule against the wishes of his political allies." Post-verdict, however, DaveScot shifted from cocksure confidence in the "good old boy" to insults. When *Time* magazine named Judge Jones one of 100 "people who shape our world," DaveScot offered some unflattering comparisons: "The magazine who made these men 'Man of the Year'—1938–Adolf Hitler; 1939–Joseph Stalin; 1942–Joseph Stalin; 1957–Nikita Krushchev; [and] 1979–Ayatullah Khomeini—now brings you Judge John Jones as a 2006 Honorable (pun intended) Mention." Dembski responded appreciatively: "Thanks, Dave, for contextualizing this milestone in our proper appreciation of important personages. . . . What a crock." (Of course, *Time*'s Person of the Year designation is not an honorific title but an assessment of the designee's influence on the year's events.)

In a December 20, 2005, post on DI's *Evolution News & Views* blog, CSC associate director John West sniped, "Judge Jones . . . wants his place in history as the judge who issued a definitive decision about intelligent design. This is an activist judge who has delusions of grandeur." DI then hastily self-published a book, *Traipsing into Evolution: Intelligent Design and the Kitzmiller vs. Dover Decision*, charging that Jones "repeatedly misrepresented both the facts and the law in his opinion, sometimes egregiously." The nastiness of DI's attacks on the judge can be seen as directly proportional to his opinion's power and accuracy.

THE AFTERMATH

Probably anticipating a future legal case, for which Jones's opinion would serve as a strong precedent and for which I might be called as an expert witness, ID creationists continue their efforts to discredit both Judge Jones and me. They employ their usual tactics: lacking scientific evidence for ID, they make things up or slander their opposition. DI's attacks on Judge Jones stem partly from his reliance on my testimony and pre-trial reports, as he indicated in his opinion: "Dr. Barbara Forrest . . . has thoroughly and exhaustively chronicled the history of ID in her book and other writings for her testimony in this case. Her testimony, and the exhibits ... admitted with it, provide a wealth of statements by ID leaders that reveal ID's religious, philosophical, and cultural content." Consequently, CSC fellow Jonathan Witt has interwoven attacks on the judge with attacks on me: "Several newspapers ... are highlighting Judge John Jones's spurious determination that intelligent design is creationism in disguise. They're accurately reporting the judge's opinion here, for his decision reads like a condensation of atheist-activist Barbara Forrest's mythological history of intelligent design. ... In following Barbara Forrest's fallacious reasoning and mythological history of intelligent design, Judge Jones has erred badly."

On April 12, 2006, presenting long-discredited pro-ID arguments at the University of Montana School of Law, CSC fellow and law professor David DeWolf complained that Jones "thought it was his job to declare what was orthodox in science ... [and] to decide theological questions." Asserting that Jones's opinion "effectively targeted" DI as "the author of this Wedge Document," DeWolf lobbed a double *ad hominem:* "Judge Jones's reliance on the Barbara Forrest testimony in the trial makes the John Birch Society look rational." He also told the audience—falsely—that "there is a letter from Barbara Forrest, one of the witnesses in the Dover trial, a letter to an editor of a journal saying basically don't publish things from people who are advocates of intelligent design." At a May 5, 2006, luncheon we both attended at the Massachusetts School of Law (MSL), I asked him publicly whether he had seen this letter. He admitted—before an audience of his legal peers—that he had not. (The luncheon was held prior to a taping of MSL's television program, *Educational Forum,* and was attended by law-school faculty and administrators.) This time he was telling the truth: I never wrote such a letter. However, DeWolf continues to promote ID using the same debunked arguments that he and his ID associates used before the trial. In an article with former DI attorney Seth Cooper, he complains that Jones's ruling was "based upon evidence and characterizations of intelligent design that have been sharply contested by leading proponents of intelligent design." Yet, like Dembski, Meyer, and Campbell, neither DeWolf nor Cooper was anywhere in sight when they had a chance to defend ID in court.

The legal defeat of ID is forcing Wedge strategists to seek new markets for their creationism and to work their conservative Christian market more thoroughly. They are peddling ID abroad: DI has added international signatories to "A Scientific Dissent from Darwinism." Even during the trial, they held an ID conference in Prague. Domestically, Dembski has been reduced to riding the coattails of conservative pundit Ann Coulter, who devoted four chapters of her latest book, *Godless,* to attacking evolution. These chapters contain the standard creationist canards, but with Coulter's recognizable stylistic stamp: "Imagine a giant raccoon passed gas and perhaps the resulting gas might have created the vast variety of life we see on Earth. And if you don't accept the giant raccoon flatulence theory for the origin of life, you must be a fundamentalist Christian nut who believes the Earth is flat. That's basically how the argument for evolution goes." Coulter credits her ability to write these chapters to "the generous tutoring of Michael Behe, David Berlinski, and William Dembski, all of whom are fabulous at translating complex ideas." Dembski acknowledges his assistance: "I'm happy to report that I was in constant correspondence with Ann regarding her chapters on Darwinism—indeed, I take all responsibility for any errors in those chapters." He has dubbed Coulter "the Wedge for the masses."

These tactics by DeWolf and Dembski highlight the bankruptcy of ID and the blustering cowardice of its leaders, who must capture support with brazen deceit and sarcastic punditry. The trial was Dembski's moment to shine, to explain on the legal record why ID is a "full scale scientific revolution," as he wrote in *The Design Revolution* (InterVarsity Press, 2004, p. 19). Instead, plaintiffs' witness Robert Pennock read to Judge Jones Dembski's statement regarding ID's revolutionary status—and then dismantled it. Ironically, Dembski had his arch-critics right where he wanted us—on the witness stand and under oath. He could have been there, implementing his strategy, helping to "squeeze the truth" out of us, "as it were." In November 2005, after the trial ended, Dembski posted on his "Design Inference" Web site a combined version of his May 11 and 16, 2005, "vise strategy" blog pages, labeled as a "Document prepared to assist the Thomas More Law Center in interrogating the ACLU's expert witnesses in the Dover case." He appended a list of "Suggested Questions," which, he wrote, "will constitute a steel trap that leave the Darwinists no room to escape." (It is worth noting that after his last question, Dembski conveniently noted that "with these questions, we don't need to get into the positive ID program—i.e., what ID is doing specifically to advance ou[r] understanding of biology. That will come out under cross-examination of our side.") But when he had an opportunity to witness firsthand how his trap would operate, he was nowhere to be found. He "escaped critical scrutiny" by quitting rather than face cross-examination. He is apparently $20,000 richer for it, however, marking yet another difference between us: whereas I served pro bono, Dembski charged $200 per hour and threatened to sue TMLC for payment for one hundred hours of work he claims to have done prior to quitting. In late June 2005, he told Canadian ID supporter Denyse O'Leary that TMLC had agreed to pay him.

After ID's dramatic, unequivocal defeat in *Kitzmiller*, Dembski's priorities remained remarkably consistent: "This galvanizes the Christian community. . . .

People I'm talking to say we're going to be raising a whole lot more funds now." If failure is that lucrative, one can only imagine how well-remunerated he and his ID colleagues would be if they could tell the truth and back up their claims about "intelligent design theory."

ACKNOWLEDGEMENT

I would like to thank Glenn Branch and Nick Matzke of the National Center for Science Education for their comments and suggestions.

NOTES

1. According to Richard Thompson, TMLC had agreed to allow Stephen Meyer, as "an officer of the Discovery Institute," to have his attorney present but would not allow Dembski and Campbell "to have attorneys, that they were going to consult with . . . and not with us." DI corroborates Thompson's statement about the offer to Meyer but says that Meyer "declined the offer because the previous actions of Thomas More had undermined his confidence in their legal judgment."

2. In addition to Rothschild, the primary attorneys were Stephen Harvey, Thomas Schmidt, and Alfred Wilcox of Pepper Hamilton; Witold "Vic" Walczak of the Pennsylvania ACLU; and Richard Katskee of Americans United for Separation of Church and State. Other attorneys who were not in court helped in pre-trial preparations.

3. A document is available on the Discovery Institute Web site, "The 'Wedge Document': So What?'" which the DI falsely claims was "originally published in 2003." But it was written in response to my book, which was published in January 2004. The document's properties show that CSC associate director John West created it on January 8, 2004, and the Internet Archive shows no such document on the DI website at the end of 2003. It does show a page with a link entitled "What Is the Wedge Document?" dated February 10, 2004, indicating that the document was on the site at that time.

4. Despite the fact that Carpenter played no real role in the trial, he wrote an article for the February 2006 *American School Board Journal* titled "Deconstructing Dover: An Expert Witness for the District Shares Lessons from 'Scopes 2005.'" In this article, he implements DI's deceptive strategy of trying to disconnect teaching ID from "teaching the controversy about evolutionary theory" (code talk for teaching ID) and offers "lessons" for school boards: "For school board members interested in 'teaching the controversy' about evolutionary theory, the lessons here are clear: Conduct the public's business in and with the public. Make sure your policies are justified with a clear and secular purpose. Neither officially advocate nor prohibit intelligent design. Seek out and pay attention to expert advice" (21–22). School boards would be well advised to ignore Carpenter's article and steer clear of teaching this nonexistent

controversy in any form. As an "Education Correspondent" for Focus on the Family, Carpenter also wrote what appear to be variants of the *ASBJ* artcle for FoF's *Citizen* magazine.

SELECTED BIBLIOGRAPHY

Davis, Percival, and Dean H. Kenyon. *Of Pandas and People: The Central Question of Biological Origins.* 2nd ed. 1993. Dallas, Texas: Haughton Publishing Company.

Kitzmiller et al. v. Dover Area School District, Memorandum Opinion. December 20, 2005. Available at www.pamd.uscourts.gov/opinions/jones/04v2688a.pdf.

Forrest, Barbara, and Paul R. Gross. *Creationism's Trojan Horse: The Wedge of Intelligent Design.* New York: Oxford University Press, 2004.

Dembski, William A. "The Vise Strategy: Squeezing the Truth Out of Darwinists." *Uncommon Descent.* Available at www.uncommondescent.com/index.php/archives/59, May 11, 2005.

Dembski, William A. "The Vise Strategy II: Essence of the Strategy." *Uncommon Descent* Available at www.uncommondescent.com/index.php/archives/72. May 16, 2005.

National Center for Science Education. "Legal Issues, Lawsuits, Documents, Trial Materials, and Updates." *Evolution Education and the Law.* Available at www2.ncseweb.org/wp/?page_id=5.

NOTE ABOUT REFERENCES

For space reasons, most of the author's original seventy endnotes, totaling the equivalent of 2,750 words, have been omitted from this published version, with her approval. Instead, four of the notes and a short selected bibliography she prepared appear at the end. An electronic version of the article with all of the author's endnotes, most with hyperlinks for opening full texts, is on CSI's Web site at www.csicop.org under "special articles." —Editor

FOUR COMMON MYTHS
ABOUT EVOLUTION

Charles Sullivan and Cameron McPherson Smith

Nearly 150 years after Charles Darwin published *On The Origin of Species*, the theory of evolution is still widely misunderstood by the general public. Evolution isn't a fringe theory, and it's not difficult to understand, yet recent surveys reveal that roughly half of Americans believe that humans were created in their present form 10,000 years ago (Brooks 2001, CBS 2004). The same number reject the concept that humans developed from earlier species of animals (National Science Board 2000).

But the evidence is clear that no species, including humans, simply "popped up." Each life form has an evolutionary history, and those histories are intricately intertwined. If we don't understand that complex evolution, we will make poor decisions about our future and that of other species. Should we genetically modify humans? How about our food crops? What effects will global warming have on human biology? None of these questions, nor many others of immediate concern to humanity, can be usefully addressed unless we understand the evolutionary process.

In examining how evolution is portrayed in the mass media, we found many problems; chief among them was the use of inaccurate expressions. In this article we examine the commonly used phrases "evolution is only a theory," "the ladder of progress," "missing links," and "only the strong survive."

These expressions are misleading at best, and simply wrong at worst. Most of these phrases have ancient roots, describing biology as it was understood centuries ago. They lead to a distorted picture of what evolution is and how it works.

EVOLUTION IS ONLY A THEORY

Have you ever heard people challenge evolution by claiming that "it's only a theory?" The Cobb County School District in Georgia did just that when it sought to put stickers on high school biology textbooks stating that, "Evolution is a theory, not a fact, regarding the origins of living things."[1] The problem with this claim rests with two different uses of the word *theory*. In popular usage the word refers to an unsubstantiated guess or assumption, as when someone theorizes that a light moving across the night sky must be an alien spaceship. When scientists use the word *theory*, however, they're referring to a logical, tested, well-supported explanation for a great variety of facts.[2] In this sense the theory of evolution is as well supported as the theory of gravitation or other explanatory models in fields such as chemistry or physics. While it's true that much of the evidence for evolution is not obtained by laboratory experiments, as in chemistry and physics, the same can also be said for geology and cosmology.

A geologist cannot travel back in time to observe first hand the formation of Earth's crust, and a cosmologist cannot witness the collapsing of a star into a black hole, but this doesn't mean that scientific theories about the nature of these phenomena are simply unsubstantiated guesses. Some scientific theories do a better job of accounting for the facts than others, and in biology there is no competing scientific theory with more explanatory power than evolution. Biologist Theodosius Dobzhansky put it best when he said, "Nothing in biology makes sense except in the light of evolution."

Many people confuse evolutionary theory with Lamarckism, named for the French naturalist Jean-Baptiste Lamarck (1744–1829). In one sense Lamarck was an evolutionist in that he favored the view that new species had evolved from ancestral species, but he was mistaken about the mechanism by which species change, and about the time required for these changes. Lamarck thought that the mechanism for biological change was the transmission to the next generation of characteristics *acquired during the life span of an individual*. His most famous example is that of the giraffe. According to Lamarck, the giraffe's ancestors had shorter necks, and they would stretch their necks to reach higher foliage in trees. Their descendants then inherited longer necks because the characteristics of these newly stretched necks of the parents were passed down to their offspring. Moreover, Lamarck thought that the evolution of a new species could occur within a few generations or even one. His position was reasonable for its time, yet it happens to be incorrect.

But acquired characteristics are not passed on.[3] If you lose your arm in an accident, your offspring will not be born with a missing arm. If you lift weights to gain muscle mass, you will not transmit larger muscles to your offspring. Jews have been practicing circumcision for hundreds of generations, yet there is no evidence that this acquired feature is biologically inherited.

The position of modern evolutionary theory (Neo-Darwinism[4]) is that some ancestors of giraffes acquired slightly longer necks through random mutation. These animals could eat food that was a little out of reach of others of their species, and so they tended to be healthier, to live longer, and have a better chance than their fellows at mating and passing on to the next generation their genes for longer necks. Many such incremental changes over a long period of time are required for a new species—or a neck as long as a giraffe's—to arise.

The evolution of giraffes or other life forms should not be thought of as a singular process. There are at least three independent processes that, when taken together, form our idea of evolution. These are replication, variation, and selection. Replication is essentially reproduction. Variation refers to the random changes—typically mutations—arising in offspring, making them different from their parents. Selection refers to the process whereby those individuals best adapted to their environment tend to be the ones that survive, passing on their genes. These three processes occur every day in nature, and it is their cumulative effect that we call evolution.

If an entirely new scientific theory with more explanatory power is formulated, then Neo-Darwinism will have to be swept aside just as Lamarckism was. Creationism and Intelligent Design don't qualify as competing scientific theories because they're not scientific. They don't offer natural explanations for biological phenomena, but rather supernatural explanations which cannot be tested scientifically. Neo-Darwinism offers a natural explanation to account for the facts of evolution, and rejects supernatural explanations.

When discussing the theory of evolution it's important to realize why it's misleading to say that evolution is *only* a theory. Evolution is indeed a theory, but it's a theory with a lot of evidence on its side, and with more explanatory power than any competing theory in biology.

THE LADDER OF PROGRESS

The word *evolution* is sometimes used to mean progress. People speak of moral evolution when discussing certain cultural changes that have been for the better, such as the increased recognition of the rights of women. Or they speak of technological evolution when comparing present-day technology with that of ancient hunter-gatherers. This sense of the word *evolution* implies a progressive development toward better or more advanced stages. It is this non-biological sense of *evolution* that influences people to think of biological evolution as involving ladder-like progress from lower to higher stages.

The idea of an evolutionary ladder of progress has its roots in Classical Greek and Medieval European concepts about the nature of the universe. The most common manifestation is known as the Great Chain of Being, which was most influential in Europe from the fifteenth through the eighteenth centuries. The basic idea of the Great Chain of Being is that God and his creation form a hierarchy which is ordered from the least perfect things or beings at the bottom of the chain to the most perfect at the top, namely, God himself. Simply put, the ranking from bottom to top is as follows: rocks or minerals, plants, animals, man, angels, God.

The Great Chain of Being scheme wasn't designed with evolution in mind since the prevailing idea of the time was that God made all existing species, in their modern forms, long ago. The Great Chain of Being is best described as a method of classification. This idea began to lose support before the Darwinian revolution, but Darwin's ideas and their refinement ultimately broke the links of the Great Chain of Being.

The modern biological understanding of evolution does not involve progress in the sense of a natural upward goal toward which all life is striving.[5]

A study of the DNA of Darwin's finches on the Galapagos Islands (Petren et al. 1999) provides a good example of why this idea of progress makes no sense in evolution. The study's findings suggest that the first finches to arrive on the islands were the Warbler finches (*Certhidea olivacea*), whose pointy beaks made them good insect eaters. A number of other finches evolved later from the Warbler finches. One of these is the *Geospiza* ground finch, whose broad beak is good for crushing seeds, and another is the *Camarhynchus* tree finch with its blunt beak, which is well adapted for tearing vegetation.[6]

Even though the seed-eating and vegetation-eating finches evolved from insect-eating finches, the former are not "more evolved" than the latter, or "higher" on some evolutionary ladder. Since finch evolution on the Galapagos Islands was driven primarily by diet, the ground finches simply became better adapted at making a living on seeds, the tree finches on vegetation, and the Warbler finches on insects.

If seeds were to become scarce on the Galapagos Islands, it's conceivable that the seed-eating finches—which are a more recent species—could become extinct, while the insect-eating finches—which have been around much longer—would continue to thrive. The concepts of "higher" and "lower" do not apply to the Galapagos finches or anywhere else in evolution. It is fitness or adaptability relative to the environment that matters. Species cannot foretell the future in order to adapt themselves deliberately to environmental changes, and if the environment changes drastically, those adaptations that were once favorable may turn out to be unfavorable.

Even though biologists reject the Great Chain of Being or any similar ladder-of-progress explanation of evolution, the idea still persists in popular culture. A more accurate analogy would be that of a bush that branches in many directions. If we think of evolution over time in this way, we're less likely to be confused by notions of progress because the branches of a bush can grow in various directions in three dimensions, and new branches can sprout off of older branches without implying that those farther from the trunk are better or more advanced than those closer to the trunk. A more recent branch that has split off from an earlier branch—like a species that has evolved from an ancestral species—does not indicate greater progress or advancement. Rather, it is simply a new and different growth on the bush, or more specifically, a new species that is sufficiently adapted to its environment to survive.

THE MISSING LINK

"Fossils May be Humans' Missing Link" reported the *Washington Post* on April 22, 1999. The story states that fossils discovered in Ethiopia "... may well be the long-sought immediate predecessor of human beings." But almost fifty years earlier, paleontologist Robert Broom published *Finding the Missing Link* (1950), about his discovery of fossil "ape men" in South African caves. And since 1950, reports of the discovery of "missing links" have been continuous. What's going on? How is it that the "missing link" has been discovered repeatedly?

The problem lies in a false metaphor. When we say "missing link," we invoke a metaphorical chain, a set of links that stretch far back in time. Each link represents a single species, a single variety of life. Because each link is connected to two other links, each is intimately connected to past and future forms. Break one link, and the pieces of the chain can be separated, and relationships lost. But find a lost link, and you can rebuild the chain, reconnect separated lengths. One potent reason for the attractiveness of this metaphor is that it allows for the drama of the quest, the search for that elusive missing link.

But the metaphor is as misleading as it is attractive. The concept that each species is a link in a great chain of life forms was largely developed in the typological age of biology, when species "fixity" (the idea that species were unchanging) was the dominant paradigm. Both John Ray (1627–1705) and Carolus Linnaeus (1707–1797), the architects of biological classification (neither of whom believed in evolution), were concerned with describing the order of living species, an order they each believed was laid out by God (Ray suggested that the divinely specified function of biting insects was to plague the wicked). But while the links of a chain are discrete, unchanging, and easily defined, groups of life forms are not.[7] We generally define a *species* as some interbreeding group that can not, or does not, productively breed with another group. But since species are not fixed (they change through time), it can be difficult to be sure where one species ends and another begins. For these reasons, many modern biologists prefer a continuum metaphor, in which shades of one life form grade into another.[8] Life is not arranged as links, but as shades. The metaphorical chain is far less substantial than it sounds.

Thus the chain metaphor is wrong. It doesn't accurately represent biology as we know it today, but as it was understood over four centuries ago. The myth persists because of convenience; it is easier to think of species as types, with discrete qualities, than as grades between one species and another. In school, we learn the specific characteristics of plants and animals; this alone is not a problem, except that we are not often exposed to the main ramification of evolution: *that those characteristics will change through time.*

Clearly, both the *Post* article and Broom's book describe the discovery of australopithecenes, African hominids[9] that lived well over three million years ago. Australopithecenes walked upright, like modern humans, but they had large, chimp-like teeth, and smallish, chimp-like brains. Australopithecenes made rudimentary stone tools that are more complex than any chimpanzee's termite-mound probe stick, but far less complex than the symmetrical tools made by early members of our genus, *Homo*. In terms of anatomy and behavior, some australopithecenes really do appear to be "half human." Additionally, it's widely believed that early *Homo* descended from some variety of late australopithecene. Broom was right after all, but so was the *Post*, a "missing link" has indeed been found. It is *Australopithecus*. But there were many varieties of *Australopithecus*, as well as *Homo*, and there is no obvious place to draw a discrete line separating a shade of late *Australopithecus* from an early grade of *Homo*. Therefore, it's more accurate to say that we have found some "grade" or "shade," rather than "the missing link."[10]

We can curb the false metaphor by changing our wording. In classes, in textbooks, in discussions with our students, and in press releases (the critical connection between academia and the general public), we have to start saying that we're looking for *a* missing link, rather than *the* missing link. Better yet, we should replace the "missing link" stock phrase with something more accurate.

ONLY THE STRONG SURVIVE

Around a million years ago, an ape so large that it's now known as *Gigantopithecus* roamed the bamboo forests of South Asia. Standing nine feet tall, weighing from 600 to 1,000 pounds, and with a bamboo-crushing jaw the size of a mailbox, this was a truly strong creature. But today, all that remains of *Gigantopithecus* are a few fossil teeth and jawbones, quietly resting in museum vaults.

If only the strong survive, how did early *Homo*—protohuman bipeds that were in the same area at the same time, and were less than half the size of *Gigantopithecus*—survive? Wouldn't any clash between these creatures result in the strapping über-ape annihilating the competition?

Yesterday's giants can be today's museum specimens. If only the strong survive, though, how is this possible? Indeed, how is it that humans are now ascendant on Earth, but, when stripped of tools and culture, we are among the most helpless of animals?

The answer, of course, is that strength can be measured in many ways. Brawn is one measure; brain is another. But this distinction is often lost in popular culture. When we say "the strong," or even "the fittest," most people immediately think of competition between individuals. These individuals, we imagine, are pitted against one another in some evolutionary arena, where they fight for survival and mates. The strongest survive, pass on their genes, and propagate their lineage. The loser, and its entire lineage, goes extinct.

But this notion of single combat in a single arena of competition is too simple. In reality, there are dozens of arenas, dozens of problems any organism must face in life. Perhaps direct competition with other individuals is one, but every day individuals are kicked from one arena to the next. If the river dries up, you're now in the Arena of Water Conservation. If the temperature suddenly drops, you're pushed into the Arena of Heat Conservation. If the properties of the vegetation you eat begin to change, you're now in the Arena of Metabolic Versatility.

In short, survival is much more complex than is implied by the single-arena concept of combat between individuals. Life forms struggle against a wide array of factors, and often against more than one factor at a time. In biology, these factors are known as *selective pressures*.

Selective pressures also change. A certain selective pressure can be particularly hard-pressing for a period, shaping the course of evolution, but later that pressure may ease, and another concern becomes primary. And since the environment is always changing, no species can ever be sure what selective pressures it will have to cope with tomorrow. Indeed, such conscious anticipation of the future is precluded for most species (could deer have anticipated the invention of guns?), and evolution is entirely reactive, shaping species according to past and present environments, but never "looking" into the future.[11]

We humans, and all life forms, exist and struggle not in any single arena, but in an immense web of selective pressures that is incomprehensibly complex and ever-changing. Survival is much more involved than simply beating down your immediate peers.

Why does the single-combatant, evolutionary arena myth persist? The answer is probably deeply intertwined with renaissance values of individualism too complex to examine here[12] but it is clearly related to nineteenth-century Social Darwinism. Social Darwinists grafted Darwin's basic ideas about biological evolution to human society and economy. To them, progress could only be made by eliminating imperfections from humanity, and this was best done by competition. That competition, neatly summarized by Herbert Spencer's term "survival of the fittest," was taken to mean the competition between individuals. It is significant that today's reality-TV programs are steeped in this metaphor, in which the concept of survival via ruthless individual competition is paramount.

The best way to curb this myth is to teach that brute strength does not guarantee long-term success. In fact, no single characteristic does. More importantly, we need to describe *why* there is no single key to long-term success, because we never know how our selective environment is going to change. For humanity, then, the only hope for success, for survival, is in remaining flexible and adaptive. Real strength is in adaptability, which comes from genetic and cognitive variation.

CONCLUSION

A picture of evolution based on the common myths we've outlined is a mosaic of confusion. It's very important to remedy this confusion, because how we think of ourselves, and every other species on Earth, is directly related to how we understand evolution. We can either see ourselves as separated from a natural world that simply serves as a theater for our evolution,[13] or as one of many coevolving species of life on Earth. The former view is more likely to persist if we continue to describe evolution using obsolete or faulty expressions. The latter view, which is accurate, will be promoted by a better use of language, and by acknowledging what we have learned about biology in the past 150 years.[14]

Solutions for conveying this accurate view must include more careful use of language and metaphor to explain exactly what evolution is, and how it happens.

NOTES

1. The entire disclaimer reads: "This textbook contains material on evolution. Evolution is a theory, not a fact, regarding the origin of living things. This material should be approached with an open mind, studied carefully, and critically considered." This led to a lawsuit, *Selman v. Cobb County School District*. On January 13, 2005, a federal judge found this policy unconstitutional.

2. See, for example, "What's Wrong with 'Theory not Fact' Resolutions." *National Center for Science Education*. December 7, 2000. Available at http://www.ncseweb.org/resources/articles/8643_whats_wrong_with_theory_not__12_7_2000.asp.

3. However, a recent study on fruit flies suggests that some genetic instructions that are not encoded in the DNA may be passed on to offspring by way of material encompassing the DNA (Lin et al., 2004).

4. Developed in the 1930s, Neo-Darwinism (also called the Modern Synthesis) integrates Darwin's theory of natural selection with the theory of genetic inheritance first proposed by Gregor Mendel and subsequently refined by later biologists.

5. Biologists disagree about whether there is an evolutionary tendency toward complexity, primarily because there is no consensus on how complexity should be defined and measured.

6. While evolution isn't "trying" to do anything at all, its results (species becoming better adapted to their ecological niches over time) can be considered progress. To understand how the concept of progress can be applied to "evolutionary arms races," see Dawkins, R., *The Blind Watchmaker*. (New York: Norton, 1986), chapter 7.

7. The species concept is introduced in Strickberger (1985: 747–756), but also see Mallet (1995) for the need to review how we define species.

8. Lions and tigers once coexisted naturally in India, and although they are outwardly very different, they can mate to create tigons or ligers. Since such hybrids were never found in nature, however, it is known that lion and tiger did not interbreed naturally. Thus, genetically, lion and tiger can be classified as one species, but behaviorally, they differed enough to be considered separate species by biologists, and in nature this difference was maintained by the animals themselves (Wilson, 1977: 7).

9. Hominids are large primates that walk upright. Those of the genus *Australopithecus* (which predate the human lineage *Homo*) are referred to as australopithecenes. They appeared over 4 million years ago. Many hominid varieties have existed, but *Homo sapiens sapiens* is the only living hominid.

10. The link metaphor also suggests that any given species is represented by only one chain, as when we see a diagram of hominids, first knuckle-walking, then hunched over in a half-stand, then upright as modern man. This depiction does not show several other bipedal hominid varieties to which we are related, such as the robust australopithecenes (appearing over 4 million years ago, disappearing about 1 million years ago) or the Neanderthals (who appear around 300,000 years ago and are extinct by c.30,000 years ago). The depiction suggests that there was one, unbroken chain, from quadruped to biped, but actually there have been bipeds that have gone extinct (as well as quadrupeds that exist today).

11. Humanity, of course, is uniquely proactive. We can imagine the future, and we prepare for it by controlling our evolution with all sorts of social and biological methods. Social methods include complex kinship and marriage rules that ensure gene flow among different populations. Biological methods include mass vaccination programs against polio and smallpox.

12. See for example Shanahan (2004), an interesting comment in Commager (1965: 82–83), and Butterfield (1965: 222–46).

13. We suggest that this view of humanity contributes to wasteful use of resources; for example, humanity has chronically overfished nearly every fishery ever discovered; see Jackson et al. (2001)

14. It's not sufficient to constantly explain away old, inaccurate expressions: we must develop new ones. What is the use in keeping old metaphors or phrases that do not point to reality? For example, we may say "The Bush of Evolution" or "The Labyrinth of Evolution," rather than "The Ladder of Evolution." A good way to find such new metaphors may be to create a Web site where anyone could suggest them, and, after a time, select new ones to start incorporating into our common speech. Perhaps poets, being familiar with the power of image and metaphor, could be helpful.

REFERENCES

Brooks, D. J. "Substantial numbers of Americans continue to doubt evolution as explanation for origins of humans." The Gallup Organization. 2001. Available online at http://www.gallup.com/poll/content/login.aspx?ci=1942.

Broom, R. *Finding the Missing Link.* London: Watts & Co., 1950.

Butterfield, H. *The Origins of Modern Science.* New York: MacMillan. 1965.

CBS News Polls. "Creationism trumps evolution." CBSNEWS.com. 2004. Available online at www.cbsnews.com/stories/2004/11/22/opinion/polls/main657083.shtml.

Commager, H. S. *The Nature and Study of History.* Columbus, Ohio: Charles E. Merrill Books, 1965.

Dobzhansky, Theodosius. "Nothing in biology makes sense except in the light of evolution." *The American Biology Teacher* 35 (1973): 125–129.

Jackson, J. et al. "Historical overfishing and the recent collapse of coastal ecosystems." *Science* 297 (2001): 629–37.

Lin, Q., Q. Chen, L. Lin, and J. Zhou. "The Promoter Targeting Sequence mediates epigenetically heritable transcription memory." *Genes & Development* 18 (2004): 2639–651.

Mallet, J. "A species definition for the modern synthesis." *Trends in Ecology and Evolution* 10 (1995): 294–99.

National Science Board. "Science and Engineering Indicators." Washington, D.C. US Government Printing Office. 2000. Available online at www.nsf.gov/sbe/srs/seind/pdf/c8/c08.pdf.

Petren, K., B. R. Grant, and P. R. Grant. "A phylogeny of Darwin's finches based on microsatellite DNA length variation." *Proceedings of the Royal Society of London* B266 (1999): 321–29.

Shanahan, T. *The Evolution of Darwinism: Selection, Adaptation and Progress in Evolutionary Biology.* New York: Cambridge University Press. 2004.

Strickberger, M. W. *Genetics.* New York: Macmillan, 1985.

Suplee, C. "Fossil find may be that of humans' immediate predecessor." *Washington Post,* April 23, 1999, pp. A3, A11.

Wilson, E. O. *Sociobiology.* Harvard, MA: The Belknap Press of Harvard University Press, 1977.

ONLY A THEORY?
Framing the Evolution/Creation Issue

David Morrison

Public opinion polls tell us that we are losing the battle to explain the nature of evolution and the central role that evolutionary concepts play in modern science. Tens of millions of Americans scoff at evolution and try to protect their children from what they consider to be a pernicious concept.

Given the overwhelming scientific support for evolution, we must be doing something wrong in discussing this issue with the public. There are several ways in which scientists and educators might enhance their effectiveness in this debate. The problems relate to framing the issues, or rather, allowing the opponents of evolution to frame them. Framing involves the selective use of language or context to trigger responses, either support or opposition. We see it in the often deceptive titles of legislation, such as a "clear skies act" or "forest renewal act" or, on the other side, a "death tax." "Pro-choice" or "pro-life" advocates are always careful to frame their position with the proper emotion-charged terms. (The subject is artfully described in George Lakoff's book *Don't Think of an Elephant.*)

As a prime example, we doom our communications efforts with many nonscientists by defending the "*theory* of evolution." *Theory* is quite simply the wrong word. Polls indicate that three quarters of Americans agreed that "evolution is commonly referred to as the theory of evolution because it has not yet been proven scientifically." Those who advocate adding "only a *theory*" disclaimers in textbooks know that to call evolution a *theory* is sufficient to undermine its acceptance.

Channeling the discussion into a debate over the "theory of evolution" is an example of framing. Since the great majority of Americans understand the word *theory* to imply uncertainty and vagueness, the name itself predisposes the answer. It is as if a criminal defendant were described by the judge and other court officials as "the murderer." Not many juries would want to let a known murderer free, no matter how the evidence was presented. The one who frames the debate often wins.

Yet many proponents of evolution seem content to argue about the "theory of evolution" and its educational role. As scientists, they were taught that a scientific theory is a systematic set of principles that has been shown to fit the facts, and has stood up against attempts to prove it false. A theory is thus the highest level of understanding, synthesizing a wide variety of observations and experiments. But that is not what the word *theory* means to 99 percent of Americans, including many scientists and educators when they are outside the classroom.

Dictionaries have noted the changing definition of this word. Older dictionaries give preference to the scientific definition and consider the use of *theory* to refer to a guess or hunch to be a form of slang. Today, the slang meaning prevails, and a theory is a belief, something taken to be true without proof, an assumption, a suggestion, a hypothesis. Similarly, *theoretical* is used as a synonym for *tentative*, an idea that has not been tested with observations.

How do we really use the term in everyday language? A theory is a hunch that a detective comes up with in a murder mystery. It is one of several competing ideas, none of them proved. Fringe theories and conspiracy theories are crazy ideas that are out of the mainstream. New medicines or changes in the tax laws may be good in *theory* but don't work in practice. Among some scientists, theorists are thought to lack solid grounding in the facts.

What about scientific usage? We don't hear much anymore about the Theory of Gravitation, or the Atomic Theory of Matter, or the Theory of Plate Tectonics. These phrases have a vaguely antique flavor. Gravitation and atoms and plate tectonics are accepted as legitimate subjects that don't need the preface "Theory of." The only two areas where "Theory of" remains in common use are Theory of Relativity and Theory of Evolution. Relativity is associated with Einstein, a genius whose work was abstract and unintelligible to laypeople. I doubt if most people realize that the principles of relativity have been tested, or that relativity has practical implications, for example, in calculating the interplanetary trajectories of spacecraft. Judge for yourselves what this association implies for "Theory of Evolution."

There is another usage that should be mentioned: *theory* as a discipline, such as organization theory, color theory, economic theory, music theory, etc. These phrases imply the existence of a knowledge base or conceptual framework, and their names are given to university courses or areas of specialization. In science there are chaos theory, cosmological theory, information theory, and—yes—evolutionary theory (as used in the title of Steve Gould's last book). This usage is, however, rarely discussed in arguments about "only a theory."

If we accept the framing that calls this topic the theory of evolution, we face a dilemma. Some people just ignore the problem and concentrate on presenting the facts of evolution. They may believe that these facts and their implications are self-evident. But the human brain does not always work that way. We have seen recent examples. The majority of voters who supported the 2004 re-election of George W. Bush told pollsters that they believed that Saddam Hussein had weapons of mass destruction and that he had been behind the September 11 terrorist attacks, despite countless news stories to the contrary. Many people believe that airplanes are more dangerous than cars, no matter what risk statistics are presented. Even after well-conducted trials showed that herbal medicines such as echinacea are ineffective, public sales have remained strong.

Alternatively, many scientists and educators recognize the public misunderstanding of the scientific term *theory* and try to explain this to the audience. In her excellent book *Evolution vs. Creationism*, Eugenie Scott devotes much of the first chapter to the scientific meaning of the terms *theory, hypothesis, falsification*, etc. However, few members of a nonscientific audience with concerns about teaching evolution to their children are ready to accept that this word, whose meaning they know perfectly well, has in fact an almost opposite definition in science. Thus before asking the audience to consider that their opinions about evolution might be wrong, we start by asking them to accept a contrary definition for a familiar word. Anyone who teaches knows how hard it is for students to unlearn things they already know and believe. So why do we accept this wholly unnecessary burden when discussing evolution? No wonder those who frame evolution as a theory often win.

We should be discussing simply evolution, the same way we might discuss plate tectonics or genetics or any other branch of science. To debate the "theory of evolution" is a trap. It is letting our opponents frame the discussion to their benefit. Once we stop defending the theory of evolution, we are also free to criticize "only a theory" disclaimers in textbooks without apology or diversionary explanations.

THE EITHER/OR FALLACY

The concept of framing has other implications for the creation-evolution debate. One that we are all familiar with is the effort to portray this as a choice between two models—godless evolution versus divinely inspired creationism. In a two-model formulation, any perceived difficulty with evolution becomes support for creationism. We should not accept this framing of the conflict. Actually, I would think that today it would be obvious that there are at least three models on the table: evolution, the young-earth creationism of biblical literalists, and the more sophisticated concept of Intelligent Design (ID), which accepts the age of the universe and the change in Earth's biota over time. It is a tribute to the discipline of the anti-evolution camp that they have avoided most public debates between the biblical literalists and ID. But we are free to exploit this split and to ask which alternative is proposed to be taught alongside evolution in science classes.

Most opponents of evolution in the schools are Christian fundamentalists, and many of them believe that evolution is a moral issue, a struggle between the forces of good and evil. Obviously proponents of evolution do not see it this way. What we want is a level playing field where we can present the facts. But as noted above, facts often lose out in a confrontation with deeply held beliefs.

To achieve a level playing field, we should avoid debating evolution in a religious context. Specifically, we should not speak on these issues in a church other than our own. In an unfamiliar church speaking to an audience of like-minded opponents of evolution, all the cards are stacked against us—not because there is anything antireligious about evolution, but because the audience *believes* there is. This situation also creates a temptation to debate religion itself, such as arguing about the "true" message of the first book of Genesis or contrasting the beliefs of Roman Catholics and fundamentalist Protestants. This is not likely to be a winning strategy for the scientist, who is again a victim of framing.

AN ISSUE OF VALUES

Many antievolutionists base their opposition not on scientific issues, but on their belief that evolution threatens their value system, specifically their family values. I don't see why we should not face this issue directly. Family values are not the monopoly of the creationist advocates. Most in these audiences will share a common interest in the education of their children, which is a fundamental American family value.

If I were speaking to an audience of parents, I would stress that evolution is central to many sciences, and that one cannot be a scientist without understanding it. I would note that students who don't understand evolution will have trouble getting into the best colleges. I would comment on the great number of scientists and engineers being graduated in "competitor" countries like China, and note that American students are not top scorers in international science tests. We can make a persuasive and nonpartisan case that the future of the nation depends on its scientific and technical literacy. Relatively few Americans will reject such arguments and state a preference for ignorance. Teaching evolution is part of a bigger issue of the competitiveness and economic well being of the nation (and of the state and local community). This is a real issue of values, for our children and their futures. And it is an issue that might appeal to lay members of school boards and textbook selection committees.

Health is another value issue. Medical research is supported by Americans across a wide political spectrum. How many people understand the role of evolution in the development of new medical treatments? Or the place played by genetics in creating new drugs? There is no more dramatic (or scary) example of evolution than the emergence of drug-resistant pathogens, as well as recent diseases such as AIDS. Newspaper stories about threatened pandemics due to mutations in avian flu or other emerging viruses can only be understood in an evolutionary context. Suppressing the study of evolution cuts off future opportunities to improve public health. Surely this argument on values has wide appeal.

BEYOND BIOLOGY

We also suffer when we accept that evolution should be debated purely in terms of biology and biology courses. In the present American public school system, these courses are already watered-down so that evolution is likely to be mentioned in only or one or two chapters or discussed in only one study unit of a biology course. It is easy to belittle a subject that seems so marginal. We should reframe this issue in terms of crosscutting ideas that affect all science. The audience should know that evolution and the concept of deep time are essential to geology and astronomy and genetics and pharmacology as well as high-school biology. It is also an opportunity to tell an audience how many people in this country are working in evolution-related jobs. If the audience comes from a traditional conservative religious background, they may have no idea that evolution is widely accepted among scientists and medical professionals and that it contributes to the livelihood of many Americans.

Many conservative Americans support competition and believe that free-market economic conditions are essential to national success. Most of them would be shocked to know that this philosophy has traditionally been known as social or economic Darwinism. Perhaps we should note that Darwinian natural selection is in many ways nature's equivalent to free-market competition. The other side of this argument addresses the belief that evolution leads to socialism and communism. Perhaps it is worth noting that Stalin's support for the anti-Darwinian biologist Trofim Lysenko set back Soviet agriculture for a generation and contributed to the starvation of millions of Russians.

Finally, we can reframe the issue in terms that do not immediately offend a conservative religious audience. The context in which most opponents fear and reject evolution is that of human origins. The scientific story of the evolution of our human ancestors is fascinating, but it also provokes the strongest resistance. My own interests, as an astrobiologist, are in microbial evolution, which is a less threatening subject. I remember a young teacher coming up to me after a lecture and saying how amazed she was that I had talked for an hour about evolution and the history of life without mentioning primate evolution or human origins. I am not suggesting that we ignore the fascinating story of hominid evolution, but I bet that there are many people who would be more receptive to evolutionary concepts if we refrained from an in-your-face challenge to their convictions about human origins or the nature of the human soul.

Another topic that is controversial is the origin of life. Creationists love to confuse origins with evolution. They frequently use criticisms of, for example, the relevance of the Miller-Urey experiment to undercut the entire concept of biological evolution. Let's not be sidetracked into the problems of the origin of life. While a great deal of research has been done to define the conditions under which life began on Earth and to understand basic biochemistry, the actual process by which living things emerged is not understood by science. I believe we can and should admit this mystery. There may be many people who will open their minds to the ideas of evolution as long as we don't claim that science has all the answers, especially about the ultimate origin of life and the meaning of being human. These topics can be explored later, when we have overcome initial emotional resistance to any form of evolution.

Above all, I hope that we can frame the evolution-creationist debate in ways that open our audience to the exciting ideas and accomplishments of science. When appropriate, we should be happy to defend teaching evolution in the context of family values and economic advantage as well as pure science. There is no reason to make this a debate about religion; we are almost sure to lose such a confrontation. But we must also understand where our audience is coming from and find ways to present the science in a non-confrontational and accessible way.

IS INTELLIGENT DESIGN CREATIONISM?

Massimo Pigliucci

Intelligent design proponents such as William Dembski claim that science is incomplete because it doesn't admit the possibility of "explaining" the world by invoking the action of a (supernatural) designer. (Dembski usually shies from overtly calling for a supernatural designer, but since natural design is already accepted as part of the explanatory framework of science—for example in the case of the Search for Extraterrestrial Intelligence—his shyness must be only a political expedient.) And yet, it is demonstrably the case that Dembski doesn't understand (or refuses to understand for psychological reasons) how partly historical sciences such as evolutionary biology actually work.

As an example, let us consider a standard objection to evolutionary theory that Dembski presents at every occasion (for example, in his "The Logical Underpinning of Intelligent Design," a chapter in Dembski and Ruse's *Debating Design*, Cambridge University Press). In discussing naturalistic scenarios for the evolution of complex biological structures, such as the bacterial flagellum, Dembski claims that the following conditions must be met: "(a) the probability of each step in the series [of evolutionary changes leading to a complex structure] can be quantified, (b) the probability at each step turns out to be reasonably large, and (c) each step constitutes an advantage to the evolving system."

Sounds reasonable, right? Not really. Any biologist (beginning with Darwin) would agree on c, though often we simply do not know enough about the structural biology and history of a given structure to be able to give a full list of steps (more on this below); *b* is rather obvious; the real problem is *a*. Dembski here sets the bar so high that he knows it cannot reasonably be met, which means of course that he wins by default. The situation is rather analogous to that of a classic creationist claim, which is why I think the logic of ID is not significantly different from the logic of creationism in general. One of the recurring objections that creationists raise to evolutionary theory is that, in their opinion, there are no intermediate fossils to make a convincing case that, say, *Homo sapiens* really did evolve from a closely related group of primates. A patient paleontologist will then produce several fossils of different species of *Australopithecus*, together with many species of pre-*sapiens Homo*, all perfectly good candidates as intermediates. "Ah!" the creationist cries in disbelief, "but now you have many more gaps to explain!"

This response is not a joke, I have actually heard plenty of creationists making it with the clear impression of having definitely trumped my materialistic delusions (and, I suspect, expecting me to kneel in front of them to accept Jesus). The fault in the logic should be obvious: the fossils produced by the paleontologist make a compelling case because they are of the appropriate morphology (they do show intermediate characteristics between the most ancient and the most recent species), and because they are found in the temporal sequence predicted by evolutionary theory. While any individual fossil does not clinch the case, the entire ensemble allows scientists to make what philosophers call an "inference to the best explanation": the idea that *Homo sapiens* evolved by successive modifications from primate ancestors is made increasingly more likely with every *Australopithecus* and pre-*sapiens Homo* fossil that is found. This, incidentally, was exactly the type of argument used by Darwin in *The Origin of Species*, which is why he famously referred to it as "one long argument" (the longer, the more convincing).

Dembski, in asking for condition *a*, is asking for the same sort of thing that standard creationists want when they demand a minute-by-minute account of evolution. He knows very well that the demand is impossible to fulfill for the same reasons that we cannot produce every single intermediate change that actually occurred during the evolution of *Homo sapiens*: evolutionary biology is a historical science, and the historical traces of many of the relevant events are forever lost to science. This doesn't mean that the inference to the best explanation isn't a powerful tool of scientific investigations. Indeed, many ID proponents themselves use precisely such a tool in claiming that the available facts "point" in the direction of an intelligent designer as the best explanation for the complexity of living organisms.

That said, sometimes biologists do get lucky enough to actually be able to show, both theoretically and by comparative anatomy, all or almost all the intermediate steps that may have occurred to bring about the evolution of a given complex structure. This is the case for the vertebrate eye, as beautifully summarized in a classic paper co-authored by legendary evolutionary biologist Ernst Mayr (Salvini-Plawen and Mayr, 1977; see also Nilsson and Pelger, 1994). We now have both a set of computer simulations showing how the complex vertebrate eye can evolve from simple photoreceptors, and a collection of currently living organisms actually displaying many of the predicted forms (all perfectly functional, which answers the classical creationist question of "what is half an eye good for?").

As the informed reader might recall, the vertebrate eye used to be the quintessential example of "irreducible complexity" and a famous warhorse of intelligent design theory at the time of William Paley (early nineteenth century). It is not by chance (shall we call it rhetorical design?) that modern ID proponents such as Michael Behe and Dembski never mention the eye and instead retreat to the depth of microscopic structures such as the bacterial flagellum. Should we be able to do for the flagellum what we did for the eye, what would the next hiding place of not-so-crypto-theists such as Behe and Dembski be? Only time will tell.

REFERENCES

Nilsson, D. E., and S. Pelger. "A pessimistic estimate of the time required for an eye to evolve." *Proceedings of the Royal Society of London* B256 (1994): 53–58.
Salvini-Plawen, L. V., and E. Mayr. "On the evolution of photoreceptors and eyes." *Evolutionary Biology* 10(1977): 207–63.

THE MEMORY WARS

Martin Gardner

In the late 1980s and early 1990s the greatest mental health scandal in North America took place. Thousands of families were cruelly ripped apart. All over the United States and Canada, previously loving adult daughters suddenly accused their fathers or other close relatives of sexually molesting them when they were young. A raft of bewildered, stricken fathers were sent to prison, some for life, by poorly informed judges and jurors. Their harsh decisions were in response to the tearful testimonies of women, most of them middle-aged, who had become convinced by a psychiatrist or social worker that they were the victims of previously forgotten pedophilia.

On what grounds were these terrible accusations made? They were (supposedly) long-repressed memories of childhood sexual abuse brought to light by self-deluded therapists using powerful suggestive techniques such as hypnotism, doses of sodium amytal (truth serum), guided imagery, dream analysis, and other dubious methods.

Of all such techniques the most worthless is hypnotism. Mesmerized patients are in a curious, little-understood state of extreme suggestibility and compliance. They will quickly pick up subtle cues about what a hypnotist wants them to say, and then say it. The notion that under hypnosis one's unconscious takes over to dredge up honest and accurate memories of a distant past event is one of the most persistent myths of psychology. There simply is no known way, short of confirming evidence, to distinguish true from false memories aroused by hypnotism or any other technique. After many sessions with a sincere but misguided therapist, false memories can become so vivid and so entrenched in a patient's mind that they will last a lifetime.

An early leader in debunking the belief that recollections of childhood traumas can be repressed for decades is the distinguished experimental psychologist Elizabeth Loftus, a professor at the University of California, at Irvine. She was awarded the prestigious $200,000 Gramemeyer Award for Psychology given annually by the University of Louisville, and also elected to the National Academy of Sciences. Her passionate book *The Myth of Repressed Memory*, written with Katherine Ketcham, has become a classic treatise on what is called the false memory syndrome (FMS). See also her article "Creating False Memories," *in Scientific American* (vol. 277, no. 3, 1997).

Another influential and tireless crusader against FMS is educator Pamela Freyd. In 1992 she established the nonprofit FMS Foundation after she and her husband Peter were falsely accused by their daughter of sexually molesting her when she was a child. Freyd continues to edit the Foundation's bimonthly newsletter, and provides information and moral support to wrongly accused parents. By 1992 more than ten thousand distressed parents had contacted the FMS Foundation for advice on how to cope with a son or daughter's charges. Today dozens of Web sites carry on the fight against FMS.

Over the past twenty years hundreds of papers and dozens of excellent books have shed light on the FMS epidemic. In addition to Loftus's book I reluctantly limit my list to three others: Eleanor Goldstein's *Confabulations: Creating False Memories, Destroying Families;* Mark Pendergrast's *Victims of Memory: Incest Accusations and Shattered Lives;* and Frederic Crews's *The Memory Wars: Freud's Legacy in Dispute.* See also three eye-opening articles on "Recovered Memory Therapy and False Memory Syndrome," in *The Skeptic's Encyclopedia of Pseudoscience* (vol. 2), edited by Michael Shermer. Psychiatrist John Hochman's article should be read by every attorney who defends a victim of the FMS. Here is his final paragraph:

Meanwhile, there is a large FMS subculture consisting of women convinced that their "recovered memories" are accurate, therapists keeping busy doing RMT [Repressed Memory Therapy], and of authors on the "recovery" lecture and talk show circuits. In addition, there are some vocal fringes of the feminist movement that cherish RMT since it is "proof" that men are dangerous and rotten, unless proven otherwise. Skeptical challenges to RMT are met by emotional rejoinders that critics are front groups for perpetrators, and make the ridiculous analogy that "some people even say the Holocaust did not happen." RMT will eventually disappear, but it will take time.

In 2001 the FMS Foundation sent a survey questionnaire to 4,400 persons who had contacted the Foundation for advice. An overwhelming number of the accusers (99 percent) were white, 93 percent were women, 86 percent were undergoing mental therapy, and 82 percent later accused their fathers of incest when they were children. Ninety-two percent said the recovery of repressed memories was the basis of their accusations.

The number of charges peaked in 1991–1992, which accounts for 34 percent of the accusations, then the rate slowly declined. By 1999–2000 the number was down to .02 percent. The decline prompted psychiatrist Paul McHugh of Johns Hopkins School of Medicine to write an optimistic article, "The End of a Delusion: The Psychiatric Memory Wars Are Over," in *The Weekly Standard* (May 26, 2003). The survey's data are analyzed and commented on in "From Refusal to Reconciliation," by McHugh, Harold Lief, Pamela Freyd, and Janet Fetkewicz, *The Journal of Nervous and Mental Disease* (August 2004).

The authors of this valuable paper distinguish three stages of accusers:

1. Refusers. Those whose beliefs about past abuse are set in concrete. They refuse all contact with anyone who does not share their convictions.

2. Returners. Those who return to their families but do not retract their charges or discuss them.

3. Retractors. Those who eventually realize that their awful memories are fabrications. They reconcile with their parents. The authors quote from a retractor's moving letter:

> I could not face the horrible thing I had done to my parents, so I had to believe the memories were true. Even though I got away from that horrible therapist, I could not go back to my entire extended family and say that I was temporarily insane and nothing had happened. It was easier for my self-esteem to pretend that I had been sexually abused by someone, and it was still my parents' fault because they should have protected me. (*FMSF Newsletter*, December 1998)

In the 1990s the FMS mania also blighted the lives of hundreds of preschool teachers and daycare personnel. Small children were taken by hysterical parents to trauma therapists, convinced that their children had been sexually exploited even though at first they could not recall such abuse. After many therapy sessions, repressed memories seemed to surface.

One of the most publicized cases involving preschool children concerned the Little Rascals daycare center in Edenton, North Carolina, a town decimated by the case. On the witness stand, brainwashed little rascals told wild, unbelievable tales. They "recalled" seeing the center's co-owner, Robert Kelly, murder babies. One child said "Mr. Bob" routinely shot children into outer space. Another lad told the court that Kelly had taken a group of youngsters aboard a ship surrounded by sharks. He threw a girl overboard. Asked if the sharks had eaten her, the boy replied no, he (the boy) jumped into the water and rescued her!

Robert Kelly was convicted on ninety-nine counts of first-degree sex offenses and sentenced to twelve consecutive life terms. It was the longest sentence in North Carolina history. Kelly spent six years in prison before an appeals court released him on $200,000 bond. Kelly's friends and coworkers, including his wife and the center's cook, got harsh sentences. A 1995 television documentary, *Innocence Lost*, left no doubt that the children had confabulated.

Many recent investigations have established how easily children can be led by inept therapists to imagine events that never happened. This was amusingly demonstrated by a simple experiment reported by Daniel Goleman in his article "Studies Reflect Suggestibility of Very Young Witnesses" (*New York Times*, June 11, 1993). A boy was falsely told he had been taken to a hospital to treat a finger injured by a mousetrap. In his first interview he denied this had happened. By the eleventh interview he not only recalled the event, but added many details. In fact, only extremely rarely are memories of traumatic events repressed until years later, only then emerging under suggestive therapy. On the contrary, it is far more common for victims to try vainly to forget a traumatic incident.

There are books defending the revival of long-repressed memories. By far the worst is *The Courage to Heal*, by Ellen Bass and Laura Davis. A bestseller in 1988, its rhetoric persuaded tens of thousands of gullible women that their mental and behavior problems were caused by forgotten childhood sex abuse, and led them to seek validation through trauma therapy.

Another book, almost as bad, is *Secret Survivors*, by trauma therapist Sue Blume. "Incest is easily the greatest underlying reason why women seek therapy," she wrote. ". . . [i]t is not unreasonable that more than half of all women are survivors of childhood sexual trauma." Both statements are, of course, preposterous.

In 1989 Holly Ramona sought treatment for bulimia. After months of therapy by a family counselor, and later by a psychiatrist, she began to get memories of being raped by her father when she was an infant. Firm believers in Freudian symbols, Holly's two therapists convinced her that she disliked mayonnaise, soup, and melted cheese because they reminded her of her father's semen. She was unable to eat a banana unless it was sliced because it resembled her father's penis. Under oath she testified that her father had forced her to perform oral sex on the family dog!

Holly's father sued the two therapists. Lenore Terr, a psychiatrist who was an expert witness at the trial, told the jury that Holly's dislike of bananas, cucumbers, and pickles confirmed her recovered memories of being forced to perform oral sex on her father. Terr has been an "expert witness" on other similar trials. Basic Books carelessly published her shameful work, *Unchained Memories: True Stories of Traumatic Memories, Lost and Found*. Happily, a California court refused to buy Terr's Freudian speculations. Holly's father won a settlement of half a million dollars.

As the FMS plague spread it took on ever more bizarre forms. Quack psychiatrists began regressing patients back to traumas in their mother's wombs. One therapist uncovered memories of traumas while a patient was stuck in a fallopian tube!

Those convinced that evil aliens kidnapped and tortured them with horrible experiments in hovering UFOs started to confirm their fears by repressed memory therapy. The most absurd of many books on recovered memories of flying-saucer abductions are by Temple University's historian David M. Jacobs, and two books by the late John E. Mack, a Harvard psychiatrist. Mack believed that the extraterrestrials are friendly, and come here from higher space dimensions. Harvard was unable to fire him because, like Jacobs, he had tenure. (See the interview with Mack in *The New York Times Magazine*, March 20, 1994.)

A more tragic application of the FMS rested on the beliefs of countless Protestant fundamentalists that the horrors of the End Times are fast approaching. Satan, aware of the Biblical prophecy that Christ will return to Earth and cast him into a lake of fire, is now on an angry rampage. He is establishing vile cults throughout the United States, Canada, and elsewhere—cults in which unspeakable rituals are performed, such as eating babies and drinking blood and urine. Dozens of shabby books about such madness have been published in spite of a thorough investigation by the FBI which concluded that, aside from the acts of pranksters, there is no evidence that Satanic cults exist here or anywhere else. In England a report by the UK Department of Health reached a similar conclusion after investigating eighty-four cases of alleged organized Satanic cults.

If revived memories of cannibalizing babies are true, thousands of Satanically mutilated infant bodies should be buried around the nation. Not one has been found. Why? Because, fundamentalists argue, the Devil is so powerful that he is able to obliterate all such evidence! For lurid accounts of bogus memories of Satanic rituals and details about false recollections, see chapters 6 and 11 in my book *Weird Water and Fuzzy Logic*.

Among a raft of books and articles debunking the myth of Satanic cults, one of the best is Lawrence Wright's "Remembering Satan" (*The New Yorker*, May 17 and 24, 1993). It is reprinted in his book with the same title. Another excellent reference is sociologist Jeffrey Victor's book *Panic: The Creation of a Contemporary Legend*.

A third crazy spinoff from the false memory wars concerns New Age psychiatrists who believe in reincarnation. Under suggestive therapy, Shirley MacLaine has recalled numerous adventures experienced in her past colorful lives. A few incarnation therapists are even using hypnotism to retrieve "recollections" of events a patient will experience in future lives!

In 1991 Geraldo Rivera introduced three trauma survivors on his talk show. One woman said she had murdered forty babies while in a Satanic cult but totally forgot about it until her memories emerged during therapy. Well-known entertainers spoke on other talk shows about their long-buried memories of pedophilia. Comedienne Roseanne Barr revealed that her parents had abused her when she was three months old! Her wild tale, vigorously denied by her dumbfounded parents, made the cover of *People* magazine.

The memory wars are slowly subsiding, but they are still far from over. There are four reasons for the decline:

1. Reversals by enlightened appellate courts of harsh, undeserved sentences—many for life—of innocent victims of FMS.

2. The gradual education of judges, jurors, attorneys, police officers, and people in the media.

3. An increasing number of "recanters," now in the hundreds, who realize how cruelly they have been misled.

4. A growing number of large settlements of malpractice lawsuits against therapists by recanters and wrongly accused relatives.

For sensational accounts of a few such actions see chapter 11, cited earlier, of my *Weird Water* book.

AIDS DENIALISM VS. SCIENCE

Nicoli Nattrass

Acquired immunodeficiency syndrome (AIDS) has killed more than twenty-five million people and remains a major threat to humankind (UNAIDS, 2006). The human immunodeficiency virus (HIV) causes AIDS by undermining the immune system, eventually resulting in death (Simon et al., 2006). Although no cure has been discovered, scientific advances have resulted in the development of antiretroviral drugs (ARVs) to prevent mother-to-child transmission of HIV (Brocklehurst, 2006) and to extend the lives of AIDS patients (Smit et al., 2006). HIV has been isolated and photographed, and its genome has been fully described. Yet a group of AIDS denialists in Australia (the so-called Perth Group) insists that HIV does not exist—recently testifying to this effect in an Australian court in defense of Andre Parenzee, an HIV-positive man charged with having unprotected sex with several women and infecting one of them with HIV. Other AIDS denialists accept the existence of HIV but, following Peter Duesberg (a molecular biologist at the University of California), believe it to be harmless. What unites them all is the unshakable belief that the existing canon of AIDS science is wrong and that AIDS deaths are caused by malnutrition, narcotics, and ARV drugs themselves.

AIDS denialists are eccentric but not irrelevant, because they campaign actively against the use of ARVs and promote the dangerous view that HIV is harmless (and some say not even sexually transmitted). South African president Thabo Mbeki took the AIDS denialists so seriously that he delayed the introduction of ARVs to prevent mother-to-child transmission of HIV and invited the leading AIDS denialists to serve on his "Presidential AIDS Advisory Panel" (Nattrass, 2007). They recommended that ARVs be avoided and that all forms of immune deficiency be treated with vitamins and "alternative" and "complementary" therapies including "massage therapy, music therapy, yoga, spiritual care, homeopathy, Indian ayurvedic medicine, light therapy and many other methods" (PAAP, 2001: 79, 86).

This leap—from the critique of mainstream biomedical science on AIDS to the promotion of unproven and unregulated alternative therapies—is a replay of the classic quack-marketing strategy of promoting belief in alternative remedies by sowing disbelief and skepticism about the medical establishment (Hurley, 2006: 216). It is thus not surprising that AIDS denialism has been used by vitamin salesmen (notably the Dr. Rath Health Foundation), self-styled alternative healers, and some traditional healers to promote their worldviews and products (Nattrass, 2007). One of South Africa's current health-policy failings is that, instead of cracking down on those making unsubstantiated health claims and creating markets for their wares, the health minister (Manto Tshabalala-Msimang) has provided cover and support for them.

AIDS denialists downplay their links with the purveyors of alternative therapies, preferring instead to characterize themselves as brave "dissidents" attempting to engage a hostile medical/industrial establishment in genuine scientific "debate." They complain that their attempts to raise questions and pose alternative hypotheses have been unjustly rejected or ignored at the cost of scientific progress itself.

Dissent and critique are, of course, central to science, but so, too, is respect for evidence and peer review. While it was intellectually respectable to dissent diametrically from mainstream views in the early days of AIDS science when relatively little was known about AIDS pathogenesis, this is no longer the case. In the 1980s, it was understandable that AIDS dissidents were uneasy about the claim that one virus could cause so many different diseases. But, once it was shown that HIV worked by undermining the immune system, thereby rendering the body vulnerable to a host of opportunistic infections, their concerns should have been put to rest. Similarly, the wealth of data on the successes of ARV treatment should have alleviated their initial worries about its overall therapeutic benefit. Thus one of the early AIDS dissident doctors, Joseph Sonnabend, had, by 2000, welcomed the lifesaving capacity of ARVs, describing them as a "wonderful blessing" (Sonnabend, 2000). However, this did not deter today's AIDS denialists, who continue to cite his dated views on their Web sites in support of their unchanged views.[1]

Given their resistance to all evidence to the contrary, today's AIDS dissidents are more aptly referred to as AIDS denialists. This stance may be attributable, in part, to a genuine misunderstanding of the science of HIV. For example, in his affidavit to the Australian court in the Parenzee case, a member of the Perth Group, Valendar Turner, testified that HIV had not been isolated because it had been identified only through the detection of reverse transcription (the process of writing RNA into DNA), an activity not unique to retroviruses (Turner, 2006: 4). In subsequent testimony for the prosecution, Robert Gallo (the discoverer of retroviruses and codiscoverer of HIV) pointed out that HIV had been identified as a retrovirus through the detection of reverse transcriptase, which is an enzyme unique to retroviruses, not the activity of reverse transcription, per se. He added that "only a fool" would mistake the two (Gallo, 2007b: 1310, 1313–14).

Misunderstanding the science of AIDS may be part of the story, but it does not explain why AIDS denialists are so hostile to and disbelieving of AIDS science. Part of the answer probably has to do with the belief that AIDS science cannot be trusted because the "scientific establishment" has been corrupted by the pharmaceutical industry (see Farber, 2006). This resonates with what Jon Cohen (2006: 1) calls "pharmanoia," or "the extreme distrust of drug research and development that's sweeping the world." John le Carré's novel (and subsequent hit Hollywood movie) *The Constant Gardener*, which provides a conspiratorial account of unethical medical trials in Africa, is a classic in this genre (Le Carré, 2001). This book was cited approvingly in a South African AIDS-denialist document coauthored by President Mbeki as being "well researched" and "illuminating" about the way the pharmaceutical industry influences academic research (Mbeki and Mokaba, 2002).

The pharmaceutical industry is, of course, far from angelic. There are documented cases where drug companies have designed trials in ways to promote sales of particular products rather than to test the best possible treatments; where clinical trials in poor countries have been unethical; where early research indicating dangerous side effects has been ignored for too long; where patent law has been abused to prevent low-cost competition; where too many resources have been spent on marketing "me-too" drugs (that is, drugs that are only marginally different from existing products) rather than investing in innovative drug development; and where unethical financial inducements have been made to doctors, researchers, and politicians (Goozner, 2004; Angell, 2005). However, what such cases suggest is that the pharmaceutical industry (and industry-funded research) needs to be carefully scrutinized and regulated. It does not imply that the entire industry and associated medical science are harmful to humans. As Cohen (2006) argues, the problem with the new pharmanoia is that it has put "Big Pharma" on a par with "Big Tobacco" and, through wild exaggeration, has turned "shades of moral grey into black."

The same applies to AIDS research, where the pharmaceutical industry has a clear incentive to fund and support those research activities most likely to generate profits in the future. This means that additional mechanisms need to be created to ensure that more risky and less profitable—but nevertheless important—areas of research, like vaccine development, are supported. It does not imply, as asserted by the AIDS denialists, that the pharmaceutical industry is funding a global conspiracy including all AIDS scientists, epidemiologists, and medical practitioners to invent a disease in order to market harmful drugs. (This tactic has also been used to great effect by Kevin Trudeau in his infomercials; see *Skeptical Inquirer*, January/February 2006, "What They Don't Want You to Know.") Aside from there being no evidence for this, the idea is incoherent, because the profit motive driving pharmaceutical companies gives them an incentive to keep people alive on chronic therapy as long as possible, not to kill them off quickly with dangerous drugs. Disrespect for AIDS scientists and physicians is a defining characteristic of AIDS denialists. Protected by a cloak of hubris—only they have the intelligence and moral courage to see the world for what it is—they portray themselves as lone, persecuted standard-bearers of the truth. As AIDS scientist John Moore (2006: 293) commented bitterly, their stance implies that "tens of thousands of health care professionals and research scientists are either too stupid to realize that HIV is not the cause of AIDS, or too venal to do anything about it for fear of losing income from the government or drug companies." Equally galling for scientists is the fact that most of the outspoken AIDS denialists are journalists or academics with no scientific training and that those who have medical qualifications have never actually worked on HIV.

In the normal course of scientific engagement, this would leave the denialists with little if any credibility. Gallo made this point very well in the Parenzee court case with regard to Turner: "Is he a virologist? Does he do experiments on AIDS?" he asked the defense attorney when presented with Turner's belief that HIV had not been isolated. "No," interjected the judge. "He's qualified in emergency medicine." "I see," replied Gallo. "I am not. Don't ever come to me if you are hurt" (Gallo, 2007b: 1272–73). In a subsequent e-mail message to the scientists and activists who run the anti-AIDS-denialist Web site www.aidstruth.org, Gallo talked of his amazement at the "mass ignorance coupled with the grandiosity of selling themselves as experts" displayed by the Perth Group, saying that "it would be like us arguing with Niels Bohr on quantum mechanics" (Gallo, 2007a).

The only active AIDS denialist with any major scientific standing is Duesberg, who is a member of the National Academy of Sciences and the first person to isolate a cancer gene.[2] But his credibility to speak on AIDS is tarnished by the fact that he has never conducted any scientific research on HIV, let alone published it in peer-reviewed scientific journals. He simply does not have any evidence to support his erroneous claim that AIDS is caused by recreational and ARV drugs rather than HIV.

Unable to convince his scientific peers, Duesberg relies on the media (including the Internet) to promote his views directly to the public. His cause was assisted substantially by *The Sunday Times* in London from 1992 to 1994, when the science editor ran many long pieces attempting to discredit AIDS science. This enabled Duesberg to achieve a form of socially constructed credibility outside of conventional scientific channels (Epstein, 1996: 105–78), which, in turn, prompted John Maddox (then editor of *Nature*) to go on the offensive and subject *The Sunday Times* during this period to regular critical review in Nature.

Largely as a consequence of Duesberg's profile, the scientific community was compelled to pay greater attention to his ideas than was warranted by their content. In 1991, Gallo (1991: 287–97) devoted ten pages of his book on discovering HIV to demolishing Duesberg's speculations. A couple of years later, Science investigated Duesberg's claims and concluded that none of them stood up to scrutiny (Cohen, 1994). Undaunted, Duesberg and his colleague David Rasnick restated their long-refuted hypotheses in a 1998 article (which was followed immediately in the same journal by a point-by-point refutation [Galea and Chermann 1998]). None of this had any impact either on Duesberg or on journalists such as Farber, who continued to promote his views, largely unchanged from the early 1990s.

AIDS scientists are understandably baffled by such conviction-driven refusal to accept the implications of the weight of evidence to the contrary. As Gallo said of Duesburg in 1988, he is "like a little dog that won't let go" (quoted in Cohen, 1994: 1644). Moore (1996) went even further, comparing Duesberg to the Black Knight from Monty Python and the Holy Grail who, after having all his limbs hacked off by his opponent, keeps on trying to fight with his teeth.

One of Duesberg's tactics is to exploit the uncertainty that is ever-present in science and demand increasingly exacting standards of "proof," and, when this is not forthcoming, proclaim fallaciously that the alternative hypothesis must be true. As Maddox observed

> Duesberg has not been asking questions or raising questions he believes should be answered, but has been making demands and implying (but sometimes saying outright) to colleagues, "Unless you can answer this, and right now, your belief that HIV causes AIDS is wrong." It is as if a person were to have told

Schrödinger in 1926, "Unless you can calculate the spectrum of lithium hydride, quantum mechanics is a pack of lies" (interestingly, that deceptively simple question is only now being answered). (Maddox, 1993: 109)

This kind of fallacious reasoning is evident among other kinds of denialists, too, such as evolution deniers who see any gap in the fossil record as proof that God must have created the world (Mooney, 2005).

Their zealous attachment to key ideas has a further consequence: the inability or refusal of AIDS denialists to weigh up risks and benefits. Thus, as soon as any toxicity can be shown for an ARV drug in any context, they conclude that the drug should not be prescribed in any situation. For example, when clinical evidence emerged that adverse events occurred among mothers on long-term Nevirapine therapy, this was seized upon by Farber (2006) to argue that Nevirapine should never be used in any circumstances—even as a single dose to prevent maternal transmission of HIV, a drug regimen that had been shown to be safe. When this error was pointed out (Gallo et al., 2006), the AIDS-denialist group "Rethinking AIDS" backed Farber's strategy on the spurious grounds that it moved "neatly" between the two trial results as part of a single argument against Nevirapine (Rethinking AIDS, 2007). They claimed, without any evidence, that both trials showed significant adverse events, when, in fact, not a single life-threatening event has ever been shown for single-dose Nevirapine.

All "debates" with AIDS denialists end up in a stalemate simply as a consequence of their refusal to play by the rules of reasonable debate. This is evident in the "rapid-responses" Web pages of the *British Medical Journal* (BMJ), where AIDS denialists such as Papadopulos-Eleopulos and Rasnick accounted for a disproportionate amount of space before the BMJ revised its rules and excluded this "shouting match of the deaf" (Butler, 2003). Typically, the denialists would paste large amounts of convoluted text into their rapid-response submissions and then argue at length with anyone who responded. After trying to engage with the denialists, Peter J. Flegg, a physician from Blackpool Victoria Hospital, finally erupted with the following:

> What is taking place on this forum is a farce, not a debate . . . . Good scientists are meant to accept new evidence and incorporate this into their hypotheses. The denialist approach is to ignore new evidence that is contradictory to their predetermined stance. After comprehensive rebuttal of any point of view, the denialist tactic is to quickly switch to a different topic. Then later, when no-one is looking, they can switch back to the original theme, hoping no-one will realise that these points were completely discredited on an earlier occasion. (Flegg, 2003)

Exactly the same tactics are evident on science blogs when AIDS denialists enter into "debate." Tara C. Smith's science blog "Aetiology" hosted several rambling and ultimately unproductive interactions with AIDS denialists—most notably Harvey Bialy (Duesberg's biographer and fellow member of Mbeki's AIDS panel). The denialists conceded nothing, not even when the case was clearly an open-and-shut one to any reader. For example, AIDS denialists persistently cite an old study by Nancy Padian that reported low HIV transmission rates between sexual partners (Padian et al., 1997) as supporting evidence for their claim that HIV cannot be transmitted sexually. When reminded that participants in the Padian study were strongly counseled to practice safe sex (which means that the study cannot be used to back the claim that HIV, per se, is difficult to transmit sexually) and when presented with evidence from other studies showing that the risk of sexual transmission can be 20 percent or higher in developing countries, the denialists simply changed the topic. This prompted Chris Noble to comment:

> Well, we seem to have drifted a long way from the *famous Padian study* which according to Harvey Bialy "*demonstrated so well that sexually transmitted HIV was a figment.*"
>
> I note that Bialy never once made a comment that was relevant to the study. These are the people that claim that HIV cannot possibly cause AIDS. You ask them for justification and they give you the "Padian study."
>
> You demonstrate that this study cannot be used to conclude that HIV is not sexually transmitted and they go all silent, bring up other studies or in Bialy's case proceed to insult everyone that doesn't worship Peter Duesberg.
>
> I predict that in the future the exact same people will again cite the "Padian study" as proof that HIV is not sexually transmitted. (Noble, 2006)

Exactly as he predicted, the denialists continued to misrepresent Padian's study (see Farber quoted in Kruglinski, 2006 and Turner, 2006: 13–14) and, even when Padian herself protested about the way that AIDS denialists have misused her work and ignored the available evidence (Padian n.d.). The denialists dismissed her piece as "info-ganda" (George, 2006).

This lack of respect for the integrity of scientists makes it very difficult for AIDS scientists to make any headway. As Brian Foley, a scientist who works with the HIV database at Los Alamos National Laboratory, commented after a long blog exchange with South African AIDS denialist Anita Allen:

> There is no such thing as "scientific debate" really. Science is about experiments, data and theories to explain the data. If Anita says "The virus has never been isolated" and I say "In fact dozens of infectious molecular clones of HIV-1 have been generated and that is as good as "isolation" gets

for retroviruses," one of us has to be lying. (Foley, 2006)

Foley's comments point to the central role of integrity and respect for expertise in science. He is saying that for Allen, who is not a scientist, to claim that HIV does not exist amounts to her accusing him of misunderstanding or lying about the vast HIV databank he has at his fingertips. For him, her refusal to accept the mountain of evidence (and his bona fides to report it) amounts to her opting to believe—and propagate—lies. As far as the scientific community is concerned, the "debate" over whether HIV causes AIDS has long been settled. As the AIDS scientists and activists who run the Web site www.aidstruth.org put it:

> For many years now, AIDS denialists have been unsuccessful in persuading credible peer-reviewed journals to accept their views on HIV/AIDS, because of their scientific implausibility and factual inaccuracies. That failure does not entitle those who disagree with the scientific consensus on a life-and-death public health issue to then attempt to confuse the general public by creating the impression that scientific controversy exists when it does not. (AIDSTruth, 2007)

Unfortunately, President Mbeki was precisely one of those who was convinced that a scientific controversy existed—and, by slowing the rollout of ARVs in the public sector for both HIV prevention and AIDS treatment, his belief resulted in the loss of many thousands of lives (Nattrass, 2007). He has also been associated with Christine Maggiore, the controversial HIV-positive American AIDS denialist who does not practice safe sex and campaigns actively against the use of ARVs (Moore and Nattrass 2006). When Maggiore was pregnant with her second child, she was featured on the cover of *Mothering* magazine with "no AZT" emblazoned across her abdomen. She did not take ARVs to prevent infecting her baby with HIV and increased the risk of transmission yet further by breastfeeding the child. Tragically, her daughter died three years later of what the Los Angeles coroner attributed to AIDS-related pneumonia (Ornstein and Costello, 2005). Maggiore, however, continues to deny that HIV had anything to do with the death, claiming instead that the child died because of an allergic reaction to an antibiotic, despite substantial evidence to the contrary (Bennett, 2006).

People in positions of authority, be they statesmen like Mbeki or parents like Maggiore, hold the lives of others in their hands. For them to reject science in favor of AIDS denialism is not only profoundly irresponsible but also tragic. But responsibility for unnecessary suffering and death rests also with the AIDS denialists who promote discredited and dangerous views and encourage people to reject scientifically tested treatments.

NOTES

1. See www.virusmyth.net/aids/index/jsonnabend.htm.

2. Another AIDS denialist with scientific credentials is Kary Mullis, who won a Nobel Prize in chemistry for inventing the polymerase chain reaction. However, he, too, has never done any scientific research on HIV or AIDS and, unlike Duesberg, is not active in the AIDS denialist movement. His autobiography (Mullis, 1998) documents his skepticism about the relationship between HIV and AIDS as well as his encounters with aliens and his belief in flying saucers and astrology.

REFERENCES

AIDSTruth. Answering AIDS denialists and AIDS lies. AIDSTruth Web site. 2007. Available at: http://www.aidstruth.org/answering-aids-denialists.php.

Angell, Marcia. *The Truth about Drug Companies: How They Deceive Us and What to Do about It.* New York: Random House, 2005.

Bennett, Nicholas. "A report on Eliza-Jane Scovill's Death, in rebuttal to that of Mohammed Al-Bayatti." 2006. Available at: http://catallarchy.net/blog/wp-content/images/A_report_on_Eliza_Ver2.pdf.

Brocklehurst, Peter. "Interventions for reducing the risk of mother-to-child transmission prevention of HIV infection." *Cochrane Database of Systematic Reviews* 2006(2). Available at: http://www.mrw.interscience.wiley.com/cochrane/clsysrev/articles/CD000102/pdf_fs.html.

Butler, Declan. "Medical Journal under Attack as Dissenters Seize AIDS Platform." *Nature* 426 (November 2003): 215,

Cohen, Jon. "The Duesberg Phenomenon." *Science* 266 (December 9 1994.): 1642–49;

"Pharmanoia: Coming to a Clinical Trial Near You." *Slate.* (February 21 2006). Available at: http://www.slate.com/id/2136721/.

Duesberg, Peter, and David Rasnick. "The AIDS Dilemma: Drug Diseases Blamed on a Passenger Virus." *Genetica* 104 (1998): 85–132.

Duesberg, Peter, Claus Koehnlein, and David Rasnick. "The Chemical Bases of the Various AIDS Epidemics: Recreational Drugs, Anti-viral Chemotherapy and Malnutrition." *Journal of Bioscience* 28, no. 4 (2003): 383–412.

Epstein, Stephen. *Impure Science: AIDS, Activism and the Politics of Knowledge.* Berkeley, CA: University of California Press, 1996.

Farber, Celia. "Out of Control: AIDS and the Corruption of Medical Science." *Harper's Magazine.* (March 2006): 37–52.

Flegg, Peter J. *Letter* (Rapid Response): HIV/AIDS—There is no "Debate." (March 3 2003.) Available at: http://www.bmj.com/cgi/eletters/326/7381/126/e#30113.

Foley, Brian. Comment following Media's Manto-bashing has undermined important nutrition message (*Akhona Cira, JournAIDS* [blog], August 11). September 3 2006.

Available at: http://www.journaids.org/blog/2006/08/11/media%e2%80%99s-manto-bashing-has-undermined-important-nutrition-message/#comments.

Galea, Pascal, and Jean-Claude Chermann. "HIV as the Cause of AIDS and Associated Diseases." *Genetica* 104 (1998): 133–42.

Gallo, Robert. *Virus Hunting: AIDS, Cancer and the Human Retrovirus: A Story of Scientific Discovery.* New York: Basic Books, 1991.

Gallo, Robert, et al. "Errors in Celia Farber's March 2006 article in Harper's Magazine." 2006. Available at: http://www.aidstruth.org/harper-farber.php#a1 and http://www.tac.org.za/Documents/ErrorsInFarberArticle.pdf.

Gallo, Robert. E-mail message to the author and others. February 11 2007. *Testimony to the Australian Court of Criminal Appeal in the Andre Parenzee case.* February 12 2007b. Available at: http://aras.ab.ca/articles/legal/Gallo-Transcript.pdf.

"George." Comment following The *Padian* waffle (Hank Barnes, *You Bet Your Life* [blog], August 9). August 10 2006. Available at: http://barnesworld.blogs.com/barnes_world/2006/08/more_on_african.html.

Goozner, Merrill. *The $800 Million Pill: The Truth behind the Cost of New Drugs.* Berkeley, CA and Los Angeles, CA: University of California Press, 2004.

Hurley, Dan. *Natural Causes: Death, Lies, and Politics in America's Vitamin and Herbal Supplement Industry.* New York: Broadway Books, 2006.

Kruglinski, Susan. "Questioning the HIV Hive Mind: Interview with Celia Farber." *Discover* 19 (October 2006). Available at: http://www.discover.com/web-exclusives/celia-farber-interview-aids/?page.

Le Carré, John. *The Constant Gardener.* New York; London; Toronto; and Sydney, Australia: Pocket Star Books, 2001.

Maddox, John. "Has Duesberg a right of reply?" *Nature* 363 (May 13 1993): 109.

Mbeki, Thabo, and Peter Mokaba. "Castro Hlongwane, Caravans, Cats, Geese, Foot and Mouth Statistics: HIV/AIDS and the Struggle for the Humanisation of the African." Circulated to ANC branches (2002): 1–132. Available at: www.virusmyth.net/aids/ data/ancdoc.htm (Note: This document was produced anonymously. However, it was circulated in the ANC by Peter Mokaba, and the document's electronic signature links it to Mbeki—and, hence, Mbeki is widely believed to be the primary author.)

Mooney, Chris. *The Republican War on Science.* New York: Basic Books, 2005.

Moore, John. "A Duesberg adieu!" *Nature* 380 (March 28 1996): 293–94.

Moore, John, and Nicoli Nattrass. "Deadly Quackery." *New York Times.* June 4 2006.

Mullis, Kary. *Dancing Naked in the Mind Field.* New York: Vintage Books, 1998.

Nattrass, Nicoli. *Mortal Combat: AIDS Denialism and the Struggle for Antiretrovirals in South Africa.* Pietermaritzburg, South Africa: University of KwaZulu-Natal Press, 2007.

Noble, Chris. Comment following Discussion of the *Padian* paper (Smith, Tara C., *Aetiology* [blog], February 23). March 1 2006. Available at: http://scienceblogs.com/aetiology/2006/02/discussion_of_the_padian_paper.php.

Ornstein, Chris, and Dan Costello. "A Mother's Denial, a Daughter's Death." *Los Angeles Times.* September 24 2005.

Padian, Nancy. "Heterosexual transmission of HIV." *AIDSTruth.org* (Web site). Available at: http://www.aidstruth.org/nancy-padian.php.

Padian, Nancy. N.d., Stephen C. Shiboski, Sarah O. Glass, and Eric Vittinghoff. "Heterosexual Transmission of Human Immunodeficiency Virus (HIV) in Northern California: Results from a Ten-Year Study." *American Journal of Epidemiology* 146, no. 4 (1997): 350–57.

Presidential AIDS Advisory Panel (PAAP). "A Synthesis Report of the Deliberations by the Panel of Experts Invited by the President of the Republic of South Africa, the Honourable Thabo Mbeki." Available at: http://www.info.gov.za/otherdocs/2001/aidspanelpdf.pdf.

Rethinking Aids. "Correcting Gallo: Rethinking AIDS responds to Harper's 'out of control' critics" (Item #6: Comparing Clinical Trial PACTG 1022 to HIVNET 012). September 27 2006.. Available at: http://www.rethinkingaids.com/GalloRebuttal/Farber-Gallo-06.html.

Simon, Viviana, David D. Ho, and Quarraisha Abdool Karim. "HIV/AIDS Epidemiology, Pathogenesis, Prevention and Treatment." *Lancet* 368 (August 5 2006): 489–504.

Smit, Colette, the CASCADE Collaboration, et al. "Effective Therapy has Altered the Spectrum of Cause Specific Mortality Following HIV seroconversion." *AIDS* 20, no. 5 (2006): 741–49.

Sonnabend, Joseph. "Honouring with Pride, 2000: Honouree: Joseph Sonnabend." Available at: http://www.amfar.org/cgi-bin/iowa/amfar/record.html?record=22.

Turner, Valendar. *Affidavit* (in the Andre Parenzee case, Australia). 2006. Available at: http://garlan.org/Cases/Parenzee/Turner-Affidavit.pdf.

UNAIDS. "AIDS Epidemic Update, December 2006." 2006. Available at: http://data.unaids.org/pub/EpiReport/2006/2006_EpiUpdate_en.pdf

GLOBAL CLIMATE CHANGE
TRIGGERED BY GLOBAL WARMING

Stuart D. Jordan

This paper will offer compelling evidence from a large body of research that global climate change caused by global warming is already underway and requires our immediate attention. The research in question appears in refereed scientific literature, and most of it reflects a broad consensus of the worldwide climatology community. The principal points of this position paper are summarized below and are considered in detail, with supporting references, in the text that follows.

Convincing evidence that Earth's climate is undergoing significant, and in some cases alarming, changes has accumulated rapidly in recent years, especially during the past three decades.

The conclusion that there is significant warming of Earth's surface is not based primarily on theoretical models, although these models do succeed in replicating the existing database with growing success. Instead, global warming is a fact confirmed by an enormous body of observations from many different sources. Indeed, the focus of research has now shifted from attempts to establish the existence of global warming to efforts to determine its causes.

Although the exact extent of harm from global warming may be difficult to predict now, it can be said with confidence that the harmful effects of global warming on climate will significantly outweigh the possible benefits.

The probability is extremely high that human-generated greenhouse gases, with carbon dioxide as the major offender, are the primary cause of well-documented global warming and climate change today.

Much can be done *now* to mitigate the effects of global warming and the associated climate change. Difficulties in addressing the problem are not caused primarily by unavailable technology, but by the lack of sufficient incentives to implement the new technologies more aggressively.

After consideration of these points, the paper will end with a brief analysis of the role of the political process in addressing these issues. No detailed recommendations will be made, but some general suggestions will be offered. This final section will argue that any solution to this major global problem will require contributions from all major elements of our society, from the academic research community to American industry. Getting the science right is the critical first step, but implementation of solutions will need more broadly based cooperation that takes economic realities *and opportunities* into account. We will end this paper on that note, expressing the view that without a determined political effort, a successful attack on climate change is unlikely to happen soon.

INTRODUCTION

Convincing evidence that Earth's climate is undergoing significant—and in some cases alarming—changes has accumulated rapidly in recent years, especially during the past three decades.

Most climatologists regard the final decade of the twentieth century as the warmest in the past millennium (Albritton et al., 2001). Recently, there have been observations suggesting that 2005 may have been the warmest year on record. Rapid temperature increases in the Arctic have already produced a significant reduction in Arctic sea ice (Fisher et al., 2006). Reduced sea ice decreases the reflection and increases the absorption of sunlight in the Arctic, an almost "runaway" process that further amplifies global warming. Recent research has demonstrated that the Earth's energy budget is out of balance, with more energy captured from the sun than is currently radiated back to space (Hansen et al., 2005). This makes global warming inevitable. This is not due to any measurable increase in the incoming solar energy, but to an increase in the amount of that energy captured and retained by Earth.

Glaciers are melting rapidly (Albritton et al., 2001), and the net loss of ice to seawater from Greenland (Chen, Wilson, and Tapley, 2006; Dowdeswell, 2006) and, more recently, from Antarctica (Alley et al., 2005; Hodgson et al., 2006;

Overpeck et al., 2006) is alarming. Measured increases in sub-ice-sheet meltwater are lubricating and accelerating the flow of this ice into the oceans. Not only does this promote mean (average) sea level rise, which has already begun, it could eventually modify current patterns of global ocean circulation, an important regulator of climate. The oceans are also a major sink for removing carbon dioxide from the atmosphere (Sarmiento and Wofsy, 1999). Because the atmospheric concentration of carbon dioxide continues to increase, a long-established equilibrium seems to have been compromised, and there is evidence that this absorption process is starting to saturate (McAvaney et al., 2001).

The effects of these major changes in the oceans, which we know have occurred in previous paleoclimates, would be catastrophic if a significant fraction of the Greenland ice sheet or, far worse, the Antarctic ice sheet were to melt. A simple calculation shows that the frozen water in the Greenland ice sheet alone would, if melted, raise the global sea level by about seven meters (about twenty-two feet). This rise would be independent of the effect of higher temperatures on warming the oceans' surface waters, which, by decreasing their density, would raise the sea level still higher, as has already begun.

While major changes in the oceans could well be the most alarming consequence of global warming for our civilization, many other known harmful effects are well documented. Some of these will be reviewed briefly below in our evaluation of the costs and possible benefits of global warming.

The conclusion that there is significant warming of Earth's surface is not based primarily on theoretical models, although these models do increasingly succeed in replicating the existing database. Instead, global warming is a fact confirmed by an enormous body of observations from many different sources. Indeed, the focus of research has now shifted from attempts to establish the existence of global warming to efforts to assess its causes.

There remains a small number of scientists who claim that the current warming trend may not reflect primarily human activity. We consider their arguments in this paper. In addition, some errors in analysis may have been made by colleagues whose position favors anthropogenic (human-caused) activity driving the changes. The work of these critics is useful, and reflects the way science proceeds and should proceed. However, even these critics no longer deny the reality of global warming and the currently accelerating climate change.

Given many serious consequences certain to follow from increased climate change, we are naturally led to ask three questions relevant to any policy initiatives formulated to address global warming:

1. What are the trade-offs between possible benefits claimed for global warming compared to the known harmful effects?
2. What are the most likely causes of the current rapid rise in the mean global surface temperature?
3. What can be done to mitigate global warming and the climate change that results?

These questions will be addressed in the following three sections of this paper.

POSSIBLE BENEFITS VERSUS KNOWN HARMFUL EFFECTS

Some of those reluctant to initiate programs to address the harmful effects of climate change frequently point out that there may also be benefits from global warming. Since some dislocation of current activities would be inevitable under any major program for change, these possible benefits should be examined and compared to the harmful effects of failing to initiate effective palliative measures. In effect, a rough benefit-to-cost assessment becomes important whenever investment of resources is involved, as is certainly the case here.

AGRICULTURE AT HIGHER LATITUDES

The most commonly cited benefit of global warming is the longer growing season that higher-latitude regions of North America and Asia would enjoy. That warmer temperatures are moving northward in the Northern Hemisphere has been well documented. However, only if rainfall and other conditions conducive to successful agriculture also march north in tandem with temperature is this scenario likely to be realized. This is currently unknown. While several current general circulation models (GCMs) of global climate all agree that global warming will induce different changes in different geographical regions, further work is needed before they can predict with confidence the details of regional changes. We already have two historically recent examples that demonstrate the fragility of allegedly rich lands awaiting development. One is the American Rocky Mountain West, constantly threatened with water shortages and clearly incapable of supporting a large population (in spite of much nineteenth-century hype to the contrary). The other is former Soviet Premier Nikita Khrushchev's "virgin lands" program for transforming central Asia into a vast breadbasket. The effort was largely a failure. In addition, a major dislocation of ecosystems can be expected as new plants and animals move into new regions and out of others. While we cannot say the end result of this process will always be harmful on balance, we can say that the ecosystems involved will all be stressed during these changes, especially if the changes are rapid. From the point of view of ecosystem health, rapid change is extremely unlikely to be beneficial.

INCREASES IN THE RATE OF PLANT GROWTH

The other benefit that is often claimed for global warming is that plants will grow more rapidly. It is further noted that a more rapid rate of plant growth, combined with a higher carbon dioxide concentration in the atmosphere, will partially offset these higher concentrations by removing more of this gas through accelerated photosynthesis. However, these possible benefits must be weighed against other possible effects of accelerated plant growth in currently fallow regions, assuming that local soil and moisture conditions already noted might allow it. This would surely lead to ever-larger human populations in potentially fragile areas, an impact hard to predict but not likely to be beneficial. There is even recent evidence that, while increasing the carbon dioxide concentration of the atmosphere appears to accelerate initial plant growth, the increased growth rate will taper off rather quickly. From these considerations, it is not obvious that there would be any net benefit at all.

DESTRUCTIVE EFFECTS OF RISING SEA LEVEL

Against these possible but still questionable benefits, we need to examine the harmful effects of the climate change that results from global warming. No one denies there will be problems, of which the most dramatic may be the rise in the mean sea level already discussed. For those living in the Washington, D.C., area, it is sobering to visit the museum of the National Academy of Sciences. A hands-on exhibit reveals how much of coastal Maryland will be flooded by the Chesapeake Bay if the mean sea level rises by 0.5 meters, by 1.0 meter, and by 1.5 meters. A rise of 0.5 meters is approximately 1.6 feet. For reference, the measured rise in the mean global sea level that has already occurred in the twentieth century is on the order of half a foot (Church et al., 2001). Since surface ocean water at current temperatures expands when the water temperature rises, global warming makes some coastal flooding inevitable, with potentially dire consequences for many highly populated areas located near sea level. Melting of glaciers and icecaps further accelerates the process.

Beyond these effects that are already taking place lies the risk of catastrophic sea level rise that substantial melting of the Greenland and Antarctic ice sheets would inevitably produce. To a complete melting of the Greenland ice sheets, leading to a sea level rise of approximately twenty-two feet, a complete melting of Antarctic ice would add an additional rise of almost 200 feet (Church et al., 2001). No one predicts that human-driven global warming will produce a complete loss of the Antarctic sea ice in the foreseeable future. Nevertheless, it is sobering to recall that, to the already noted accelerating melt of the Greenland ice sheet, recent evidence shows that snowfall in the interior of Antarctica, formerly cited to compensate for ice-sheet loss from coastal regions there, is *not* increasing (Monaghan et al., 2006). The significant losses from coastal Antarctic ice sheets means there is a net transfer of ice into sea water at high southern latitudes as well (Cook et al., 2005). Even a partial loss of Antarctic ice will result in a significant rise in sea level, with resulting inundation of coastal regions.

THE SPECTER OF CATASTROPHIC CLIMATE CHANGE

There is another potentially catastrophic problem that could result from the impact of global warming on the world's oceans. This is a possible change in the global oceanic conveyor belt, which governs the many surface currents and coupled deep-water flows that are known to have important impacts on major regional climates. The best known and best studied of these is the Gulf Stream, which plays an important role in maintaining temperate conditions in northern Europe. Not all climate scientists agree on how much, if any, reduction in Gulf Stream flow has already begun, or on how important this warm current is to maintaining temperate conditions in northern Europe (Hátán et al., 2005; Kerr, 2005; Seager, 2006). However, some reduction of this flow—called the thermohaline circulation from the driving force—is inevitable if enough less-salty, lower-density meltwater is introduced into the North Atlantic. This is exactly what would result from rapid melting of the Greenland ice sheet, supplemented by further melting of the Arctic sea ice. It is also universally accepted that some cooling in northern Europe would then become inevitable (Stocker et al., 2001).

ONE EXAMPLE OF CATASTROPHIC CLIMATE CHANGE

What makes the well-studied period known as the "Younger Dryas" so disturbing is what it tells us about a sudden, devastating cold period that occurred after Earth began to emerge from the last Ice Age. The Younger Dryas is a period of approximately one thousand years duration, centered around 12,000 B.C. During the inception of the Younger Dryas, the mean global surface temperature—known from sea-core sediments taken from many ocean basins around the globe—dropped over ten degrees C in a few decades. Furthermore, while the geological evidence is still under investigation, a large-scale picture of the Younger Dryas is emerging. As the Laurentian Ice Sheet that formerly covered much of North America was receding, a huge freshwater lake, Lake Agassiz, formed west of what is today Lake Superior. As long as the frigid waters of this lake drained through the current Mississippi Valley into the Gulf of Mexico, the situation was stable. Then, rather abruptly, geological processes in southern Canada appear to have blocked this southern outlet, releasing an enormous flow of cold water through the St. Lawrence Valley into the North Atlantic (Broecker 2006). The effect on suppressing the thermohaline-driven circulation of the global oceanic conveyor belt was sufficiently strong to plunge much of Earth into a new period of drastically colder temperatures for about one millennium. A detailed account of the Younger Dryas, and of how ice-core data and other data can be used to confirm this picture, as well as other climate changes that have occurred since Earth emerged from the last major ice age, is presented with reference to many original sources in the National Academy of Sciences book *Climate Crash* (Cox, 2005).

CLIMATE AS A SEMICHAOTIC SYSTEM SUSCEPTIBLE TO SIGNIFICANT SHIFTS

We cannot assert that the Younger Dryas scenario would necessarily be repeated if global warming causes much of the Greenland ice sheet to melt. There are too many other factors that would need to be considered. One such factor is the rate of melting, since the release of cold water from Lake Agassiz was apparently quite precipitous. However, the Younger Dryas does tell us that dramatic climate change can occur very quickly if some large-scale process triggers a dramatic shift in a parameter the climate depends upon, pushing the climatic system beyond a critical point for stability. Climate is a highly nonlinear, semichaotic system. These systems can respond in hard-to-predict ways to a number of triggering processes that, under critical conditions, dramatically shift them from one equilibrium state to another, a property that mathematicians have already demonstrated in complex nonlinear systems. While this is a highly technical point, it is an extremely important one to grasp for understanding large-scale climate processes. A good description for an intelligent layperson can be found in *Climate Crash*.

The trigger for the Younger Dryas was the eruption of a massive cold water flow into the North Atlantic. A climate crash sharing many common features with the Younger Dryas has occurred even more recently during the allegedly stable Holocene, around 8200 B.C. (Ellison, Chapman, and Hall, 2006). The evidence for this more recent event was obtained from sea bottom sediments that, thanks to new techniques, now offer promise of providing further confirmation of their global character back through the most recent ice age (Nicholson, et al., 2006). The next triggering event could be different. It is difficult to predict if persistent increases in carbon dioxide and other greenhouse gases might induce a sudden major climate shift in the foreseeable future. While the probability would seem to be small today, no one can rule out such a future change with certainty. On timescales relevant to shifts in global equilibria, the rate of change of the mean global surface temperature over the past three decades is sobering.

LESS OBVIOUS HARMFUL EFFECTS OF CLIMATE CHANGE

If we assume that the climate system is sufficiently stable so that no Younger Dryas catastrophe awaits us before corrective measures can be taken, there are still many less harmful effects we can expect that are already underway as a result of global warming. Consider the effect of a modest sea level rise on migrating insectivorous songbirds returning to the United States from tropical America in the spring. Having lost roughly half of their body weight flying over water, these birds are critically dependent on food sources in low-lying coastal regions along the northern Gulf of Mexico. If the swampy areas of southern Louisiana alone were flooded by sea water, the reduction in the population of these birds could be drastic. Agricultural experts estimate that the insects eaten by these birds save American agriculture billions of dollars annually. Addressing bird migration raises an even subtler issue. Advocates of wind power to address global warming are often criticized because large windmills kill birds and bats. However, windmills can be defended to bird lovers by considering the net effect on birds and bats of rising sea levels. The number of migrating birds who perish from this new energy technology may be small compared to the number who will perish from starvation in migration, if these southern coastal areas are flooded by sea-level rise. Of course, there are many other often-overlooked examples of economic hardships associated with sea-level rise. Recently, managers of several American insurance companies announced that they may no longer insure low-lying coastal properties, citing not Hurricane Katrina but instead rising sea level.

Many of the deleterious effects of global warming on climate are not immediately obvious. Others include the increase in pathogens in densely populated mid-latitude regions; documented evidence that storms are likely to become more violent (Emanuel, 2006; Mann and Emanuel, 2006); increasing severity of forest fires (Westerling et al., 2006); the further amplification of global warming from a decrease in Earth's reflection of sunlight resulting from lesser ice and snow cover, especially in the Arctic; and many more subtle effects.

Global warming is inducing instabilities in the Earth's climate that we already know have many harmful effects. What we do not know yet is *how* serious the situation might become. The possibilities range from bad and costly to fix—such as saving many valuable coastal areas from flooding and fighting new and more pervasive diseases—to the much less probable but potentially catastrophic effects of a sudden, significant climate shift that could be difficult to reverse.

Although the exact extent of harm from global warming may be difficult to predict now, it can be said with confidence that the harmful effects of global warming on climate will significantly outweigh the possible benefits.

However, we have yet to address the critical question of what is primarily responsible for the current warming trend. Is it our own anthropogenic activity? Or is it instead, as some have claimed, natural fluctuations that we are helpless to influence and therefore must simply stoically accept and ride out?

LIKELY CAUSES OF THE CURRENT RAPID RISE IN GLOBAL SURFACE TEMPERATURE

Showing that the effects of global warming and climate change are likely to be far more harmful than beneficial is only the first step we must take toward a solution. The question of the causes must now be addressed. Only if the causes can be addressed by determined action on our part is there any hope of mitigating or eliminating the harmful effects. Consequently, we need to examine climate fluctuations due to natural processes over which we may have no control. These have been proposed as the primary drivers of the current rapid global temperature rise. This section examines the arguments for these natural processes, and also for the human (anthropogenic) activities that many scientists believe are the primary causes.

TWO IMPORTANT DEFINITIONS

It is useful to introduce precise definitions of two terms that appear throughout this paper. The *mean global surface temperature* is defined as the average of the air temperature measured at the land surface and of the surface water temperature measured over large bodies of water, statistically weighted by the fact that more measurements are available in some equal-area grid "points" than in others. Since the advent of the space age, these temperatures can now be uniformly measured from satellite instruments over the entire globe. By comparing the results of the more recent space observations with current measurements made in a more traditional way from the ground, or from ships and anchored sensors on the oceans, it has been possible to determine the accuracy of earlier ground-based and sea-based observations, which proved to be surprisingly good in most cases.

The other term we use frequently here is *climate*. Most people realize that climate and weather differ mainly in the timescales involved. The timescale for climate used by many contemporary climatologists is of the order of thirty years, while weather is a daily phenomenon that we know can change abruptly within hours. The recent advent of paleoclimatology based on deep ice-core studies has somewhat shaken the idea of gradual climate change. As discussed above, we now know that Earth's global climate can become rapidly unstable under certain conditions. Indeed, changes of over ten degrees C in one or two decades have occurred many times over the past 400,000 years (Folland et al., 2001). This has now been extended to 600,000 years and is discussed further in the just-released IPCC-2006 Science Report. It is only somewhat reassuring that ice-core studies suggest that the global climate system has probably been more stable over the past 10,000 years (the Holocene) than at any time over the 600,000–year record. Yet even during this relatively stable Holocene epoch, there is evidence that regional climate change was an important factor in the rise and fall of several early historical civilizations (Cox, 2005). Perhaps the best way to define *global climate* is the average of the global weather, averaged over a time interval appropriate to the rate at which the climate is changing. That is the viewpoint we have adopted in this paper, using a definition of an inherently dynamic phenomenon that emphasizes our primary concern, which is *climate change*.

With this background we now examine the question of what is likely to be mainly responsible for the current rapid rise in the mean global surface temperature. We begin by discussing what is almost certainly *not* causing it.

MIGHT THE SUN BE DRIVING GLOBAL WARMING?

We noted earlier that the sun is not responsible for the current rapid global temperature rise. (We emphasize here that these remarks refer to the last several decades. The role of the sun over longer epochs remains an area of active research.) Three hypotheses have been advanced for solar driving of the current rapid temperature rise.

One speculates that there is a small but persistent rise in the solar flux at the minimum phase of solar sunspot activity, when the small periodic increase in flux due to this activity is absent. So far, measurements from space lack the accuracy to establish the precise value, but they do put an upper limit on any possible increase, which is far too small to explain the observed terrestrial temperature rise. In addition, there are other solar observers who infer from other sensitive data that there is no secular increase in this flux at all, over the last three decades of measurements (Hudson, 2004).

A second solar hypothesis is that the tiny energy increase in the ultraviolet radiation at the peak of the sunspot cycle produces enough change in the upper layers of Earth's atmosphere that, properly coupled through planetary waves with the weather-producing lower layers, it will exert a sufficient effect on the lower atmosphere (the troposphere) to produce cyclic and increasing global warming. This hypothesis demands that there be a significant and persistent increase in the level of solar activity, hence ultraviolet radiation, over recent solar cycles, and that a strong periodic signature of the solar cycle be seen in this process. Neither has been observed (North, Wu, and Stevens, 2004), nor is there any evidence to date for a coupling mechanism nearly strong enough to support this view.

The final hypothesis involves the modulation of intergalactic cosmic rays by the solar-cycle-dependent interplanetary magnetic field. Reductions in this magnetic field during the minimum of solar activity increases cosmic-ray penetration into Earth's atmosphere. It was postulated that, during this period of increased cosmic ray penetration, cloud formation would be enhanced through the "cloud chamber" effect well known to physicists in the high-energy laboratory. An early, surprisingly high positive correlation of increased cloud formation with enhanced cosmic-ray flux stimulated interest in this possibility, but further work with a more extensive database has negated it. This was not surprising, given a large number of other problems with this hypothesis too technical to review here.

In summary, there is no evidence that the sun is responsible for current global warming, and a great deal of evidence to suggest otherwise. Solar forcing of the current global warming is extremely unlikely (Ramaswamy et al., 2001).

INTERNAL MODES OF THE CLIMATE SYSTEM

Another possibility that has been tentatively proposed is that some long-term quasi-periodic process confined to Earth itself is responsible. To date, no one has provided a plausible, testable mechanism for such Earth-bound forcing. While there is speculation that a very large-scale phenomenon such as the El Niño Southern Oscillation (ENSO) might play a role in quasi-periodic global warming and climate change, there is no convincing evidence to support this. In the assessment of the IPCC-2001 Science Report, the likelihood that internal modes of the climate system are causing global warming and influencing the current climate change range from unlikely to extremely unlikely (Stocker et al., 2001).

THE 'URBAN HEAT-ISLAND' EFFECT

S. Fred Singer has suggested that the urban heat-island effect may be responsible for global warming (Singer, 2002). Because large urban areas are warmer by several degrees than their surroundings, Singer suggests that, when these effects are averaged into a computation of the global temperature, with proper allowance for the higher density of records in the urban areas, the increase in the global temperature can be explained. This is not correct. The work of a number of scientists that has appeared in the refereed literature and which takes into account the relative density of observing stations shows that, while the urban heat-island effect is indeed large in major urban centers, its effect on the mean global surface temperature is almost negligible (Folland et al., 2001).

ARE CLIMATE SCIENTISTS BIASED TOWARD ANTHROPOGENIC FORCING?

In contrast to the legitimate research that raised the possibility that either the sun or internal modes of the climate system might be responsible for global warming and climate change, there is a final suggestion one occasionally hears that reflects a serious ignorance of how science proceeds. This is the speculation that the work of the great majority of climate scientists who support anthropogenic forcing of global warming and climate change is motivated primarily by a desire to get more support for their research. Thus, it is further suggested, they are introducing a strong bias into the way they report their results.

In reply to this, it is worth noting that, aside from proving a new hypothesis, there is nothing that will enhance a scientist's reputation more than disproving one that has gained some notoriety. Generally speaking, the better the scientist, the more likely personal pride will motivate him or her to proceed within the "rules." The idea that there is an unwritten self-serving understanding among climate scientists to falsely exaggerate the implications of their work for financial gain reflects widespread ignorance of how science proceeds.

EARTH'S ORBITAL DYNAMICS

While the Milankovich effect, named after its proponent, is now regarded as the likely cause of the ice ages, these changes in the eccentricity of Earth's orbit around the sun and in Earth's spin-axis orientation occur over far too long a duration to explain the relatively rapid global warming now occurring (Morell, 2004). Also, since we know that the next major changes due to these processes will be to enter a new cold period, the present trends are of the wrong kind to be explained by Earth's dynamics. To date, no convincing case can be made that global warming is caused by natural processes over which we may have no control in the foreseeable future.

Since the case for natural forcing of the current rapid global temperature rise is weak, we must ask what human activities are likely to be major causes. Here a near consensus has arisen among most climatologists that human-rated greenhouse gases are the main culprit. While the scientific arguments supporting this position are now known to most knowledgeable people, it is useful to repeat them here in the interest of presenting a complete contemporary picture of global warming and climate change.

THE GREENHOUSE EFFECT

There is complete agreement among research scientists whose work appears in the refereed literature that the "greenhouse effect" is a real physical process that is contributing to global warming. The greenhouse effect occurs because the sun's relatively short-wavelength radiation from the hot solar surface passes through Earth's atmosphere and reaches ground level to be absorbed by

Earth, where the surface temperature is much lower than the sun's. Much of the longer-wavelength infrared radiation emitted by the cooler Earth is absorbed by Earth's atmosphere. Thus, a part of the sun's radiant energy is trapped by Earth's atmosphere. If the amount of incoming solar radiation that is absorbed and trapped exceeds the amount that is reflected and re-radiated into space, there is an imbalance, and Earth's temperature will begin to rise. The excess amount will warm the atmosphere until a higher equilibrium temperature is reached, at which the total rate of energy loss from Earth to space equals the amount that is retained by Earth.

Indeed, thanks to the greenhouse effect, our atmosphere has had equilibrium temperatures (within a limited range, determined by many different processes) sufficient to sustain, ultimately, human life. If Earth had no atmosphere, the mean global surface temperature would be approximately fifty degrees centigrade colder!

The problem of the current global warming occurs because of the aforementioned net imbalance in Earth's energy budget, with more solar energy being retained than is radiated back into space. This is due to the increase in greenhouse gases produced by human activity since the beginning of the Industrial Revolution, a process that continues today with an ever-increasing atmospheric concentration of the culprit gases (Prentice et al., 2001; Sarmiento and Gruber, 2002). This situation makes global warming and some associated climate changes inevitable.

THE DOMINANT ROLE OF CARBON DIOXIDE TODAY

Greenhouse gases produce this imbalance when their concentration in the atmosphere is increased. The role of the most-offending greenhouse gas, carbon dioxide, will illustrate the process. Because carbon dioxide gas is a major by-product of burning the fossil fuels that power much of our industry and transportation, its production and release into the atmosphere continue to increase today. Carbon dioxide is an efficient absorber of the infrared radiation emitted by Earth, and increasing its concentration in our atmosphere increases the rate at which Earth's infrared radiation is trapped. Moreover, by warming the atmosphere above the previous equilibrium temperature, more water vapor is produced, which is the most effective greenhouse gas of all (Soden et al., 2005). The warming associated with these processes is further amplified by deforestation, which reduces the capacity of the land to absorb atmospheric carbon dioxide. Since the land is the only other major natural sink for this greenhouse gas, along with the oceans, deforestation contributes significantly to global warming (Sarmiento and Wofsy, 1999).

Other warming processes are set in motion by those stimulated by an increased concentration of atmospheric carbon dioxide. Some of these could produce "runaway" phenomena until they go to completion. In the first part of this article, we cited the effect of reducing the bright reflecting ice cover in the Arctic, leading to greater absorption of solar radiation by the less reflecting water surface, thus inducing still more melting. In this case, the process could continue until the Arctic Ocean is entirely ice-free. Methane is an even more efficient greenhouse gas than carbon dioxide. Melting of the permafrost at high latitudes could release large quantities of methane into the atmosphere, and Arctic regions are warming most rapidly of all. While the consequences of methane release are still being debated in the scientific literature, this and other processes triggered by continued global warming are now under investigation, as their cumulative effect could be extremely harmful (Prather et al., 2001).

Once carbon dioxide gas is released into our atmosphere, it stays there for well over a century. This means that even if we could freeze the release of these greenhouse gases at the current rate of production their concentration in our atmosphere will continue to increase for at least several decades. This fact, more than any other, may be the reason some climatologists are genuinely alarmed, especially in light of the rapid industrialization of China and India with their huge populations yearning for better lives, coupled with the apparent reluctance of some influential circles in our own country to take global warming and climate change seriously.

Finally, there is one more disturbing result emerging from research on carbon dioxide in the Earth's atmosphere. As its atmospheric concentration increases, so does its partial pressure over the oceans and the ability of the surface layers of the ocean to absorb it. Current research suggests that the oceans are already beginning to saturate in their ability to take up atmospheric carbon dioxide at the current rate (Sarmiento and Wofsy, 1999). As noted above, these investigators have also shown that the oceans are one of the two most effective sinks for removing carbon dioxide from the atmosphere, along with bioprocesses on land.

Thus, from almost every point of view, the increasing atmospheric concentration of carbon dioxide is considered by most climatologists to be the greatest problem we face in addressing global warming and global climate change today.

THE RELIABILITY OF CLIMATE MODELS AND THEIR PREDICTIONS

All major computer models that calculate a mean global surface temperature in response to the input of many physical processes agree on the importance of greenhouse gases, especially carbon dioxide, in driving global warming and climate change. However, since these models alone permit us to predict future climate change, we need to ask how reliable their predictions are. The best of these models, the GCMs (general circulation models), are still often criticized for many reasons, even with the advent of faster computers and a rapid growth of sophisticated modeling. Some of this criticism is undoubtedly justified, which is why this paper began by emphasizing that evidence for current global warming is based primarily on an enormous amount of hard data—carefully made observations taken over the entire globe—and not primarily on models. In addition, there may be no climate modeler anywhere today who would claim that there exists a current model that has a sufficiently fine grid to predict detailed local climate patterns over the globe as the climate changes.

However, some relatively recent problems with climate models have been at least partially solved, removing some of the most common criticisms. Several of the GCMs have now been refined to the point of replicating the past temperature history of Earth remarkably well, when their output is compared to either direct temperature measurements or reasonable proxies for estimating past climate conditions. An excellent review of climate models, as of the year 2000 C.E., is given in McAvaney et al. (2001) and was updated in the IPCC-2006 Science Report.

One of the problems that current research is beginning to address is the influence of aerosols on how much global warming occurs. The IPCC-2001 Science Report (Penner et al., 2001; Ramaswamy et al., 2001) discusses aerosols extensively and notes that their role in global warming—outside of the tropospheric cooling effect of large volcanic eruptions—is much less understood than that of greenhouse gases. This becomes significant when we consider that recent research suggests a larger role for aerosols in general and that this role is likely to contribute to a net *cooling* (Bréon, 2006). During the middle of the twentieth century, the mean global surface temperature failed to increase, contrary to what the effect of increasing greenhouse-gas emission led many scientists to expect. However, industrial aerosols were also increasing at a rapid rate during the same period, until they became strongly regulated due to concerns about "acid rain." Inclusion of strong aerosol cooling would clearly tend to counterbalance the normal temperature increase from greenhouse gases alone. Additional work showed that a decrease in the amount of solar radiation reaching Earth's surface, called *global dimming*, coinciding approximately with the aforementioned flat temperature profile, ended in the late 1980s, and now we are experiencing *global brightening* along with a rapid global temperature rise (Wild et al., 2005). (It is important to note that this is not an increase in the solar flux impinging on the top of our atmosphere but only on the amount that reaches the ground.) If we consider these two recent developments together, much of the apparent mystery of the flat temperature profile in the mid-twentieth century appears to go away. Current work on models is exploring this possibility, as our understanding of aerosols improves.

Until recently, another puzzle for GCM modelers was a disagreement between their model predictions and what appeared to be reliable observations showing that temperatures in the troposphere did not increase during the current period of a rapid rise in the mean global surface temperature. The models predicted the two temperatures should march together. However, a serious error was found in the calibration of the satellite data used for the tropospheric temperature measurements. When this error was corrected, it brought the two temperatures much closer together (Mears and Wentz, 2005). Confidence in the models correspondingly increased.

While further work is required to improve climate models, it is clear that they are improving rapidly. Future climate predictions, while still only valid within a range, are also improving. A benchmark for the sensitivity of different models to the actual climate has been to compare their outputs by assuming a doubling of the atmospheric concentration of carbon dioxide since the beginning of the Industrial Revolution, a doubling that is now almost universally expected to occur. Between 1995 and the publication of the IPCC-2001 Science Report, there was little increase in the computed range of predicted global-temperature increases under this assumption, and the range was from 1.5 to 4.5 degrees centigrade (1.0 degree centigrade equals 1.8 degrees Fahrenheit). However, when the results of the improved, and generally considered best, three US GCMs were compared in 2002, the range of predicted global temperature increases had been narrowed to 2.5–3.0 degrees centigrade (Kerr, 2004). A rise of 2.5 degrees centigrade may not seem like much, but when we consider the climate changes that have already occurred in response to the measured increase in the twentieth century of 0.6 degrees centigrade (with an estimated uncertainty of 0.2 degrees centigrade), it becomes disturbing.

SUMMARY

No scientist to date has made a strong case (one supported by a large number of his colleagues who publish in the refereed journals) for any observation(s) or mechanism(s) that can explain the current rapid global warming trend by invoking natural causes. Arguments for solar forcing, for forcing by internal modes of the climate system (natural processes that operate within the Earth system itself), and for the urban heat-island effect have either failed to offer hard evidence or have been completely discredited. Nor do Earth's long-duration, quasi-periodic, dynamical motions explain the current rapid temperature rise.

In contrast, global warming forced by a growing atmospheric concentration of greenhouse gases, especially carbon dioxide, is based on sound science that refers to a mechanism that is well understood and universally accepted by the scientific community. Models that incorporate this and many other known processes support this conclusion, and the models themselves, while still in need of improvement, are becoming increasingly reliable for making global predictions.

The probability is extremely high that human-generated greenhouse gases, with carbon dioxide as the major offender, are the primary cause of well-documented global warming and climate change today.

ACKNOWLEDGMENTS

The author thanks Drew Shindell, of the Goddard Institute for Space Studies, for pointing out that there remain unsolved problems in the geology of southern Canada that could modify the current picture of what caused the eruption of floodwater know as the Younger Dryas. He also thanks Judith Pap, of the Goddard Space Flight Center, for reading the manuscript carefully and offering suggestions for clarification based on her expertise in assessing solar irradiance variation. Finally, appreciation is due to Ronald A. Lindsay, staff member and legal advisor to the Center for Inquiry (and now its CEO), for ensuring both readability for an intelligent but not necessarily scientifically trained reader, as well as uniformity of format for the collection of papers of which this is but one.

A Position Paper from the Center for Inquiry Office of Public Policy, Washington, D.C. Reviewing Committee: Paul Kurtz, PhD, Thomas W. Flynn, Ronald A. Lindsay, JD, PhD, Toni Van Pelt—December 2006. The full version of this report is on the CFI/DC web site at http://www.cfidc.org/opp/jordan.html.

REFERENCES

Albritton, D. L., et al. "Technical summary." In *Climate Change 2001: The Scientific Basis.* New York: Cambridge University Press, 2001.

Alley, R. B., et al. "Ice Sheet and Sea-Level Changes." *Science* 310 (2005): 456.

Bréon, F.-M. "How do Aerosols Affect Cloudiness and Climate?" *Science* 313 (2006): 623.

Broecker, W. "Was the Younger Dryas Triggered by a Flood?" *Science* 312 (2006): 1147.

Chen, J. L., C. R. Wilson, and B. D. Tapley. "Satellite Gravity Measurements Confirm Accelerated Melting of Greenland Ice Sheet." *Science* 313 (2006): 1958.

Church, J. A., et al. "Changes in Sea Level." In *Climate Change 2001: The Scientific Basis.* New York: Cambridge University Press, 2001.

Cook, A. J., et al. "Retreating Glacier Fronts on the Antarctic Peninsula over the Past Half-Century." *Science* 308 (2005): 541.

Cox, J. D. *Climate Crash.* Washington, D.C.: Joseph Henry Press, 2005.

Dowdeswell, J. A. "The Greenland Ice Sheet and Global Sea Level Rise." *Science* 311 (2006): 963.

Ellison, C. R. W., M. R. Chapman, and I. R. Hall. "Surface and Deep Ocean Interactions during the Cold Climate Event 8200 Years Ago." *Science* 312 (2006): 1929.

Emanuel, K. "Hurricanes: Tempests in a Greenhouse." *Physics Today.* (August 2006): 74.

Fisher, D., et al. "Natural Variability of Arctic Sea Ice over the Holocene." *EOS* (Transactions of the American Geophysical Union) 87, no. 28 (2006): 273.

Folland, C. K., and T. R. Karl. "Observed Climate Variability and Change." In *Climate Change 2001, The Scientific Basis.* New York: Cambridge University Press, 2001. (See especially boxed summary, p.106).

Hansen, J., et al. "Earth's Energy Imbalance: Confirmation and Implications." *Science* 308 (2005): 1431.

Hátán, H., et al. "Influence of the Atlantic Subpolar Gyre on the Thermohaline Circulation." *Science* 309 (2005): 1841.

Hodgson, D. A., et al. "Examining Holocene Stability of Antarctic Peninsula Ice Sheets." *EOS* 87, no. 31 (2006): 305.

Hudson, H. "Solar Energy Flux Variations." In *Solar Variability and Its Effect on Climate Change*, edited by J. Pap, and P. Fox. *American Geophysical Union Monograph #141.* 2004.

Kerr, R. A. "Confronting the Bogeyman of the Climate System." *Science* 310 (2005): 432.

MacAvaney, B. J., et al. "Model Evaluation." In *Climate Change 2001: The Scientific Basis.* New York: Cambridge University Press, 2001.

Mann, E., and K. Emanuel. "Atlantic Hurricane Trends Linked to Climate Change." *EOS* 87 no. 24 (2006): 233.

Mears, C. A., and E. J. Wentz. "The Effect of Diurnal Correction on Satellite-Derived Lower Tropospheric Temperature." *Science* 309 (2005): 1548.

Monaghan, A. J., et al. "Insignificant Changes in Arctic Snowfall since the International Geophysical Year." *Science* 313 (2006): 827.

Morell, V. "Now What?" *National Geographic* 56. (September 2004). (See especially the excellent illustrated description of the Milankovich effect on climate, p. 64.)

Nicholson, C. et al. "Santa Barbara Basin Study Extends Global Climate Record." *EOS* 87, no. 21 (2006): 205.

North, G. R., Q. Wu, and M. J. Stevens. "Detecting the 11-Year Solar Cycle in the Surface Temperature Field." In *Solar Variability and Its Effect on Climate Change.* Edited by J. Pap and P. Fox. *American Geophysical Union Monograph #141.*

Overpeck, J. T., et al. "Arctic System on Trajectory to New Seasonally Ice-Free State." *EOS* 86, no. 34 (2005): 309.

Overpeck, J. T., et al. "Paleoclimatic Evidence for Future Ice-Sheet Instability and Rapid Sea-level Rise." *Science* 311 (2006): 1747.

Penner, J. E., et al. "Aerosols, Their Direct and Indirect Effects." In *Climate Change 2001: The Scientific Basis.* New York: Cambridge University Press, 2001.

Prather, M., et al. "Atmospheric Chemistry and Greenhouse Gases." In *Climate Change 2001: The Scientific Basis.* New York: Cambridge University Press, 2001.

Prentice, I. C., et al. "The Carbon Cycle and Atmospheric Carbon Dioxide." In *Climate Change 2001: The Scientific Basis.* New York: Cambridge University Press, 2001.

Ramaswamy, V., et al. "Radiative Forcing of Climate Change." In *Climate Change 2001: The Scientific Basis.* New York: Cambridge University Press, 2001.

Sarmiento, J. L., and S. C. Wofsy. "A U.S. Carbon Cycle Science Plan, prepared at the request of the U.S. Global Change Research Program." 1999.

Sarmiento, J. L., and N. Gruber. "Sinks for Anthropogenic Carbon." *Physics Today* (August 2002): 30.

Seager, R. "The Source of Europe's Mild Climate." *American Scientist* 94 (2006): 334.

Singer, S. F. *The Kyoto Protocol Is Not Backed by Science.* Arlington, VA, 2002: The Science and Environmental Policy Project. (See especially discussion of the urban heat-island effect, p. 10.)

Soden, B. J., et al. "The Radiative Signature of Upper Tropospheric Moistening." *Science* 310 (2005): 841.

Stocker, T. F., et al. "Physical Climate Processes and Feedbacks." In *Climate Change 2001: The Scientific Basis.* New York: Cambridge University Press, 2001.

Westerling, A. L., et al. "Warming and Earlier Spring Increase Western Forest Wildfire Activity." *Science* 313 (2006): 940

Wild, M., et al. "From Dimming to Brightening: Decadal Changes in Solar Radiation at the Earth's Surface." *Science* 308 (2005): 847.

THE GLOBAL WARMING DEBATE
Science and Scientists in a Democracy

Stuart D. Jordan

As the author of the previous chapter of this volume on global warming and climate change, I offer some concluding thoughts on this topic, which enjoys an emerging consensus among an overwhelming majority of researchers working in the field but remains controversial among some in the general public. Because the issue deals with future activities of important sectors of the American economy, it is not surprising that when the *Skeptical Inquirer* originally published the same material, as two articles, it received an unusually large number of responses to it. They speak for themselves, as do the follow-up comments of the critics and the supporters published in the magazine's September/October 2007 issue. Since my response to the critics appears in that issue as well, I will not review that dialogue again. However, the exchange did raise important questions about the role of science (and of scientists) in a democracy, and it is this topic I wish to address here.

It is important that the public have a reasonable understanding of what science is and of what science can and cannot do in helping to settle issues that eventually reach the public square and thus take on a political dimension. Most people understand that science is a process for seeking the truth about how the natural order works. It is the process itself, not the results of applying it, that lies at the heart of science. Fewer people may realize that this process virtually guarantees the integrity of science in the long run even if individual scientists make mistakes, as all occasionally do, or if a (very) rare individual is actually dishonest and falsifies data. This guaranty results not from any intrinsic moral superiority of scientists themselves, but from the fact that research examined by scientific colleagues in the most prestigious medium, the refereed publications, is quickly subjected to ruthless examination for any errors. Those who detect an error often gain as much credit for their scrutiny as those whose work survives it. Scientists who deliberately avoid this scrutiny by publishing their work in less respected media are understandably and properly given less credence for their efforts. History has demonstrated convincingly that the latter work is much more likely to contain serious errors.

Science does not offer certainty. The results of modern science are typically presented in the language of statistics and probabilities. This is especially true of scientific studies of complex phenomena, of which climate science is an excellent example, even though these phenomena remain rooted in the basic laws of nature. Nevertheless, the existence of "uncertainty" has led some individuals less familiar with science to interpret any uncertainty as evidence for "a major scientific controversy" even when there is none. Thus the general public is vulnerable to the claim that a major scientific dispute over climate science is underway between two equally large and well-qualified groups of scientists, when this is simply not so. Often this false claim is made by those who wish to discourage action to address the problems associated with climate change. There are certainly a few scientists of integrity who remain skeptical of the current near consensus, but the interested reader might consider the language of some of the critics and investigate their sources.

The real issue at stake today is what to do in light of what science has uncovered. Here there is a real controversy. One group favors action in response to the alarming evidence that global warming is definitely occurring, most likely driven by anthropogenic greenhouse gases, while the other side opposes this view for reasons ranging from a few still-unresolved scientific questions to concerns of a more economic and political nature. Typically, the latter group is dominated by those fearing change of the industrial status quo, and they tend to be more vehement in advocating their position.

This leads into two final issues needing comment. The first acknowledges the importance of addressing the economic dislocations and economic opportunities that will result from actions to mitigate the effects of global warming. There is an understandable—not always unwise—human tendency to want to continue with the familiar. This produces a natural inclination to oppose change unless it becomes disastrous not to do so, which can lead to overlooking the many—in this case economic—opportunities associated with pursuing more climate-friendly and eco-friendly technologies. These include many technologies already available, with others undergoing current development that could be accelerated if proper economic incentives were provided. Interested readers can find examples, which I was unable to include in my articles due to space constraints, at www.cfidc.org/opp/jordan.html. The final question I wish to address is what the responsibilities of scientists are in a democracy that, de facto, provides much of the funding for their research. Many working scientists would prefer to have little to do with the political process, yet there is no denying that most scientists today receive much of their support from governments. Performing one's research with integrity is obviously part of the answer, but is it the full answer? Some would say yes and defend this position by noting that distracting a competent researcher from his or her research is likely to reduce scientific productivity. As one who has both performed and managed research, I agree with this position under most circumstances. However, if major public policies depend on science for their proper formulation, as is true of climate science today, a strong case can be made that it becomes the duty of the scientist to inform the public and the political establishment of the best science available on the issue, especially when there are others exerting a major effort to suppress consideration of it. A historical example was the effort of the atomic scientists following World War II to inform the public of the unprecedented power and appalling destructiveness of nuclear weapons. A growing number of climate scientists, and others in related fields, are engaging in a similar educational effort today. I believe this effort serves the public well, and that it should continue.

A SKEPTICAL LOOK AT SEPTEMBER 11TH
How We Can Defeat Terrorism by Reacting to It More Rationally

Clark R. Chapman and Alan W. Harris

uman beings might be expected to value each life, and each death, equally. We each face numerous hazards—war, disease, homicide, accidents, natural disasters—before succumbing to "natural" death. Some premature deaths shock us far more than others. Contrasting with the 2,800 fatalities in the World Trade Center (WTC) on September 11, 2001 (9/11), we barely remember the 20,000 Indian earthquake victims earlier in 2001. Here, we argue that the disproportionate reaction to 9/11 was as damaging as the direct destruction of lives and property. Americans can mitigate future terrorism by learning to respond more objectively to future malicious acts. We do not question the visceral fears and responsible precautions taken during the hours and days following 9/11, when there might have been even worse attacks. But our nation's priorities remain radically torqued toward homeland defense and fighting terrorism at the expense of objectively greater societal needs. As we obsessively and excessively beef up internal security and try to dismantle terrorist groups worldwide, Americans actually feed the terrorists' purposes.

Every month, including September 2001, the US highway death toll exceeds fatalities in the WTC, Pentagon, and four downed airliners combined. Just like the New York City firefighters and restaurant workers, September's auto crash victims each had families, friends, critical job responsibilities, and valued positions in their churches and communities. Their surviving children, also, were left without one parent, with shattered lives, and much poorer than the 9/11 victims' families, who were showered with 1.5 million dollars, per fatality, from the federal government alone. The 9/11 victims died from malicious terrorism, arguably compounded by poor intelligence, sloppy airport security, and other failed procedures we imagined were protecting us. While few of September's auto deaths resulted from malice, neither were they "natural" deaths: most also resulted from individual, corporate, and societal choices about road safety engineering, enforcement of driving-while-drunk laws, safe car design, and so on.

A LACK OF BALANCE

Why does 9/11 remain our focus rather than the equally vast carnage on the nation's highways or Indian earthquake victims? Some say, "Oh, it was a natural disaster and nothing could be done, while 9/11 was a malicious attack." Yet better housing in India could have saved thousands. As for malice, where is our concern for the 15,000 Americans who die annually by homicide? Apparently, the death toll doesn't matter, not if people die all at once, not even if they die by malicious intent. We focus on 9/11, of course, because these attacks were terroristic and were indelibly imprinted on our consciousness by round-the-clock news coverage. Our apprehension was then amplified when just a half dozen people died by anthrax. Citizens apparently support the nation's sudden, massive shift in priorities since 9/11. Here, we ask "Why?"

Suppose we had reacted to 9/11 as we did to that September's auto deaths. That wouldn't have lessened the destroyed property, lost lives and livelihoods, and personal bereavement of family and associates of the WTC victims. But no billions would have been needed to prop up airlines. Local charities wouldn't have suffered as donations were redirected to New York City. Congress might have enacted prescription drug benefits, as it was poised to do before 9/11. Battalions of National Guardsmen needn't have left their jobs to provide a visible "presence" in airports. The nation might not have slipped into recession, with resulting losses to businesses, workers, and consumers alike. And the FBI might still be focusing on rampant white-collar crime (think Enron) rather than on terrorism. While some modest measures (like strengthening cockpit doors) were easy to implement, may have inhibited some "copycat" crimes, and might even lessen future terrorism, we believe that much of the expensive effort is ineffective, too costly to sustain, or wholly irrelevant.

Some leaders got it right when they implored Americans after 9/11 to return to their daily routines, for otherwise "the terrorists will win." Unfortunately, such exhortations seemed aimed at rescuing the travel industry rather than articulating a broad vision of how to respond to terrorism. We advocate that most of us more fully "return to normal life." We suggest that the economic and emotional damage unleashed by 9/11, which touched the lives of all Americans, resulted mostly from our *own* reactions to 9/11 and the anthrax scare, rather than from the objective damage. We recognize that our assertion may seem inappropriate to some readers, and we are under no illusion that natural human reactions to the televised terrorism could have been wholly averted and redirected. We, too, gaped in horror at images of crashing airplanes and we contributed to WTC victims as well. But from within the skeptical community there could emerge a more objective, rational alternative to post-9/11. Citizens could learn to react more constructively to future terrorism and to balance the terrorist threat against other national priorities. It could be just as important to combat our emotional vulnerability to terrorism as to attack Al Qaeda.

Terrorism, by design, evokes disproportionate responses to antisocial acts by a malicious few. By minimizing our negative reactions, we might contribute to undermining terrorists' goals as effectively as by waging war on them or by mounting homeland defenses. We do *not* "blame the victims" for the terrorists' actions. Rather, we seek that we citizens, the future targets of terrorism, be empowered. As Franklin D. Roosevelt famously said, "The only thing we have to fear is fear itself." We can help ensure that terrorists don't win if we can minimize our fears and react more constructively to future terrorism. We don't suggest that this option is easy or will suffice alone. It may not even be possible. But human beings often best succeed by being rational when their emotions, however tenacious and innate, have let them down.

DEATH AND STATISTICS

It is a maxim that one needless or untimely death is one too many. So 20,000 victims should be 20,000 times worse. But our minds don't work that way. Given the national outpouring of grief triggered by the initially estimated 6,500 WTC deaths, one might have expected celebration in late October when it was realized that fewer than half that many had died. But there were no headlines like "3,000 WTC Victims Are Alive After All!" The good news was virtually ignored. Weeks later, many—including Defense Secretary Donald Rumsfeld—continued to speak of "over 5,000 deaths" on 9/11.

To researchers in risk perception, this is natural human behavior. We are evolved from primitive nomads and cave dwellers who never knew, personally, more than the few hundred people in their locales. Until just a few generations ago, news from other lands arrived sporadically via sailors; most people lived and died within a few miles of where they were born. Tragedies invariably concerned a known, nearby person. With the globalization of communication, the world—not just our local valley—has entered our consciousness. But our brains haven't evolved to relate, personally, to each of 6 billion people. Only when the media singles out someone—perhaps an "average layperson" or maybe a tragic exception like JonBenet Ramsey—do our hearts and minds connect.

When an airliner crashes, and reporters focus on a despairing victims' spouse or on the last cellular phone words of a doomed traveler, our brains don't think statistically. We imagine ourselves in that airplane seat, or driving to the airport counseling center when our loved one's plane is reported missing. Actually, 30,000 US commercial flights occur each day. In 2001, except for September 11th and November 12th (when an airliner crashed in Queens, New York, killing more passengers and crew than in the four 9/11 crashes combined), no scheduled US commercial air trips resulted in a single passenger fatality. Indeed, worldwide airline accidents in 2001—including 9/11—killed fewer passengers than during an average year. But statistics can't compete with images of emergency workers combing a crash site for body parts with red lights flashing. We are gripped by fear as though the tragedy happened in our own neighborhood, and another might soon happen again.

Some responses to 9/11 were rational. Soon after jumbo jets were used as flying bombs, workers in landmark skyscrapers might reasonably have feared that their building could be next. With radical Muslims preaching that Americans must be killed, it might behoove us to avoid conspicuous or symbolic gatherings like Times Square on New Year's Eve or the Super Bowl. Surely disaster managers must plug security loopholes that could permit thousands or millions more to be killed. But when police chiefs of countless middle American communities beef up security for *their* anonymous buildings, and search fans entering hundreds of sports fields to watch games of little note, official reactions to terrorism have run amok. To imagine that Al Qaeda's next target might be the stadium in, say, Ames, Iowa, is far-fetched indeed.

FINITE RESOURCES, INFINITE ALARM

Americans' WTC fears only grew when six people died from mailed anthrax. Postal officials patiently explained that public risks were minimal. But millions donned gloves to open their mail or gingerly threw out unopened mail; post offices rejected letters lacking return addresses; urgent mail was embargoed; and for weeks the national dialog centered on one of the least hazards we face.

An NPR radio host asked the Postmaster General if the whole US Postal System might be shut down, despite expert opinion that—in a world faced with diabetes, salmonella poisoning, and AIDS—anthrax will remain (even as a biological weapon) a bit player as a cause of death. Its sole potency is in the context of terrorism: if, by mailing lethal powder to someone, the news media choose to broadcast hysteria into every home so that the very future of our postal system is questioned, then the terrorist has deployed a powerful weapon indeed. But his power would be negated if we were to react to the anthrax in proportion to its modest potential for harm.

Research on risk perception has shown that our reactions to hazards don't match the numerical odds. We fear events (like airliner crashes) that kill many at once much more than those that kill one at a time (like car accidents). We fear being harmed unknowingly (by carcinogens) far more than by things we feel we control ourselves (driving or smoking). We fear unfamiliar technologies (nuclear power) and terrorism far more than prosaic hazards (household falls). Such disproportionate attitudes shape our actions as public citizens. Governments accordingly spend vastly more per life saved to mitigate highly feared hazards (for example, on aircraft safety) than on "everyday" risks (like food poisoning). Risk analysts commonly accept, with neutral objectivity, the disparity between lay perceptions and expert risk statistics. Sometimes it is justifiable to go beyond raw statistics. Depending on our values, we might be more concerned about unfair deaths beyond an individual's control than self-inflicted harm. We might worry more about deaths of children than of elderly people with limited life expectancies. We might dread lingering, painful deaths more than sudden ones. We might be more troubled about "needless" deaths, with no compensating offsets, than about fatalities in the name of a larger good (like those of soldiers or police). Or, in all these cases, we might not.

Why should terrorism command our exceptional attention? That the 9/11 terrorists maliciously attacked the symbolic *and actual* seats of our economic and military power (WTC/Wall Street and the Pentagon) *should* concern us if we truly think that future attacks might destroy our society. But who believes that? Government responses seem directed mostly at stopping future similar attacks . . . which returns us full circle to the question: why should that have become our primary national goal, at the sacrifice of tens of billions of dollars, of some of our civil liberties, of our traveling convenience, and of many of our pre-9/11 priorities?

Instead of rationally apportioning funds to the worst or most unfair societal predicaments, homeland security budgets soar. Nearly every airport administrator, city emergency management director, mayor, legislator, school district supervisor, tourist attraction manager, and plant operations foreman felt compelled after 9/11 to "cover their asses" by visibly enhancing their facility's security. Superfluous barricades were erected, search equipment purchased, and guards hired. Postage rates and delivery delays increase as envelopes are searched for anthrax. Even the governor of West Virginia announced a "West Virginia Watch" program; while some vigilance in that state does no harm, it is unlikely that Wheeling is high on Osama bin Laden's target list.

Meanwhile, programs unrelated to "homeland security" suffer. Finite medical resources were diverted to comforting people that their flu symptoms weren't anthrax . . . or testing to see if they were. Charitable funds that would have nurtured the homeless flowed, instead, to wealthy families of deceased Wall Street traders. Funds for education and pollution control go instead to "securing" public buildings and events. Billions of extra tax dollars are spent on military operations in Pakistan and Afghanistan rather than on enhancing American productivity. If we truly believe in "life, liberty and the pursuit of happiness" and that each life is precious, we must resist selfish forces that would take advantage of our fears and squander our energies and fiscal resources on overblown security enhancements.

Many say that spending for extra security can do no harm. But there *is* ha

rm when politicians act on views, like those of a New Yorker who in 2002 disparaged complaints about airport queues, saying, "I hope that they will be inconvenienced, and will always be inconvenienced, because we should never forget the 5,000 [*sic*] who died." "Inconvenience" sounds innocuous, but it means lost time, lost money, lost productivity, as well as increased frustration and cynicism. Disproportionate expenditures on marginal security efforts take attention, time, and resources away from other more productive enterprises. Moreover, our civil liberties are eroded by the *involuntary* nature of our "sacrifices." When a person irrationally fears crowded elevators and takes the stairs instead, only that person suffers the inconvenience of their personal response. But when everyone, fearful or not, is forced to suffer because of the fears of others, then such measures become tyrannical: we should expect rational deliberation and justifications by our leaders before accepting them. But in the aftermath of 9/11, tens of billions of dollars were immediately reallocated with little public debate. Skeptics might well question our society's acquiescence to popular hysteria and proactively challenge our leaders to balance the expenditures of our resources.

MISPERCEPTIONS OF RISK

Consider some misperceptions of risk. Many news headlines just before 9/11 concerned shark attacks and the disappearance of Chandra Levy, an extreme distortion of serious societal issues (only ten people annually are killed by sharks worldwide). We can laugh at, or bemoan, the triviality of the media. But such stories reflect our own illogical concerns. If, in allocating funds among different hazards, we deliberately *choose* to value the lives of Manhattan skyscraper office workers, postal employees, or airline frequent flyers more than we value the lives of agricultural workers or miners, it is a conscious, informed choice. But it is rarely objectivity that informs such choices. In order to help laypeople and leaders to put our options into perspective, skeptics, teachers, and journalists alike have a responsibility to put the objective past and potential threats from terrorism into contexts that ordinary people can relate to.

Let's compare 9/11 with other past and potential causes of mass death. Note that we generally can't compare prevention costs with lives saved; at best, we can compare expenditures with lives *not* saved. For example, we can compare the cost of air traffic control with midair collision fatalities, but we can only guess at the toll without any such air traffic control.

- 9/11 deaths are similar to monthly US traffic fatalities. Whatever total private/public funds are spent annually, per life saved, on improved highway and motor vehicle safety, alcohol-while-driving prevention efforts, etc., it hardly approaches homeland security budgets.
- The 9/11 fatalities were several to ten times fewer than annual deaths from falls (in the home or workplace), or from suicide, or from homicide. One can question the effectiveness of specific safety programs, counseling efforts, or laws; but, clearly, comparatively paltry sums are spent on programs that would further reduce falls, suicides, and murders.
- In autumn 2001, the Center for Disease Control (CDC) predicted that 20,000 Americans would die from complications of influenza during the then-upcoming winter, most of which could be prevented if susceptible people were vaccinated. The CDC advisory was typically buried inside newspapers whose banner headlines dealt with the anthrax attacks, which killed just a few people.
- Twice as many people died in the worst US flood (stemming from the 1900 Galveston hurricane) as at the WTC. Floods and earthquakes are major killers abroad (each of ten disasters killed over 10,000 people, and a few over 100,000, during the last three decades, chiefly in Asia) but are minor killers in modern America. Hurricane Andrew did great physical damage even though fatalities were few. What are sensible expenditures for research in meteorology and seismology, for mandatory enhancement of building codes and redevelopment, and for other measures that would mitigate natural disasters?

- The 9/11 fatalities are just 1.5 percent of those in the nation's worst epidemic (half a million died from flu in 1918), and also just 1.5 percent of the *annual* US cancer fatalities. We have waged a "war" on cancer, at the expense of research on other less feared but deadly diseases; this war's success is equivocal (five-year survivability after detection is up, but so are cancer death rates—though mainly due to decades-old changes in smoking habits). Where should "homeland security" expenditures rank against medical expenditures?
- Impacts by kilometer-sized asteroids are extremely rare, but one could send civilization into a new Dark Age. The annualized American fatality rate is about 5 percent of the WTC fatalities, although such a cosmic impact has only 1/100th of 1 percent chance of happening during the twenty-first century. Just a couple million dollars are now spent annually to search for threatening asteroids. Should we spend many billions to build a planetary defense shield, which would statistically be in proportion to what we now spend on homeland security and the war on terrorism? Might the threat to our civilization's very existence raise the stakes above even the terrorist threat?

To us, these comparisons suggest that the nation's post-9/11 expenditures have been lopsidedly large, and that a balanced approach would "give back" some funds to reduce deaths from falls, suicide, murder, highway accidents, natural disasters (including even asteroid impacts), malnutrition, and preventable or curable diseases . . . and give back our civil liberties, and just the plain pleasures of life, such as the arts and humanities, exploration, and national parks. And if truly effective means to end wars could be found, they would be especially worthy of funds, given the death toll from twentieth century wars. Before homeland security becomes dominated by vested bureaucracies and constituencies, there may yet be time to question its dominant role in our priorities.

We advocate shifting toward objective cost-benefit analyses and equitable evaluation of the relative costs of saving human lives. Of course, subjective judgments have some validity beyond strict adherence to numerical odds. But we need a national dialog to address these issues dispassionately so that future governmental decisions can eschew immediate, impulsive reactions. Individual skeptics, in our own lives, can exemplify sensible choices. Among the many dumb things we should avoid (smoking, driving without a seatbelt, or letting kids play with firearms), we must also avoid driving instead of flying, acquiescing uncomplainingly to ineffective searches at local buildings and events, and generally yielding to the new "homeland security" mania. Clear thinking about risks, rather than saying that "any improvement in security is worth it," can reduce our societal vulnerability to terrorism.

One constructive antidote to post-9/11 trauma is to enhance the information available and to foster sound appreciation, evaluation, and use of the information. Life is inherently risky, unpredictable, and subject to things we cannot know . . . but there are things we do know and can understand. Rather than scaring people about sharks, serial killers, and anthrax, the mass media could help people understand the real risks in their everyday environments and activities. Educational institutions should help students develop critical skills necessary to make rational choices. While avoiding intrusions into personal liberties, government could nevertheless collect and assess statistical data in those arenas (like air travel) where potential dangers lurk, concentrating protective efforts and law enforcement where it is most efficacious.

To conclude, we suggest that most homeland security expenditures, which in the zero-sum budget game are diverted from other vital purposes, are terribly expensive and disproportionate to competing needs for preventing other causes of death and misery in our society. While prudent, focused improvements in security are called for, the sheer costs of most security initiatives greatly distort the way we address the many threats to our individual and collective well-being. Our greatest vulnerability to terrorism is the persisting, irrational fear of terrorism that has gripped our country. We must start behaving like the informed, reasoning beings we profess to be.

ACKNOWLEDGMENTS

The authors are grateful to many friends and colleagues, espeially David Morrison, for comments and criticisms that helped us to frame these issues.

STRONG RESPONSE TO TERRORISM NOT A SYMPTOM OF FALLACIOUS STATISTICAL REASONING OR HUMAN COGNITIVE LIMITATIONS

Steven Pinker

C lark R. Chapman and Alan W. Harris, in the previous chapter, are right to question the costs in money, opportunities, and civil liberties of many of the policies adopted in response to the September 11 attacks. And they are right to call attention to the vulnerability of the human mind to fallacies in statistical reasoning, as in people's overestimation of the dangers posed by air travel, shark attacks, and trace levels of carcinogens. But they are not correct in saying that the responses to the attacks are consequences of fallacious statistical reasoning. The classic experiments by Paul Slovic, Amos Tversky, and Daniel Kahneman demonstrating those fallacies presupposes a number of conditions that are not met by the events of September 11, 2001.

First, since every event is unique, estimating risk requires one to define some class of events to be treated as equivalent, and then to compare the frequency of those events with the number of opportunities for such events to occur. For a singular event like the September 11 attacks, the equivalence class could be defined in many ways. If it is defined as "airplanes crashed into buildings," then the probability of the event multiplied by the number of deaths per event may indeed be smaller than other risks we tolerate. (Even then, one could question Clark and Harris's characterization of the casualty rates for events like September 11, because if a few parameters had been different—the hour of the day, the time available for people to escape before the towers collapsed, the success of the passenger mutiny over Pennsylvania—the death toll could have been far higher.) But if one defines the class as "acts designed to inflict as many American deaths as possible"—which could include nuclear bombs simultaneously set off in New York, Los Angeles, and Chicago—then the multiplication gives a very different result, and taking expensive measures to prevent such events is not necessarily irrational. Similarly, one gets very different risk estimates for the class "anthrax attacks" (probably small) and the class "biological attacks, including smallpox" (possibly catastrophic).

In general, it is fairly straightforward to define an equivalence class for events with physical definitions such as plane crashes, shark attacks, and lung cancer deaths. But it is not at all straightforward to define the equivalence class for events such as terrorist attacks, which are limited only by the ideology, ingenuity, and resources of the perpetrators. Prior to September 11, 2001, people had little reason to estimate that the equivalence class "terrorist attack" included massive destruction of American lives and landmarks brought about by well-funded suicidal fanatics exploiting hitherto unrecognized vulnerabilities of a technologically advanced democracy. The terrorist attacks provide new information relevant to estimating those unknowns.

Second, a probability estimate is specific to an interval of time in which the causal structure of the world remains unchanged. If the world has changed, all bets are off. If I notice that a nefarious character has just tampered with a slot machine, then ignoring the published odds is not fallacious. Or to take an example from the psychologist Gerd Gigerenzer, it would not be irrational to keep one's child out of a river that had no previous fatalities after hearing that a neighbor's child was attacked there by a crocodile that morning: there was no crocodile in the river before then, but now there is. For this reason one cannot use the rate of major terrorist attacks in, say, the past ten years to estimate the rate in the next ten years. Wahabism and anti-Americanism may be more widespread, nuclear weapons more available, copycats more emboldened, and so on. Because of these uncertainties, anyone who claims to have calculated the mathematically correct probability that a horrendous terrorist attack will take place in the next year would be talking through his hat.

There is a third reason that terrorist attacks cannot be equated with the kinds of risks that people have been shown to treat irrationally. Nonhuman causes of deaths (such as sharks, airplane part failures, and carcinogens) don't take into account how people react to them. Human causes of deaths (such as terrorists) do. Bin Laden thought that American society was so decadent and spiritually bankrupt that a few easily inflicted humiliating blows would lead to its collapse. A public response of defiance and solidarity, and the implementation of extensive preventive security measures, could change such calculations in the minds of future terrorists. Similarly, if we calibrated our response to the anthrax attacks by cost/benefit comparisons to other risks, future bioterrorists could be emboldened to inflict exactly as many deaths as we decided we could endure. But pulling out all the stops to combat this new kind of threat, even if seemingly irrational on narrow actuarial grounds in the short run, could deter perpetrators in the long run, who would have to factor this determination into their own calculations. Another way of putting it is that dealing with terrorists is a problem in game theory, not just a problem in risk estimation.

I don't disagree with Chapman and Harris's opposition to some of the measures taken by the Bush administration and other authorities. But it is not correct to call the strong response to the September 11 attacks a symptom of fallacious statistical reasoning or human cognitive limitations.

AFTERWORD

I was ambivalent about allowing this note to be reprinted. I wrote it not long after the September 11, 2001, attacks, when the initial response of Western governments to terrorism and repression by Islamist extremists was justifiable. Since that time the United States engaged in a muddled and ill-advised invasion of Iraq that had nothing to do with deterring terrorism, and a program of so-called security measures at home that are nothing short of idiotic. A few examples: color-coded nationwide security alerts; surrounding tiny rural post offices with concrete barriers; entangling foreign visitors and immigrants with pointless red tape; confiscating nail clippers at airports; prohibiting the use of cell phones at airport baggage carrousels; and blocking off public walkways near federal buildings. At the same time, programs for what should be the top priority in preventing terrorist damage—tracking down loose nukes—are underfunded.

The current American response to terrorism *is* a product of numerous fallacies, and if I had known how it would turn out, I would not have written the note the way I did. I'm allowing it to be reprinted because I think it lays out some important issues in how we should evaluate our responses to terrorism. It's a shame that the current policies were not framed more rationally and honestly.

THE ANTI-VACCINATION MOVEMENT

Steven Novella

Michelle Cedillo has autism, which her parents believe is the result of her childhood vaccines. In June 2007 they had the opportunity, along with eight other families, to make their case to the Autism Omnibus—a US Court of Federal Claims that was presided over by three "special masters" appointed for the purpose. These nine cases are the first test cases that will likely determine the fate of 4,800 other claims made over the past eight years for compensation for injuries allegedly due to childhood vaccines.

Vaccines are one of the most successful programs in modern health care, reducing, and in some cases, even eliminating serious infectious diseases. Public support for the vaccination program remains strong, especially in the United States where vaccination rates are currently at an all-time high of >95 percent (CDC, 2004). Yet, despite a long history of safety and effectiveness, vaccines have always had their critics: some parents and a tiny fringe of doctors question whether vaccinating children is worth what they perceive as the risks. In recent years, the anti-vaccination movement, largely based on poor science and fear-mongering, has become more vocal and even hostile (Hughes, 2007).

Of course, vaccines are not without risk (no medical intervention is), although the benefits far outweigh those risks. Because vaccines are somewhat compulsory in the United States—although opting out is increasingly easy—

a National Vaccine Injury Compensation Program was established to stream-line the process for compensation for those who are injured due to vaccines (USDOJ, 2007). It is this program to which the Cedillo family—and 4,800 others—are applying for compensation.

In the last decade, the anti-vaccine movement, which includes those who blame the MMR (mumps-measles-rubella) vaccine for autism, has largely merged with those who warn that mercury toxicity is the cause of many of the ills that plague mankind. The two groups have come together over the issue of thimerosal, a mercury-based preservative in some vaccines. They believe that it was the use of thimerosal in childhood vaccines that led to the apparent autism epidemic beginning in the 1990s.

Autism is a complex neurological disorder that typically manifests in the first few years of life and primarily involves a deficiency of typical social skills and behavior. In the 1990s, the number of autism diagnoses significantly increased, from between one and three to about fifteen cases per ten thousand, although the true incidence is probably between thirty and sixty per ten thousand (Rutter, 2005). During this same period, the number of vaccines given in the routine childhood schedule also increased. This led some to assume, or at least speculate, causation from correlation—perhaps the vaccines or something in them created this "epidemic" of autism.

We can now say, from multiple independent lines of evidence, that vaccines do not cause autism. For one thing, the autism "epidemic" probably does not represent a true increase in the disorder, but rather an artifact of expanding the diagnosis (now referred to as autism spectrum disorder, ASD) and increased surveillance (Taylor, 2006).

In 1998, researcher Andrew Wakefield and some of his colleagues published a study in the prestigious English medical journal *Lancet* that claimed to show a connection between the MMR vaccine and autism (Wakefield, 1998). Wakefield's theory was that the MMR vaccine, which contains a live virus, can cause a chronic measles infection in susceptible children. This in turn leads to gastrointestinal disturbances, including what he calls a "leaky gut" syndrome, which then allows for certain toxins and chemicals, like those from bread and dairy that are normally broken down by the gut, to enter the bloodstream where they can access and damage the developing brain.

Although the study was small and the evidence was considered preliminary, this article sparked a firestorm. As a result of the study and the media coverage that followed (and continues to this day), MMR compliance in Great Britain plummeted, resulting in a surge of preventable disease (Friederichs, 2006).

Subsequent to the seminal article in the *Lancet*, many follow-up studies were performed testing the autism-MMR vaccine correlation. As the follow-up studies began to be published, however, it became increasingly clear that there was no link between MMR and autism. For example, a study in the *British Medical Journal* found that autism rates continued to climb in areas where MMR vaccination rates were not increasing (Taylor, 1999). Another study found no association with MMR and autism or GI (gastrointestinal) disorders (Taylor, 2002). Other studies showed no difference in the diagnosis rate of autism either before or after the MMR vaccine was administered (Honda, 2005), or between vaccinated and unvaccinated children (Madsen, 2002). Most recently, a study found that there was no decrease in autism rates following removal of the MMR vaccine in Japan (Honda, 2005).

In 2001, the Institute of Medicine (IOM) reviewed all of the MMR-autism data available to date and concluded that there was no association and essentially closed the case (IOM, 2001)—a conclusion confirmed by still later studies, such as the Honda study in Japan cited above.

If Wakefield had simply been wrong in his preliminary findings, he would be innocent of any wrongdoing—scientists are not faulted if their early findings are not later vindicated. However, in May 2004, ten of Wakefield's co-authors on his original paper withdrew their support for its conclusions. The editors of *Lancet* also announced that they withdrew their endorsement of the paper and cited as part of the reason an undisclosed potential conflict of interest for Wakefield, namely that at the time of its publication he was conducting research for a group of parents of autistic children seeking to sue for damages from MMR vaccine producers (Lancet, 2004).

It gets worse. Investigative reporter Brian Deer has uncovered greater depths to Wakefield's apparent malfeasance. Wakefield had applied for patents for an MMR vaccine substitute and treatments for his alleged MMR vaccine-induced gut disorder (Deer, 2007). So, not only was he allegedly paid by lawyers to cast doubt on the MMR vaccine, but he stood to personally gain from the outcome of his research.

Further, during the Cedillo case testimony, Stephen Bustin, a world expert in the polymerase chain reaction (PCR), testified that the lab Wakefield used to obtain the results for his original paper was contaminated with measles virus RNA. It was therefore likely, Bustin implied, that the PCR used by Wakefield was detecting this contamination and not evidence for measles infection in the guts of children with autism who had been vaccinated, as Wakefield claimed. And finally, Nicholas Chadwick testified that the measles RNA Wakefield found matched the laboratory contamination and did not match either any naturally occurring strain or the strain used in the MMR vaccine—a fact of which he had informed Wakefield (USCFC, 2007).

All of this, plus other allegations still coming out, has caused Britain's

General Medical Council to call Wakefield before its "Fitness to Practise" panel for review of his alleged professional misconduct (GMC, 2007).

Believers in the MMR-autism hypothesis dismiss the findings of the larger and more powerful epidemiological studies that contradict a link. Instead, they have turned Andrew Wakefield into a martyr, dismissing the evidence of his wrongdoing as a conspiracy against him designed to hide the true cause of autism from the public. Wakefield is unrepentant and maintains his innocence (Gorski, 2007).

With the MMR-autism hypothesis scientifically dead, attention soon shifted to thimerosal, a mercury-based preservative found in some childhood vaccines (although not the MMR vaccine). There is little doubt, and no controversy, that mercury, the major component of thimerosal, is a powerful neurotoxin, or poison to the brain. However, toxicity is always a matter of dose. Everything becomes toxic in a high enough dose; even too much water or vitamin C can kill you. So the real question is whether the amount of mercury given to children in vaccines containing thimerosal was enough to cause neurological damage.

Proponents of the mercury hypothesis argue that the ethylmercury found in thimerosal was given in doses exceeding Environmental Protection Agency limits. This load of mercury should be considered with prenatal vaccine loads possibly given to mothers, and to other environmental sources of mercury, such as seafood. Furthermore, underweight or premature infants received a higher dose by weight than larger children. Some children, they argue, may have a specific inability to metabolize mercury, and perhaps these are the children who become autistic.

Fear over thimerosal and autism was given a huge boost by journalist David Kirby with his book *Evidence of Harm* (Kirby, 2005). Kirby tells the clichéd tale of courageous families searching for help for their sick children and facing a blind medical establishment and a federal government rife with corruption from corporate dollars. Kirby echoes the core claim that as the childhood vaccine schedule increased in the 1990s, leading to an increased cumulative dose of thimerosal, autism diagnoses skyrocketed.

In the end, *Evidence of Harm* is an example of terrible reporting that grossly misrepresents the science and the relevant institutions. As bad as Kirby's position was in 2005, in the last two years the evidence has been piling up that thimerosal does not cause autism. Rather than adjusting his claims to the evidence, Kirby has held fast to his claims, which has made him a hero alongside Wakefield of the mercury-autism-connection crowd as he has squandered his credibility.

There have now been a number of epidemiological and ecological studies that have all shown no correlation between thimerosal and autism (Parker,

2004; and Doja, 2006). I have already mentioned that the current consensus holds that there is no real autism epidemic, just an artifact of how the diagnosis is made. If there's no epidemic, there's no reason to look for a correlation between thimerosal and autism. This has been backed up by the Institute of Medicine, which has also reviewed all the available evidence (both epidemiological and toxicological) and concluded that the evidence does not support the conclusion that thimerosal causes autism (IOM, 2004).

Especially damning for the thimerosal hypothesis are the recent studies that clearly demonstrate that early detection of autism is possible long before the diagnosis is officially made. Part of the belief that vaccines may cause autism is driven by the anecdotal observation by many parents that their children were normal until after they were vaccinated—autism is typically diagnosed around age two or three. However, more careful observations indicate that signs of autism are present much earlier, even before twelve months of age, before exposure to thimerosal (Mitchell, 2006). In fact, autism expert Eric Fombonne testified in the Autism Omnibus hearings that Michelle Cedillo displayed early signs of autism clearly visibly on family video taken prior to her receiving the MMR vaccine (USCFC, 2007).

Meanwhile, evidence is accumulating that autism is largely a genetic disorder (Szatmari, 2007). This by itself does not rule out an environmental factor, but it is telling that genetic research in autism has proven so fruitful.

Mercury alarmists, in the face of this negative evidence, have been looking for rationalizations. Some have argued that the thimerosal in prenatal vaccines may be to blame, but recent evidence has shown a negative correlation there as well (Miles, 2007).

What we have are the makings of a solid scientific consensus. Multiple independent lines of evidence all point in the same direction: vaccines in general, and thimerosal in particular, do not cause autism, which rather likely has its roots in genetics. Furthermore, true autism rates are probably static and not rising.

The only researchers who are publishing data that contradicts this consensus are the father-and-son team of Mark and David Geier. They have looked at the same data and concluded that thimerosal does correlate with autism. However, the hammer of peer-review has come down on their methods and declared them fatally flawed, thus rendering their conclusions invalid or uninterpretable (Parker, 2004). Also, like Wakefield, their reputations are far from clean. They have made something of a career out of testifying for lawyers and families claiming that vaccines caused their child's autism, even though the Geiers' testimony is often excluded on the basis that they lack the proper expertise (Goldacre, 2007). The Geiers were not even called as experts in the Autism Omnibus hearings.

The Geiers are now undertaking an ethically suspect study in which they are administering chelation therapy to children with autism in conjunction with powerful hormonal therapy allegedly designed to reduce testosterone levels. Chelation therapy removes mercury, and so it is dependent upon the mercury hypothesis, which is all but disproved. Moreover, there is no clinical evidence for the efficacy of chelation therapy. The treatment is far from benign and is even associated with occasional deaths (Brown, 2006).

With the scientific evidence so solidly against the mercury hypothesis of autism, proponents maintain their belief largely through the generous application of conspiracy thinking. The conspiracy claim has been made the loudest by Robert F. Kennedy Jr. in two conspiracy-mongering articles: "Deadly Immunity" published on Salon.com in 2005 (Kennedy, 2005), and more recently "Attack on Mothers" (Kennedy, 2007). In these articles, RFK Jr. completely misrepresents and selectively quotes the scientific evidence, dismisses inconvenient evidence as fraudulent, accuses the government, doctors, and the pharmaceutical industry of conspiring to neurologically damage America's children, and accuses scientists who are skeptical of the mercury claims of attacking the mothers of children with autism.

Despite the lack of evidence for any safety concern, the FDA decided to remove all thimerosal from childhood vaccines, and by 2002 no new childhood vaccines with thimerosal were being sold in the United States. This was not an admission of prior error, as some mercury proponents claimed; instead, the FDA was playing it safe by minimizing human exposure to mercury wherever possible. The move was also likely calculated to maintain public confidence in vaccines.

This created the opportunity to have the ultimate test of the thimerosal autism hypothesis. If rising thimerosal doses in the 1990s led to increasing rates of autism diagnosis, then the removal of thimerosal should be followed within a few years by a similar drop in new autism diagnoses. If, on the other hand, thimerosal did not cause autism, then the incidence of new diagnoses should continue to increase and eventually level off at or near the true rate of incidence. In 2005, I personally interviewed David Kirby on the topic, and we both agreed that this would be a fair test of our respective positions. Also, in an e-mail to science blogger "Citizen Cain," Kirby wrote, "If the total number of 3–5 year olds in the California DDS [Department of Developmental Services] system has not declined by 2007, that would deal a severe blow to the autism-thimerosal hypothesis" (Cain, 2005).

Well, five years after the removal of thimerosal, autism diagnosis rates have continued to increase (IDIC, 2007). That is the final nail in the coffin in the thimerosal-vaccine-autism hypothesis. The believers, however, are in full rationalization mode. David Kirby and others have charged that although no new vaccines with thimerosal were sold after 2001, there was no recall, so pediatricians may have had a stockpile of thimerosal-laden vaccines—even though a published inspection of 447 pediatric clinics and offices found only 1.9 percent of relevant vaccines still had thimerosal by February 2002, a tiny fraction that was either exchanged, used, or expired soon after (CDCP/ACIP, 2002).

Those who argue for the link have put forth increasingly desperate notions. Kirby has argued that mercury from cremations was increasing environmental mercury toxicity and offsetting the decrease in mercury from thimerosal. The Geiers simply reinterpreted the data using bad statistics to create the illusion of a downward trend where none exists (Geier, 2006). Robert Kennedy Jr. dodges the issue altogether by asking for more studies, despite the fact that the evidence he asks for already exists. He just doesn't like the answer. Kennedy and others also point to dubious evidence, such as the myth that the Amish do not vaccinate and do not get autism. Both of these claims are not true, and the data RFK Jr. refers to is nothing more than a very unscientific phone survey (Leitch, 2007).

The Autism Omnibus hearings have concluded, and while we await the decision, I am optimistic that science and reason will win the day. Just as shown in the 2005 Dover trial of intelligent design where the full body of scientific evidence was given a thorough airing in court and subjected to rules of evidence and the critical eyes of experienced judges, science tends to win out over nonsense. By all accounts, the lawyers for those claiming that vaccines caused their children's autism put on pathetic performances with transparently shoddy science, while the other side marshaled genuine experts and put forth an impressive case.

But the stakes are high, and not just for the 4,800 families. If the petitioners win these test cases despite the evidence, it will open the floodgates for the rest of the 4,800 petitioners. This will likely bankrupt the Vaccine Injury Compensation Program and will also risk our vaccine infrastructure. Pharmaceutical companies will be reluctant to subject themselves to the liability of selling vaccines if even the truth cannot protect them from lawsuits.

Thimerosal still exists as a necessary preservative in multi-shot vaccines outside the United States, especially in poor third world countries that cannot afford stockpiles of single-shot vaccines. Anti-thimerosal hysteria therefore also threatens the health of children in poor countries.

And of course a victory for the anti-vaccination activists would undermine public confidence in what is arguably the single most effective public health measure devised by modern science. This decrease in confidence will lead, as it has before, to declining compliance and an increase in infectious disease.

The forces of irrationality are arrayed on this issue. There are conspiracy theorists, well-meaning but misguided citizen groups who are becoming increasingly desperate and hostile, irresponsible journalists, and ethically compromised or incompetent scientists. The science itself is complex, making it difficult for the average person to sift through all the misdirection and misinformation. Standing against all this is simple respect for scientific integrity and the dedication to follow the evidence wherever it leads.

Right now the evidence leads to the firm conclusion that vaccines do not cause autism. Yet, if history is any guide, the myth that they do cause autism will likely endure even in the face of increasing contradictory evidence.

REFERENCES

Brown, M. J., T. Willis, B. Omalu, and R. Leiker. "Deaths resulting from hypocalcemia after administration of edetate disodium: 2003–2005." *Pediatrics.* 118 no. 2 (2006): e534–36.

Centers for Disease Control. *MMWR Weekly* 53, no. 44 (November 12 2004): 1041–44. Available at http://www.cdc.gov/mmwr/preview/mmwrhtml/mm5344a4.htm.

Centers for Disease Control and Prevention Advisory Committee on Immunization. "Practice Records of the meeting held on February 20–21, 2002, Atlanta Marriott North Central Hotel." Available at http://www.kevinleitch.co.uk/grabit/acip-min-feb.pdf.

Citizen Cain. "Slouching Toward Truth—Autism and Mercury," November 30 2005. Available at http://citizencain.blogspot.com/2005/11/slouching-toward-truth-autism-and_30.html.

Deer, B. "Andrew Wakefield & the MMR scare: part 2." 2007. Available at http://briandeer.com/wakefield-deer.htm.

Doja, A., and W. Roberts. "Immunizations and Autism: A Review of the Literature." *Canadian Journal of Neurological Sciences* 33, no. 4 (2006): 341–46.

Friederichs, V., J. C. Cameron, and C. Robertson. "Impact of Adverse Publicity on MMR Vaccine Uptake: A Population Based Analysis of Vaccine Uptake Records for One Million Children, Born 1987–2004." *Archives of Diseases of Children* 200691, no. 6 (2006): 465–68. Epub April 25 2006.

Geier, D. A., and M. R. Geier. An Assessment of Downward Trends in Neurodevelopmental Disorders in the United States following Removal of Thimerosal from Childhood Vaccines. *Medical Science Monitor* 12, no. 6 (2006): CR231–9. Epub May 29 2006.

General Medical Council. July 16 2007. Available at http://www.gmcpressoffice.org.uk/apps/news/events/index.php?month=7&year=2007&submit=Submit.

Goldacre, B. "Opinions from the Medical Fringe Should Come With a Health Warning." *Guardian,* Saturday, February 24 2007. Available at http://www.guardian.co.uk/science/2007/feb/24/badscience.uknews.

Gorski, D. "Andrew Wakefield: The Galileo Gambit Writ Large in The Observer." *Respectful Insolence* July 9 2007. Available at http://scienceblogs.com/insolence/2007/07/andrew_wakefield_the_galileo_gambit_writ.php.

Honda, H., Y. Shimizu, and M. Rutter. "No Effect of MMR Withdrawal on the Incidence of Autism: A Total Population Study." *Journal of Child Psychology and Psychiatry* 46, no. 6 (2005): 572–79.

Hughes, V. "Mercury Rising." *Nature Medicine* 13, no. 8 (2007): 896–97. Epub August 31 2007.

Infectious Diseases and Immunization Committee, Canadian Paediatric Society (CPS). "Autistic Spectrum Disorder: No Causal Relationship with Vaccines." *Paediatrics & Child Health* 12, no. 5 (2007): 393–95. Available at http://www.cps.ca/english/statements/ID/pidnote_jun07.htm.

Institute of Medicine. 2001. Immunization Safety Review: Measles-Mumps-Rubella Vaccine and Autism. April 23. Available at www.iom.edu/CMS/3793/4705/4715.aspx.

Institute of Medicine. "Immunization Safety Review: Vaccines and Autism." May 17 2004.. Available at http://www.iom.edu/CMS/3793/4705/20155.aspx.

Kennedy, R. F. "Deadly Immunity." *Salon.com.* June 16 2005. Available at http://dir.salon.com/story/news/feature/2005/06/16/thimerosal/index3.html?pn=1.

———. Attack on Mothers. *Huffington Post.* June 19 2007. Available at www.huffingtonpost.com/robert-f-kennedy-jr/attack-on-mothers_b_52894.html.

Kirby, David. *Evidence of Harm: Mercury in Vaccines and the Autism Epidemic: A Medical Controversy* New York: St. Martin's Press, 2005.

Lancet Editors. *Lancet* 363(2004): 9411.

Leitch, K. "Autism amongst the Amish." *Left Brain/Right Brain* 22 (2007) Available at http://www.kevinleitch.co.uk/wp/?p=5353.

Madsen, K. M., et al. "A Population-based Study of Measles, Mumps, and Rubella Vaccination and Autism." *New England Journal of Medicine* 347, no. 19 (2002): 1477–1482.

Miles, J. H., and T. N. Takahashi. "Lack of Association between Rh Status, Rh Immune Globulin in Pregnancy and Autism." *American Journal of Medical Genetics, Part A1* 143, no. 13 (2007): 1397–407.

Mitchell, S., et al. "Early Language and Communication Development of Infants Later Diagnosed with Autism Spectrum Disorder." *Journal of Developmental and Behavioral Pediatrics* 27, no. 2 [Suppl] (2006): S69–78.

Parker, S. K., B. Schwartz, J. Todd, and L. K. Pickering. "Thimerosal-containing Vaccines and Autistic Spectrum Disorder: A Critical Review of Published Original Data." *Pediatrics* 114, no. 3 (2004): 793–804.

Rutter, M. "Incidence of Autism Spectrum Disorders: Changes over Time and Their Meaning." *Acta Paediatrica* 94, no. 1 (2005): 2–15.

Szatmari, P., et. al. "Mapping Autism Risk Loci Using Genetic Linkage and Chromosomal Rearrangements." *Nature Genetics* 39 (2007): 319–28.

Taylor, B. "Vaccines and the Changing Epidemiology of Autism." *Child Care, Health, and Development* 32, no. 5(2006): 511–19.

Taylor, B., E. Miller, et al. Autism and Measles, Mumps, and Rubella Vaccine: No Epidemiological Evidence for a Causal Association. *Lancet* 12, no. 353[9169] (1999): 2026–29.

Taylor, B., E. Miller, R. Lingam, et al. "Measles, Mumps, and Rubella Vaccination and Bowel Problems or Developmental Regression in Children with Autism: Population Study." *British Medical Journal* 16, no. 324[7334](2002): 393–96.

United States Court of Federal Claims. *Cedillo v. Secretary of Health and Human Services,* Transcript of Day 6. June 18, 2007. Available at ftp://autism.uscfc.uscourts .gov/autism/transcripts/day06.pdf.

United States Court of Federal Claims. *Cedillo v. Secretary of Health and Human Services,* Transcript of Day 8. June 20, 2007. Available at ftp://autism.uscfc.uscourts .gov/autism/transcripts/day08.pdf.

USDOJ. "About the National Vaccine Injury Compensation Program." Available at www.usdoj.gov/civil/torts/const/vicp/about.htm.

Wakefield, A. J., et al. "Ileal-lymphoid-nodular Hyperplasia, Non-specific Colitis, and Pervasive Developmental Disorder in Children." *Lancet* 351[9103](1998): 637–41.

VACCINE SAFETY
Vaccines Are One of Public Health's Great Accomplishments

Richard G. Judelsohn

O ver the past decade, the public has been presented with a large amount of information about the safety of vaccines. Among the reasons for this interest is the widespread success of routine, universal immunization of infants and children, beginning in the 1940s. Previously common, dangerous, handicapping, potentially fatal diseases (vaccine-preventable diseases) have been wiped out with this policy. As the last century drew to a close, immunization was declared the greatest public health achievement in the United States in the twentieth century.

The list of licensed and recommended vaccines has been growing, and not just for infants and children. There are now schedules from professional societies, such as theAmerican Academy of Family Physicians (AAFP) and the American College of Obstetricians and Gynecologists (ACOG), and public agencies (like the US Centers for Disease Control and Prevention–CDC and most state health departments) that indicate what vaccines should be given and when for adolescents, adults, and specific vulnerable populations.

The considerable focus on vaccines and their safety in our information-overloaded society is not surprising, with a surplus of articles in magazines, books, parenting guides, and on the Internet, and stories on radio and television. While these occasionally highlight the benefits of immunization, "No

One Got Sick or Died from a Vaccine-Preventable Disease Today" is not a very exciting story, so more often the emphasis in the media is on speculation that a vaccine caused a health problem. Furthermore, the widespread availability of litigation and liberal tort in the United States has encouraged lawsuits claiming harm from vaccines. Finally, it's human nature to assume cause and effect when something bad happens, so a vaccination is an attractive target when administered before the onset of a medical condition.

Unfortunately, most of the public receives a lot of health information from lay sources rather than their physicians. Professional knowledge of immunization is grounded in science—microbiology, immunology, epidemiology, and statistics. Vaccines are licensed by the US Federal Drug Administration (FDA) only when proven safe and effective. Recommendations for use are promulgated by committees of scientific experts composed of academics, clinicians, and other caregivers who are passionately devoted to our citizens' health and safety. The committees' conclusions, and the rationale for them, are shared with practicing physicians, who are the most reliable source of information for patients. This process is the foundation that has led to the conclusion that licensed vaccines are safe, and the fears that vaccines are harmful are unfounded.

Nevertheless, to address these unfounded fears, these and other groups of scientific experts have undertaken investigations to determine possible relationships between vaccines and autism, asthma, diabetes, multiple sclerosis, SIDS, and other diseases. *No studies have yet established a causal link between vaccines and these diseases.* For example:

> • Does hepatitis B vaccine cause SIDS (sudden infant death syndrome)? Looking at the numbers of doses of the former administered and cases of the latter, one would conclude the opposite, that hepatitis B vaccine prevents SIDS, since 90 percent of US children have received hepatitis B vaccine, and SIDS cases have dropped dramatically in the past decade (probably due to the American Academy of Pediatrics [AAP] recommendation that infants sleep on their backs).
>
> • Does MMR vaccine cause autism? This question received extraordinary attention after it was raised in an article in *The Lancet* in 1998, by Dr. Andrew Wakefield and colleagues. The co-authors and *The Lancet* both have since retracted the article and its conclusions, and Wakefield is currently on trial in the U.K. for conflict of interest at the time of its publication. (He was on retainer from lawyers suing for vaccine damages.) More important, an Institute of Medicine (IOM) expert panel evaluated the issue and concluded that the evidence favored rejection of a connection between autism and MMR vaccine. Fourteen epidemiologic studies have been performed, all demonstrating the absence of a relationship between increased rates of

autism and frequency of use of MMR vaccine. It is unfortunate that the speculation of a relationship between MMR vaccine and autism has resulted in the occurrence of vaccine-preventable diseases (especially measles) in children whose parents refused to allow them to receive the vaccine and has diverted attention from research into the real causes of autism, which has been shown to have prenatal origins.

• Is thimerosal a cause of neurologic abnormalities, including autism? The preservative thimerosal, consisting of ethyl mercury, was used in multi-dose vaccine vials. At present, most infancy and childhood vaccines are supplied in single-dose vials, and all such routine vaccines are thimerosal-free. Studies to answer this question, including five epidemiologic surveys, came to the same conclusion as the MMR vaccine–autism analyses, that there is not a relationship. A pivotal study at the University of Rochester quantifying thimerosal in childhood vaccines stated that administration of vaccines containing thimerosal does not seem to raise blood concentrations of mercury above safe levels in infants.

Many of us recall that only two generations ago we had schoolmates who limped or had withered arms due to the paralytic polio they were infected with. That disease is now extinct in the United States because of the universal use of polio vaccine. During my training, I cared for children made deaf from measles, infants blind and retarded from rubella, and those who died from bacteria like pneumococcus and meningococcus. With vaccination, those conditions no longer occur. As a physician in my early years of practice, the threat of infection with bacteria called Haemophilus influenza type B (Hib) loomed large for my patients and their families, the outcomes of brain damage or death being distinct possibilities. A vaccine was invented, adopted as policy, and given to US infants and children. I'm pleased to say I no longer worry about Hib infection.

Despite scientific proof and a long track record of vaccine safety, we see public policy based on junk beliefs, misinformation, fear, and mass hysteria. In 2006, a number of legislative bodies passed, and executives signed, bills prohibiting use of vaccines containing thimerosal. From a practical perspective, these restrictions mean little, since all but a few influenza vaccines do not contain thimerosal. But such policies send a bad message: that the vaccines that have virtually eradicated many diseases, constituting one of the greatest public health accomplishments of the past century, are dangerous. Furthermore, these policies denigrate our informed medical and scientific communities. This is a disservice to our citizens and endangers us all.

WARNING: ANIMAL EXTREMISTS ARE DANGEROUS TO YOUR HEALTH

P. Michael Conn and James V. Parker

For years we have laughed at the antics of people in some of the more extreme segments of the animal rights movement—groups like People for the Ethical Treatment of Animals (PETA). They put up billboards encouraging children to drink beer instead of milk and vilify fast food chains for cooking veggie burgers on the same grill as meat. They even wrote to Oklahoma City bomber Timothy McVeigh urging him to stop the killing at his dinner plate and to request a vegetarian dinner for his last meal. All this sure gets the media's attention and sometimes even a chuckle from the public. Well, maybe its time to stop laughing.

We may choose to ignore the poor taste of the animal rights movement in equating the Holocaust of World War II with the raising of broiler chickens or the "enslavement" of circus animals with the slavery of African-Americans in the United States. But consider this curious candor from one animal rights leader: "The life of an ant and that of my child should be granted equal consideration" (Fox, 1992). What does *that* mean?

Can we ignore this statement from PETA co-founder Alex Pacheco: "Arson, property destruction, burglary, and threat are 'acceptable crimes' when used for the animal cause" (Activist Cash.com, 2008a)?

FBI special agent David Szady, referring to Earth Liberation Front, one extremist group of the animal rights movement, said, "Make no mistake about it, by any sense or definition [this] is a domestic terrorism group" (Hemphill, 2003). Animal rightists are domestic terrorists?

In the short term and for most of us, there is no reason for the jitters. We are not the scientists who use animal models to unlock secrets of physiology that may improve our health. So far, they have been the primary targets of animal extremists' wrath—people like the two Oregon researchers whose homes and cars were vandalized in December 2007. There is no indication that the extremists will, any time soon, go after you and me for eating a hamburger, keeping a pet, taking meds, or using a pacemaker.

The inconvenient truth is that in the long term, and for all of us, there is cause for concern. The agenda of extreme animal rightists is crystal clear: end the use of all animals as food, clothing, pets, and subjects of medical research. Yet we live longer and healthier lives due to vaccinations, better drugs, and improved information about nutrition and disease prevention—longer lives are the result of animal research.

Noting the impact of these extremists on the nation's health agenda, famed heart surgeon and 2007 congressional gold medal winner Michael DeBakey said, "It is the American public who will decide whether we must tell hundreds of thousands of victims of heart attacks, cancer, AIDS, and other dread diseases that the rights of animals supersede a patient's right to relief from suffering and premature death."

Clarifying definitions will provide a good basis for discussion.

The term *animal welfare* refers to the idea that humans have a responsibility to care for animals and look out for their well-being. Because seeking animal welfare is in line with what is noblest in human nature, it is sometimes called "acting humanely." Most reasonable people agree with this. Researchers reflect these values in subscribing to high-quality care for animals, something codified into law as the Animal Welfare Act. Federal regulations are in place to minimize pain and suffering in research. At the authors' place of employment, the Oregon National Primate Research Center, animals live longer lives than their counterparts in the wild, owing to high-quality food and excellent veterinary care. One sad truth is that our animals get better medical care and nutrition than do many children in the United States.

Animal rights, sometimes used as shorthand for any concern for animals, really means the belief that animals, like humans, possess some inalienable rights. It is our view that while animals do not have such rights—rights and responsibilities are correlative, and animals are unable to take responsibility for their actions—it is our duty as humans and ethical researchers to care for them humanely, just as we care for our pets.

Animal extremists portray themselves as engaged in a "David against Goliath" struggle on behalf of animals, but are they the true animal welfarists? Hardly! In 2006 alone, PETA killed 2,981 dogs, cats, puppies, kittens, and other animals—an astonishing 97 percent of the animals left in their care, according to the group's own records supplied to the Virginia Department of Agriculture and Consumer Services (2006). For comparison, the Virginia Society for the Protection of Animals (which operates in Norfolk, Virginia, as does PETA) euthanized less than 2.5 percent of the 1,404 animals placed with them in 2006. While PETA collects tens of millions in donations by claiming to advocate for the welfare of animals, the group has actually killed 17,400 pets since 1998 (Center for Consumer Freedom, 2008).

PETA's most recently available tax filing (according to Guidestar.org) lists nearly $30 million in income from contributions, gifts, and grants offered by individuals who may believe that it is actually an animal welfare organization that helps strays.

Many of its donors are also unaware that PETA has provided cash to individuals who publicly engaged in a terrorist agenda. A few examples were provided by *Lewiston Morning Tribune* (Idaho) writer Michael Costello. "PETA donated $45,200 to . . . ALF [Animal Liberation Front] terrorist Rodney Coronado's legal defense. (Coronado was convicted in connection with an arson attack at Michigan State University that caused $125,000 worth of damage and destroyed thirty-two years of research data. On December 14, 2007, in a Federal Court in San Diego, he entered a guilty plea to one count of distribution of information related to the assembly of explosives and other charges.) They also . . . 'loaned' Coronado's father $25,000 dollars [*sic*], which to our knowledge, has not been repaid. In 1999, PETA gave $2,000 to David Wilson, a national 'ALF spokesperson.' . . . And sure enough, PETA has contributed to the ALF's sister organization; according to its own IRS filing, in 2000 PETA openly donated $1,500 to the Earth Liberation Front. The Federal Bureau of Investigation (FBI) calls ELF 'the largest and most active US-based terrorist group'" (Costello, 2003).

PETA probably doesn't want its donors to know that. Instead, it directs outrage toward "vivisectors" falsely accused of cruelty to animals to incense donors into reaching more deeply into their pockets.

"We are complete press sluts," the PETA leadership has claimed (ActivistCash.com, 2008b). On that single issue, we agree with PETA.

Although PETA is careful not to openly embrace the assaults, vandalism, and threats perpetrated by some groups, it does not oppose such violence either. Speaking of one animal extremist group whose leaders have been convicted of animal terrorism, PETA president Ingrid Newkirk said, "More power to SHAC [Stop Huntingdon Animal Cruelty] if they can get someone's attention" (ActivistCash.com, 2008b).

The Animal Liberation Front can speak for itself, however. Says one of its leaders, Tim Daley, "In a war you have to take up arms and people will get killed, and I can support that kind of action by gasoline bombing and bombs under cars, and probably at a later stage, the shooting of vivisectors on their doorsteps. It's a war, and there's no other way you can stop vivisectors" (Lovitz, 2007). Jerry Vlasak, head of the ALF Press Office, is equally candid, "I don't think you'd have to kill—assassinate—too many [doctors involved with animal testing]. I think for 5 lives, 10 lives, 15 human lives, we could save a million, 2 million, 10 million non-human lives" (McDonald, 2007).

To "sell" their story, animal extremists rely on the lack of public awareness of tight federal regulation on animal research—random inspections of facilities by the US Department of Agriculture for compliance to the rigorous standards of the Animal Welfare Act.

Animal extremists wrongfully claim that data obtained from animal research cannot be extrapolated to drug development for humans. A recent survey of 150 drug compounds from twelve international pharmaceutical companies found that animal testing *had* significant predictive power to detect most—not all, admittedly—of 221 human toxic events caused by those drugs (Olson et al., 2000).

Animals are important not just in testing for efficacy and safety of drugs but also to the basic research that leads to medical advances. Ironically, animal extremists were decrying the uselessness of our Center's basic investigations in primate stem cell biology on the very day in November 2007 that one of our scientists announced the first cloning of stem cells from non-embryonic primate tissue, subsequently hailed by *Time* magazine as the top discovery of 2007.

Animal extremists often show willful naïveté in considering human health needs. Smallpox, malaria, and polio have been nearly eradicated from much of the world—you no longer see wards of people confined to "iron lungs." Animal research is inextricably tied to improved human health. The first recognition of diabetes as a disease and the explanation of its cause, as well as its first treatment and early management, came directly from animal research conducted in universities. Improvements in treatments for this disease continue to come from these same sources. While these accomplishments are tributes to animal research, extremists fail to recognize that antibiotic-resistant tuberculosis, AIDS, diabetes, and heart disease still need the attention of researchers, who in turn need ethical animal research to advance their studies.

What are we to do when those very researchers are targeted for harassment and violence?

In 2006, members of ALF declared that they left a Molotov Cocktail outside the Bel Air home of Dr. Lynn Fairbanks, the Director of the Center for Primate Neuroethology at UCLA's Neuropsychiatric Institute. Actually, the explosive device was placed on the porch of the faculty member's seventy-year-old neighbor. Fortunately, the timing device failed (Editors, 2006).

About one year later, a group calling themselves the Animal Liberation Brigade claimed responsibility for placing a lighted incendiary device next to a car parked at the home of Dr. Arthur Rosenbaum, who is chief of Pediatric Ophthalmology at UCLA's Jules Stein Eye Institute. Authorities described the event as "domestic terrorism." The delivery address was correct this time, but fortunately for Dr. Rosenbaum and his neighbors, the device did not ignite due to ineptness on the part of the "activists" (McDonald, 2007). Police noted that the device had the potential to create great harm.

Another talented researcher, Dr. Dario Ringach, ultimately gave in to animal extremists, promising to stop his research on monkeys in exchange for cessation of harassment of his family, including his young children. "You win," he e-mailed them (Epstein, 2006).

It is worth mentioning that Ringach's, Rosenbaum's, and Fairbanks's research all were humanely conducted and met federal standards.

As you can see, we are not talking about peaceful protests here. As the editors of *Nature Neuroscience* (Editors, 2006) put it, "Over several years, the researchers have been subjected to a campaign of harassment that included demonstrations at their homes and pamphlets distributed to their neighbors, as well as threatening phone calls and emails. Elsewhere, targets of similar protests have had abuse shouted through bullhorns or painted on their homes or cars, doorbells rung repeatedly, and windows smashed or doors broken down while family members were in the house. Animal-rights Web sites post the names of scientists' spouses and children, along with their ages and schools."

According to the Foundation for Biomedical Research (2006), a handful of illegal acts by animal extremist groups in 1994 had risen to a hundred such attacks ten years later. Society for Neuroscience members reported more attacks in the first six months of 2007 than in the five-year period from 1999 to 2003, prompting that organization to release, in February 2008, the document "Best Practices for Protecting Researchers and Research: Recommendations for Universities and Institutions." It calls on universities and research institutions to "ensure the ability of researchers to conduct their research in a safe environment" and to bear the primary burden of "maintaining the fundamental principles of academic freedom." It urges support for government efforts worldwide "to combat these anti-research campaigns."

Even though these and similar events send a chilling message to researchers and young people considering the field of biomedical research, they are poorly reported in the general media. The public doesn't hear about the impact this has on students viewing research as a potential career—or those already active in the field. Nor does it hear how the loss of talented researchers threatens creation of the new knowledge needed to devise cures.

Former University of Iowa President (now of Cornell University) David Skorton worries that researchers and students are being scared off by attacks from animal rights extremists. ALF, which took credit for break-ins and destruction at the University of Iowa, distributed the home addresses of researchers who conduct animal research to animal activists. "Publicizing this personal information was blatant intimidation," Skorton pointed out, adding that because of safety worries, "numerous researchers are even concerned about allowing their children to play in their own yards." He acknowledged that the cost of such intimidation is difficult to nail down, but he believed it "could be measured by many, many lives" that might not be saved by medical advances (Lederman, 2005).

His words echoed those of Richard Bianco, vice president for research at the University of Minnesota, where an attack by vandals in 1999 caused more than $2 million in damage. "The financial aspect is the least of our problems. . . . The hardest thing is people see this and don't want to go into science," he said. "Why would they go into science when they can have their work threatened like that?" (Agri News, 2005).

Senator Orrin Hatch understood. "When research laboratories and university researchers are targeted and attacked, the ones who lose most are those who are living with a disease or who are watching a loved one struggling with a devastating illness" (Davidson, 2004).

Because it seeks to stop ethical medical research, animal extremism is bad for our health. There are several steps the public can take to help reduce this threat to public health and good science.

We should be very careful in our giving to ensure that our contributions don't wind up aiding those who use the weapons of intimidation and violence. At the same time, we want to support organizations with proven records of caring for animals or of providing humane education that enhances the care received by laboratory animals.

If we have scientists who are neighbors, we can offer to organize a neighborhood watch and volunteer to speak to the media about how we have benefited from animal research if their homes are vandalized.

While mentioning the importance of speaking out, we can contact the local organization or university supporting research (a list, by state, is available at www.statesforbiomed.org) and offer to testify about what animal research has meant to someone in our family. When our kids come home from school with animal rights literature that denigrates animal research, we can contact their teachers to ask that they invite a researcher or veterinarian from a local university or research center to visit the class or even take the students on a tour of their facility.

It is because we thought it was time to sound the alarm that we wrote *The Animal Research War*[1] describing what we think the public needs to know about this quiet war—"quiet," because it is seldom reported in the news. We want to tell people about the battle zone that we, as animal researchers, live in every day. We also want to communicate the benefits of animal research, past and potential, as well as the compassion with which researchers care for laboratory animals. If this war is lost, it is all who struggle with disease—that means all of us, sooner or later—who will bear the burden.

NOTE

1. The authors have written *The Animal Research War* (Macmillan/Palgrave, 2008), a personal account of what it is like to be terrorized, an analysis of the effect of animal extremists on the world's scientists, and the way in which the public and legal system is changing its views on animals. The book traces the evolution of the animal rights movement, profiles its leadership, and reveals the remarkable value of the research enterprise.

REFERENCES

ActivistCash.com. *Alex Pacheco Biography.* 2008a. Available online at http://www
.activistcash.com/biography.cfm/bid/1459.
————. *Ingrid Newkirk Quotes.* 2008b. Available online at http://www.activistcash.com/
biography_quotes.cfm/bid/456.
Agri News. Attacks on Animal Research Labs Carry Heavy Costs." June 28 2005. Available online at http://webstar.postbulletin.com/agrinews/285455810602578.bsp.
Center for Consumer Freedom. *PETAkillsAnimals.com.* 2008. Available online at http://petakillsanimals.com/index.cfm.
Costello, Michael. "Zero tolerance for PETA." *Lewiston Morning Tribune,* October 10 2003. Available online at http://michaelcostello.blogspot.com/2003_10_01_archive.html.
Davidson, Lee. "Hatch Flays Animal-rights 'Terrorists.'" (Salt Lake City). May 19 2004. Available online at http://findarticles.com/p/articles/mi_qn4188/is_20040519/ai_n11459837.
Editors. "Fighting Animal Rights Terrorism." *Nature Neuroscience* 9 (2006): 1195. Available online at http://www.nature.com/neuro/journal/v9/n10/full/nn1006-1195.html.

Epstein, David. "Throwing in the Towel." *Inside Higher Education.* August 22 2006. Available online at http://www.insidehighered.com/news/2006/08/22/animal.

Foundation for Biomedical Research. "Illegal Incidents Report: A 25 Year History of Illegal Activities by Eco and Animal Extremists." 2006. Available online at http://www.fbresearch.org/AnimalActivism/IllegalIncidents/IllegalIncidentsRe port.pdf.

Fox, Michael W. *Inhumane Society: The American Way of Exploiting Animals.* New York: St. Martin's Press, 1992.

Hemphill, Kendal. *Domestic Terrorists.* 2003. Available online at www.kingsnake.com/ wths/clergyman.htm.

Lederman, Doug. "Animal Rights and Eco-terrorism." *Inside Higher Education.* May 19 2005. Available online at http://www.insidehighered.com/news/2005/05/19/animal.

Lovitz, Dora. "Animal Lovers and Tree Huggers Are the New Cold-blooded Criminals?" *Journal of Animal Law* 3 (2007): 81. Available online at http://www .animallaw.info/journals/jo_pdf/Journal%20of%20Animal%20Law%20Vol%203.pdf.

McDonald, Patrick Range. Monkey Madness at UCLA. August 8 2007. Available online at www.laweekly.com/news/news/monkey-madness-at-ucla/16986/.

Olson, H., et al. "Concordance of the Toxicity of Pharmaceuticals in Humans and in Animals." *Regulatory Toxicology and Pharmacology* 32 (2000): 56–67.

Society for the Study of Neuroscience "Best Practices for Protecting Researchers and Research: Recommendations for Universities and Institutions." 2008. Available online at http://www.sfn.org/skins/main/pdf/gpa/Best_Practices_for_Protecting.pdf.

Virginia Department of Agriculture. "Animal Reporting Online: People for the Ethical Treatment of Animals" 2006. Available online at http://www.virginia .gov/vdacs_ar/cgi-bin/Vdacs_search.cgi?link_select=facility&form =fac_select&fac_num=157&year=2006.

PREDATOR PANIC:
A Closer Look

Benjamin Radford

"Protect the children." Over the years that mantra has been applied to countless real and perceived threats. America has scrambled to protect its children from a wide variety of dangers including school shooters, cyberbullying, violent video games, snipers, Satanic Ritual Abuse, pornography, the Internet, and drugs.

Hundreds of millions of taxpayer dollars have been spent protecting children from one threat or other, often with little concern for how expensive or effective the remedies are—or how serious the threat actually is in the first place. So it is with America's latest panic: sexual predators.

According to lawmakers and near-daily news reports, sexual predators lurk everywhere: in parks, at schools, in the malls—even in children's bedrooms, through the Internet. A few rare (but high-profile) incidents have spawned an unprecedented deluge of new laws enacted in response to the public's fear. Every state has notification laws to alert communities about former sex offenders. Many states have banned sex offenders from living in certain areas, and are tracking them using satellite technology. Other states have gone even further; state emergency leaders in Florida and Texas, for example, are developing plans to route convicted sex offenders away from public emergency shelters during hurricanes. "We don't want them in the same shelters as others," said Texas Homeland Security Director Steve McCraw. (How exactly thousands of desperate and homeless storm victims are to be identified, screened, and routed in an emergency is unclear.)

AN EPIDEMIC?

To many people, sex offenders pose a serious and growing threat—especially on the Internet. Attorney General Alberto Gonzales has made them a top priority this year, launching raids and arrest sweeps. According to Senate Majority Leader Bill Frist, "the danger to teens is high." On the April 18, 2005, *CBS Evening News* broadcast, correspondent Jim Acosta reported that "when a child is missing, chances are good it was a convicted sex offender." (Acosta is incorrect: If a child goes missing, a convicted sex offender is among the least likely explanations, far behind runaways, family abductions, and the child being lost or injured.) On his NBC series "To Catch a Predator," *Dateline* reporter Chris Hansen claimed that "the scope of the problem is immense," and "seems to be getting worse." Hansen claimed that Web predators are "a national epidemic," while Alberto Gonzales stated that there are 50,000 potential child predators online.

Sex offenders are clearly a real threat, and commit horrific crimes. Those who prey on children are dangerous, but how common are they? How great is the danger? After all, there are many dangers in the world—from lightning to Mad Cow Disease to school shootings—that are genuine but very remote. Let's examine some widely repeated claims about the threat posed by sex offenders.

ONE IN FIVE?

According to a May 3, 2006, ABC News report, "One in five children is now approached by online predators." This alarming statistic is commonly cited in news stories about prevalence of Internet predators, but the factoid is simply wrong. The "one in five statistic" can be traced back to a 2001 Department of Justice study issued by the National Center for Missing and Exploited Children ("The Youth Internet Safety Survey") that asked 1,501 American teens between 10 and 17 about their online experiences. Anyone bothering to actually read the report will find a very different picture. Among the study's conclusions: "Almost one in five (19 percent) . . . received an unwanted sexual solicitation in the past year." (A "sexual solicitation" is defined as a "request to engage in sexual activities or sexual talk or give personal sexual information that were unwanted or, whether wanted or not, made by an adult." Using this definition, one teen asking another teen if her or she is a virgin—or got lucky with a recent date—could be considered "sexual solicitation.") Not a single one of the reported solicitations led to any actual sexual contact or assault. Furthermore, almost half of the "sexual solicitations" came not from "predators" or adults but from other teens—in many cases the equivalent of teen flirting. When the study examined the type of Internet "solicitation" parents are most concerned about (namely someone who asked to meet the teen somewhere, called the teen on the telephone, or sent gifts), the number drops from "one in five" to just 3 percent.

This is a far cry from an epidemic of children being "approached by online predators." As the study noted, "The problem highlighted in this survey is not just adult males trolling for sex. Much of the offending behavior comes from other youth [and] from females." Furthermore, "Most young people seem to know what to do to deflect these sexual 'come-ons.'" The reality is far less grave than the ubiquitous "one in five" statistic suggests.

RECIDIVISM REVISITED

Much of the concern over sex offenders stems from the perception that if they have committed one sex offense, they are almost certain to commit more. This is the reason given for why sex offenders (instead of, say, murderers or armed robbers) should be monitored and separated from the public once released from prison. While it's true that serial sex offenders (like serial killers) are by definition likely to strike again, the reality is that very few sex offenders commit further sex crimes.

The high recidivism rate among sex offenders is repeated so often that it is accepted as truth, but in fact recent studies show that the recidivism rates for sex offenses is not unusually high. According to a US Bureau of Justice Statistics study ("Recidivism of Sex Offenders Released from Prison in 1994"), just five percent of sex offenders followed for three years after their release from prison in 1994 were arrested for another sex crime. A study released in 2003 by the Bureau of Justice Statistics found that within three years, 3.3 percent of the released child molesters were arrested again for committing another sex crime against a child. Three to five percent is hardly a high repeat offender rate.

In the largest and most comprehensive study ever done of prison recidivism, the Justice Department found that sex offenders were in fact *less* likely to reoffend than other criminals. The 2003 study of nearly 10,000 men convicted of rape, sexual assault, and child molestation found that sex offenders had a re-arrest rate 25 percent lower than for all other criminals. Part of the reason is that serial sex offenders—those who pose the greatest threat—rarely get released from prison, and the ones who do are unlikely to re-offend. If released sex offenders are in fact no more likely to re-offend than murderers or armed robbers, there seems little justification for the public's fear and the monitoring laws targeting them. (Studies also suggest that sex offenders living near schools or playgrounds are no more likely to commit a sex crime than those living elsewhere.)

While the abduction, rape, and killing of children by strangers is very, very rare, such incidents receive a lot of media coverage, leading the public to overestimate how common these cases are. (See John Ruscio's article "Risky Business: Vividness, Availability, and the Media Paradox" in the March/April 2000 *Skeptical Inquirer.*)

WHY THE HYSTERIA?

There are several reasons for the hysteria and fear surrounding sexual predators. The predator panic is largely fueled by the news media. News stories emphasize the dangers of Internet predators, convicted sex offenders, pedophiles, and child abductions. The *Today Show*, for example, ran a series of misleading and poorly designed hidden camera "tests" to see if strangers would help a child being abducted.[1]

Dateline NBC teamed up with a group called Perverted Justice to lure potential online predators to a house with hidden cameras. The program's ratings were so high that it spawned six follow-up "To Catch a Predator" specials. While the many men captured on film supposedly showing up to meet teens for sex is disturbing, questions have been raised about Perverted Justice's methods and accuracy. (For example, the predators are often found in unmoderated chatrooms frequented by those looking for casual sex—hardly places where most children spend their time.) Nor is it surprising that out of over a hundred million Internet users, a fraction of a percentage might be caught in such a sting.

Because there is little hard data on how widespread the problem of Internet predators is, journalists often resort to sensationalism, cobbling a few anecdotes and interviews together into a trend while glossing over data suggesting that the problem may not be as widespread as they claim. But good journalism requires that personal stories—no matter how emotional and compelling—must be balanced with facts and context. Much of the news coverage about sexual predation is not so much wrong as incomplete, lacking perspective.

MORAL PANICS

The news media's tendency toward alarmism only partly explains the concern. America is in the grip of a moral panic over sexual predators, and has been for many months. A *moral panic* is a sociological term describing a social reaction to a false or exaggerated threat to social values by moral deviants. (For more on moral panics, see Erich Goode and Nachman Ben-Yehuda's 1994 book *Moral Panics: The Social Construction of Deviance*.)

In a discussion of moral panics, sociologist Robert Bartholomew points out that a defining characteristic of the panics is that the "concern about the threat posed by moral deviants and their numerical abundance is far greater than can be objectively verified, despite unsubstantiated claims to the contrary." Furthermore, according to Goode and Ben-Yehuda, during a moral panic "most of the figures cited by moral panic 'claims-makers' are wildly exaggerated."

Indeed, we see exactly this trend in the panic over sexual predators. News stories invariably exaggerate the true extent of sexual predation on the Internet; the magnitude of the danger to children, and the likelihood that sexual predators will strike. (As it turns out, Attorney General Gonzales had taken his 50,000 Web predator statistic not from any government study or report, but from NBC's *Dateline* TV show. *Dateline*, in turn, had broadcast the number several times without checking its accuracy. In an interview on NPR's *On the Media* program, Hansen admitted that he had no source for the statistic, and stated that "It was attributed to, you know, law enforcement, as an estimate, and it was talked about as sort of an extrapolated number.") According to *Wall Street Journal* writer Carl Bialik, journalists "often will use dubious numbers to advance that goal [of protecting children] . . . one of the reasons that this is allowed to happen is that there isn't really a natural critic. . . . Nobody really wants to go on the record saying, 'It turns out this really isn't a big problem.'"

PANICKY LAWS

Besides needlessly scaring children and the public, there is a danger to this quasi-fabricated, scare-of-the-week reportage: misleading news stories influence lawmakers, who in turn react with genuine (and voter-friendly) moral outrage. Because nearly any measure intended (or claimed) to protect children will be popular and largely unopposed, politicians trip over themselves in the rush to endorse new laws that "protect the children."

Politicians, child advocates, and journalists denounce current sex offender laws as ineffective and flawed, yet are rarely able to articulate exactly why new laws are needed. Instead, they cite each news story about a kidnapped child or Web predator as proof that more laws are needed, as if sex crimes would cease if only the penalties were harsher, or enough people were monitored. Yet the fact that rare crimes continue to be committed does not necessarily imply that current laws against those crimes are inadequate. By that standard, any law is ineffective if someone violates that law. We don't assume that existing laws against murder are ineffective simply because murders continue to be committed.

In July 2006, teen abduction victim Elizabeth Smart and child advocate John Walsh (whose murdered son Adam spawned *America's Most Wanted*) were instrumental in helping pass the most extensive national sex offender bill in history. According to the bill's sponsor, Senator Orrin Hatch (R-Utah), Smart's 2002 "abduction by a convicted sex offender" might have been prevented had his bill been law. "I don't want to see others go through what I had to go through," said Smart. "This bill should go through without a thought." Yet bills passed without thought rarely make good laws. In fact, a closer look at the cases of Elizabeth Smart and Adam Walsh demonstrate why sex offender registries *do not* protect children. Like most people who abduct children, Smart's kidnapper, Brian David Mitchell, was not a convicted sex offender. Nor was Adam Walsh abducted by a sex offender. Apparently unable to find a vocal advocate for a child who had actually been abducted by a convicted sex offender, Hatch used Smart and Walsh to promote an agenda that had nothing to do with the circumstances of their abductions. The two high-profile abductions (neither by sex offenders) were somehow claimed to demonstrate the urgent need for tighter restrictions on sex offenders. Hatch's bill, signed by President Bush on July 27, 2006, will likely have little effect in protecting America's children.

The last high-profile government effort to prevent Internet predation occurred in December 2002, when President Bush signed the Dot-Kids Implementation and Efficiency Act into law, creating a special safe Internet "neighborhood" for children. Elliot Noss, president of Internet address registrar Tucows Inc., correctly predicted that the domain had "absolutely zero" chance of being effective. The ".kids.us" domain is now a largely ignored Internet footnote that has done little or nothing to protect children.

TRAGIC MISDIRECTION

The issue is not whether children need to be protected; of course they do. The issues are whether the danger to them is great, and whether the measures proposed will ensure their safety. While some efforts—such as longer sentences for repeat offenders—are well reasoned and likely to be effective, those focused on separating sex offenders from the public are of little value because they are based on a faulty premise. Simply knowing where a released sex offender lives—or is at any given moment—does not ensure that he or she won't be near potential victims. Since relatively few sexual assaults are committed by released sex offenders, the concern over the danger is wildly disproportionate to the real threat. Efforts to protect children are well intentioned, but legislation should be based on facts and reasoned argument instead of fear in the midst of a national moral panic.

The tragic irony is that the panic over sex offenders distracts the public from the real danger, a far greater threat to children than sexual predators: parental abuse and neglect. The vast majority of crimes against children are committed not by released sex offenders but instead by the victim's own family, church clergy, and family friends. According to a 2003 report by the Department of Human Services, hundreds of thousands of children are abused and neglected each year by their parents and caregivers, and more than 1,500 American children died from that abuse in 2003—most of the victims under four years old. That is more than *four children killed per day*—not by convicted sexual offenders or Internet predators, but by those entrusted to care for them. According to the National Center for Missing and Exploited Children, "danger to children is greater from someone they or their family knows than from a stranger."

If journalists, child advocates, and lawmakers are serious about wanting to protect children, they should turn from the burning matchbook in front of them to face the blazing forest fire behind them. The resources allocated to tracking ex-felons who are unlikely to re-offend could be much more effectively spent on preventing child abuse in the home and hiring more social workers.

Eventually this predator panic will subside and some new threat will take its place. Expensive, ineffective, and unworkable laws will be left in its wake when the panic passes. And no one is protecting America from that.

NOTE

1. For more on this, see my article "Stranger Danger: 'Shocking' TV Test Flawed" at www.mediamythmakers.com/sgi-bin/mediamythmakers.cgi.

THE CASE FOR DECENTRALIZED GENERATION OF ELECTRICITY

Thomas R. Casten and Brennan Downes

Electricity was originally generated at remote hydroelectric dams or by burning coal in the city centers, delivering electricity to nearby buildings and recycling the waste heat to make steam to heat the same buildings. Rural houses had no access to power. Over time, coal plants grew in size, facing pressure to locate far from population because of their pollution. Transmission wires carried the electricity many miles to users with a 10 to 15 percent loss, a difficult but tolerable situation. Because it is not practical to transmit waste heat over long distances, the heat was vented. There was no good technology available for clean, local generation, so the wasted heat was a tradeoff for cleaner air in the cities. Eventually a huge grid was developed and the power industry built all-new generation in remote areas, far from users. All plants were specially designed and built on site, creating economies of scale. It cost less per unit of generation to build large plants than to build smaller plants. These conditions prevailed from 1910 through 1960, and everyone in the power industry and government came to assume that remote, central generation was optimal, that it would deliver power at the lowest cost versus other alternatives.

However, technology has improved and natural gas distribution now blankets the country. By 1970, mass-produced engines and turbines cost less per unit of capacity than large plants, and the emissions have been steadily reduced. These smaller engines and gas turbines are good neighbors, and can be located next to users in the middle of population centers. Furthermore, the previously wasted heat can be recycled from these decentralized generation plants to displace boiler fuel and essentially cut the fuel for electric generation in half, compared to remote or central generation of the same power.

But the industry had ossified. Electric monopolies were allowed to charge rates to give a fair return on capital employed. To prevent excessive or monopoly profits, the utilities have long been required to pass 100 percent of any gain in efficiency to the users. This leaves utilities with no financial incentive to adopt new technologies and build decentralized generation that recycles heat. In fact, such local generation erodes the rationale for continued monopoly protection—if one can make cheap power at every factory or high-rise apartment house, why should society limit competition?

Congress tried to open competition a little bit in 1978, and some independent power companies began to develop on-site generation wherever they could find ways around the monopoly regulation. One author (Casten) was one of those early pioneers, working to develop more efficient decentralized generation since 1975.

This article summarizes extensive research into the economically optimal way to build new power generation in each of the past 30 years, given then available technology, capital costs, and fuel prices, and concludes that the continuing near-universal acceptance of the "central generation paradigm" is wrong. The result is a skeptical look at the world's largest industry—the electric power industry—with surprising conclusions.

Power industry regulations largely derive from the unquestioned belief that central generation is optimal. However we believe the conventional "central generation paradigm" is based on last century's technology. Meeting the world's growing appetite for electric power with conventional central generation will severely tax capital markets, fossil fuel markets, and the global environment. The International Energy Agency's (IEA) 2002 World Energy Outlook Reference Case—based on present policies—presents a frightening view of the next thirty years.[1] The Reference Case says world energy demand will grow by two-thirds, with fossil fuels meeting 90 percent of the increase. World electrical demand doubles, requiring construction of nearly 5,000 gigawatts of new generating capacity, equivalent to adding six times current US electric generating capacity. The generation alone will cost $4.2 trillion, plus transmission and distribution (T&D) costs of $6.6 trillion (2004 US dollars). Under this projection, global carbon dioxide emissions increase by 70 percent; see figure 1.

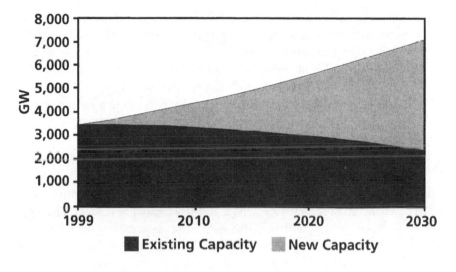

Figure 1. World installed electricity generation capacity.

The Reference Case assumes that the energy policies of each government in 2002 continue without change, a modest evolution of technology, and continued reliance on central generation of electric power, which is consistent with most existing policies and regulations. The IEA projections assume that central generation is the optimal approach, given today's technology.

The IEA report is silent on the need for (or capital cost of) new T&D, even though existing T&D is far from adequate. There were 105 reported grid failures in the United States between 2000 and 2003, and eleven of those outages affected more than a half million people.[2] US consumers paid $272 billion for electricity in 2003,[3] plus power outage costs, estimated between $80 billion and $123 billion per year. Outages thus add 29 percent to 45 percent to the cost of US power.[4] The T&D situation is worse in developing countries, where 1.6 billion people lack any access to electric power and many others are limited to a few hours of service per day. Satisfying expected load growth with central generation will clearly require at least comparable construction of T&D capacity.

Close examination of past power industry options and choices suggests that load growth can be met with just over half the fossil fuel and pollution associated with conventional central generation. *We had better get this world energy expansion right.* Consider these points:

- The power industry has not deployed optimal technology over the past thirty years.
- The universally accepted "Central Generation Paradigm" prevents optimal energy decisions.

- Decentralized generation (DG), using the same technologies used by remote central generation, significantly improves every key outcome from power generation.
- Meeting global load growth with decentralized energy can save $5 trillion of capital, lower the cost of incremental power by 35–40 percent, and reduce CO_2 emissions by 50 percent versus the IEA central generation dominated reference case.

A BRIEF HISTORY OF ELECTRIC GENERATION

Figure 2 shows that US net electric efficiency peaked in about 1910, when nearly all generation recycled waste heat and was located near users. That efficiency dropped to 33 percent over the next fifty years as the power industry moved to electric-only central generation. Industry efficiency has not improved in four decades. Technology improved, enabling conversion of fuel to electricity to rise from 7 percent at commercial inception to 33 percent by 1960. The best electric-only technology now converts more than 50 percent of the fuel to power, but the industry's average efficiency has not improved in forty-three years. No other industry wastes two-thirds of its raw material; no other industry has stagnant efficiency; no other industry gets less productivity per unit output in 2004 than it did in 1904.

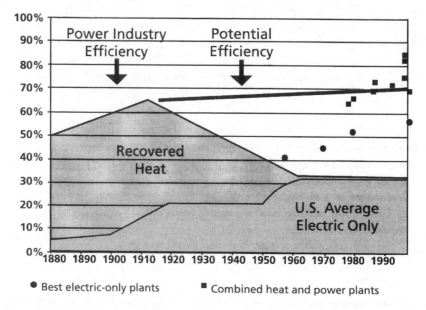

Figure 2. US electricity generating efficiency, 1880 to present.

Early generating technology converted 7 to 20 percent of the fuel to electricity, making electric-only production quite expensive. To reduce fuel costs, energy entrepreneurs, including Thomas Edison, built generating plants near thermal users and recycled waste heat, increasing net electric efficiency to as much as 75 percent. A second wave of technical progress after World War II drove electric-only efficiencies to 33 percent (after distribution losses) and increased individual plant size to between 500 and 1,000 megawatts. Central or remote generation of electricity only, while still wasting two-thirds of the input energy, became the standard. Buttressed by monopoly protection, utilities fought competing on-site generation and, by 1970, replaced all but 3 to 4 percent of local generation, ending waste heat recycling. Government regulations, developed over the first 90 years of commercial electricity, institutionalized central generation.

The third wave of technical progress should have reversed the central generation trend. Modern power plants emit only 1 to 2 percent as much nitrogen oxides as 1970 plants, come in all sizes, burn all fuels, and are good neighbors. Many technical advances make local or distributed generation technically and economically feasible and enable society to return to energy recycling, displacing boiler fuel and doubling net electric efficiency. However, protected from competition and rewarded by obsolete rules, the power industry continues to build remote plants and ignores opportunities to recycle energy.

The squares in figure 2 represent the alternative to central or remote generation. These are actual plants employing central plant generation technologies that are located near users. These combined heat and power (CHP) plants deploy the best modern electric-only technology and achieve 65 to 97 percent net electrical efficiency by recycling normally wasted heat and by avoiding transmission and distribution losses. United States Energy Information Agency (EIA) records show 931 distributed generation plants with 72,800 megawatts of capacity, about 8.1 percent of US generation. These plants demonstrate the technical and economic feasibility of doubling US electricity efficiency.

Nevertheless, the United States and world power industry ignores—and indeed actively fights against—distributed generation. Conventional central generation plants dump two-thirds of their energy into lakes, rivers, and cooling towers, while factories and commercial facilities burn more fuel just to produce heat, which is then thrown away. We believe the power industry has not made wise or efficient choices—we will set out to test this thesis with data.

A FLAWED WORLDWIDE HEAT & POWER SYSTEM

To determine whether the power industry made optimal choices, we analyzed EIA data on all 5,242 reported generation plants, separating plants built by firms with monopoly-protected territories and plants built by independent power producers. We calculated what price per KWh would be required for each of four central generation technologies, built in each year, to provide a fair return on capital.[5]

We also analyzed distributed generation (DG) technology choices. Several clarifications are necessary:

- Distributed generation is any electric generating plant located next to users.
- DG is not a new concept. Edison built his first commercial electric plant near Wall Street in lower Manhattan, and he recycled energy to heat surrounding buildings.
- DG plants employ all of the technologies that are used in central generation.
- DG plant capacities range from a few kilowatts to several hundred megawatts, depending on the users' needs. We have installed 40-kilowatt backpressure steam turbines in office buildings that recycle steam pressure drop, and managed a 200-megawatt coal-fired CHP plant serving Kodak's world headquarters in Rochester, New York.
- DG can use renewable energy, but not every renewable energy plant is DG. Solar photovoltaic panels on individual buildings or local windmills are distributed generation, while large hydro and wind farms are central generation requiring transmission and distribution (T&D).
- DG uses all fuels, including nuclear. Modern naval vessels generate power with nuclear reactors and then recycle waste heat to displace boiler fuel.

Power generated near users avoids the need for T&D. We have assumed each kilowatt of new DG will require net T&D investment equal to only 10 percent of a kilowatt, for backup services.[6] We assume DG plants require a 50 percent higher average cost of capital (12 percent versus 8 percent) due to risks and transaction costs. Industrial companies that install DG see power generation as a non-core activity and demand 35 to 50 percent rates of return, but this analysis focuses only on power companies' cost of capital.

Figures 3a and 3b depict our findings. The gray bars show the average price of power to all US consumers in each year in both figures. Figure 3a looks at coal and recycled industrial waste energy and shows the retail price per megawatt-hour (MWh) needed to fully fund new plants using four power generation technologies. The highest cost line, ranging from $60 to $83 per MWh, is for a new central generation with coal. This would have slightly reduced the average cost of delivered power through most of the period. However, had the coal plants been built at or near large users of thermal energy, as depicted by the line with circles, such plants would have required only $40 to $60 per MWh, saving $20 or more per MWh. The two lowest lines show the delivered price of power from recycling industrial waste energy (with square box) and converting steam pressure drop into power (with triangles). Recycling waste energy would have saved $35 to $65 per MWh throughout the period.

Figure 3b depicts the same average retail prices per megawatt-hour through the period, but looks at the retail price per MWh needed to fully fund new natural gas plants, and then compares this again with the prices needed to fund waste energy recycling. The cost of producing power with natural gas in a central or remote station, depicted by the top solid line, varied a great deal over the period, but was nearly always higher than the retail average price. Gas turbine manufacturers steadily improved the efficiency of the turbines, but natural gas prices rose sharply prior to 1984 and then fell through 1999. New gas plants built locally, near thermal users, could recycle the exhaust heat to displace boiler fuel and required $30 to $40 per MWh less than using the same technology (gas turbines) and same fuel (natural gas) in remote plants that cannot recycle the waste heat. The lower two lines depict the lower required costs from recycling waste energy and are identical to the lines in figure 3a, but added to show the comparison with gas options.

Thermal plants generate steam by burning fossil fuel in boilers. The steam then drives condensing steam turbines. Thermal generation technology matured in the mid-fifties, achieving maximum electric-only efficiency of 38 to 40 percent, before line losses. Over the entire period, new central oil and gas thermal plants

Energy Source: Coal

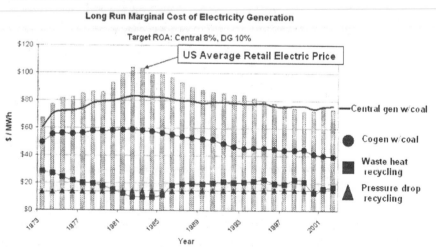

Figure 3a Long-term US marginal cost of coal and recycled energy generation options.

Energy Source: Natural Gas

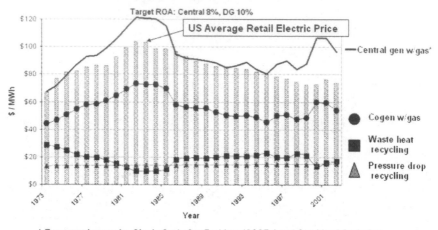

* For natural gas units, Single Cycle Gas Turbines (SCGTs) and Combined Cycle Gas Turbines (CCGTs), marginal costs have been averaged together for both central and distributed generation cases.

Figure 3b. Long-term US marginal cost of natural gas and recycled energy electric generation options.

required prices well above average retail. Gas turbines use a different cycle; the technology improved dramatically over the period. Simple cycle gas turbine plants required prices similar to those of gas-fired thermal plants until 1985–1990, when improving turbine efficiency reduced fuel and lowered required prices. New coal plants could sell power for below average retail prices each year until 1998. However, environmental rules blocked coal plants in many states.

Combined cycle gas turbine plants (CCGTs) are the same gas turbines described above, but the plants also make steam with the turbine exhaust to drive a second power generation cycle—a condensing steam turbine. The first commercial applications of CCGTs were in 1974. These plants cost less to build than an oil and gas thermal plant and initially achieved 40 percent efficiency, which rose to 55 percent by 1995.

DISTRIBUTED GENERATION RECYCLES ENERGY TO REDUCE COSTS

Distributed generation, or DG, involves building the energy-generation technologies near thermal users and recycling normally wasted heat. There is enormous economic value in recycling energy. Burning coal in combined heat and power plants saves $11 to $27 per MWh versus burning coal in new central plants. Simple cycle gas turbine plants built near users save $25 to $60 per MWh versus the same technology producing only electricity. Building combined cycle gas turbine plants near users and recycling waste heat saves even more money, reducing required costs by $25 per MWh versus the same technology built remote from users.

The lowest-cost power avoids burning any extra fossil fuel by recycling waste energy from process industries. Process industries use fossil fuel or electricity to transform raw materials and then discard energy in three forms including hot exhaust gas, flare gas, and pressure drop. Local "bottoming cycle" generation can recycle this waste into heat and/or power.

POWER INDUSTRY CHOICES FOR NEW CAPACITY

An ideal approach would build all possible plants requiring the lowest retail price per megawatt-hour first and then build plants with the next lowest needed retail price, etc.

To determine whether the electric power industry made optimal choices, we analyzed all power plants built since 1973. The new generation built in each two-year period by monopolies, which we defined as any utility with a protected distribution territory, is seen in figure 4. Monopoly utilities include investor-owned utilities, cooperatives, municipal utilities, and state and federally owned utilities. They collectively built 435,000 megawatts of new generation, but ignored energy recycling, even though it was always the cheapest option. They continued to build oil and gas thermal plants long after CCGT plants were a cheaper central option. Monopoly utilities were slow to make optimal choices among central plant technologies and completely ignored the more cost-effective distributed use of the same technologies.

Figure 5 shows the 175,000 MW of new generation built by independent power producers (IPP's) since 1973. Most new IPP plants were distributed generation and/or combined cycle plants until the last four years. The price spikes of 1998–2000 apparently induced IPP companies to install simple cycle gas turbines for peaking. Prior to 1978 passage of the Public Utility Policy Regulatory Policy Act (PURPA) it was illegal to build generation as a third party. Between 1978 and the law change in 1992, IPPs were allowed to build qualifying facilities—those that recycled at least 10 percent of the fuel's energy for heat use, or utilized certain waste fuels. After 1992, IPPs could legally build remote electric-only generation plants.

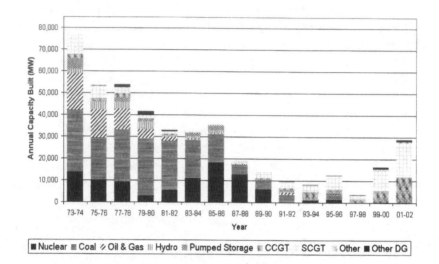

Figure 4. Annual US utility additions of electricity generating capacity by technology, 1973–2002.

For another view of industry choices, we divided plants built since 1973 into those recycling and not recycling energy. Generating plants that recycle energy must be near thermal users or near sources of industrial waste energy. Figure 6 shows that only 1.2 percent—or 5,000 of the 435,000 megawatts of new genera-

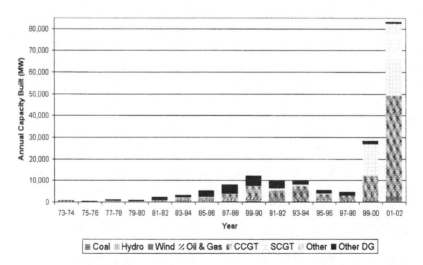

Figure 5. Total US independent power producers utility additions of electronic generating capacity by technology, 1973–2002.

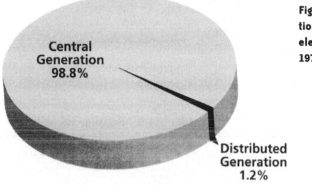

Figure 6. Total generation capacity built by US electric utilities, 1973–2002.

tion built by monopolies over the thirty-year period—recycled energy. We doubt that these choices would be profitable in a competitive marketplace.

Independent power producers built 34 percent of their total capacity as DG plants, at or near users. Figure 7 depicts the mix of central and distributed power built by IPPs since 1978.

Finally, we estimated the potential generation from the least-cost options—those plants that recycle industrial process waste energy. EPA aerometric data and other industry analyses suggest that US industrial waste energy would power 40,000 to 100,000 megawatts with no incremental fossil fuel and no incremental pollution.[7] However, EIA plant data show only 2,200 megawatts of recycled industrial energy capacity, 2.2 to 5 percent of the potential.[8]

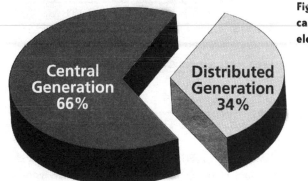

Figure 7. Generation capacity built by US electric IPPs, 1973–2002.

It seems clear that the power industry has made poor choices that have increased cost and decreased efficiency. These data show that utilities eschewed least-cost generating technologies, effectively increasing prices to all customers.

MEETING EXPECTED US LOAD GROWTH WITH LOCAL GENERATION

Our colleagues built a model to determine the best way to satisfy projected load growth for any nation over the next two decades.[9] The model incorporates relevant factors for central and distributed electric generation technologies, including projected improvements in cost, efficiency, and availability of each technology. The model assumes new central generation will require 100 percent new transmission and distribution and new decentralized generation will require new T&D equal to 10 percent of added generating capacity. The model assumes 9 percent line losses for central power, equal to US losses for 2002, and 2 percent net line losses for DG power.

Although the future surely includes some mix of central and decentralized generation, the model calculates the extreme cases of meeting all load growth with central generation, or meeting all growth with decentralized generation. Local generation that recycles energy improves every important outcome versus full reliance on central generation. Figure 8 compares the extreme cases. Full reliance on DG for expected US load growth would avoid $326 billion in capital by 2020, reduce incremental power costs by $53 billion, NO_x by 58 percent, and SO_2 by 94 percent. Full DG lowers carbon dioxide emissions by 49 percent versus total reliance on new central generation.

Impact of Generating 2020 Load Growth with Central or Decentralized Generation				
	100 percent CG	100 percent DG	Savings	Percent Change
Total Capital Cost Capacity + T&D) Billions of Dollars	$831	$504	$326	39 percent
2020 Incremental Power Cost Billions of Dollars	$145	$92	$53	36 percent
Emissions from New Load Thousand Metric Tonnes				
NO_x	288	122	166	58 percent
SO_2	333	19	314	94 percent
PM10	22	12	9	43 percent
Million Metric Tonnes CO_2	776	394	381	49 percent

Figure 8. Decentralized generation as a percentage of total US generation.

EXTRAPOLATING US ANALYSIS TO THE WORLD

We lack the data to run the US model for the world, but have taken the percentage savings to be directionally correct and applied them to the IEA load growth projections through 2030. Detailed analysis by others will undoubtedly refine the estimates, and there will be some mix of central and decentralized generation. The analysis shows the extreme cases to provide guidance.

Figure 9 shows expected world load growth with conventional central plants that convert 100 units of fuel into 67 units of wasted energy and 33 units of delivered power. The text at the bottom reflects IEA's projected capital cost for 4,800 gigawatts of new generation, totaling $4.2 trillion. The International Energy Agency was silent on T&D, so we used estimates made for the United States Department of Energy on the all-in cost per kW of new transmission to forecast $6.6 trillion cost for new wires and transformers. Assuming US average line losses (which are significantly lower than developing country line losses), 9 percent of the capacity will be lost, leaving 4,368 gigawatts delivered to users. To achieve the IEA Reference Case with central generation, the world must invest $10.8 trillion capital, roughly $2,500 per kW of delivered capacity.

Meeting IEA Reference Case load growth with decentralized generation will lower the need for redundant generation. An analysis by the Carnegie

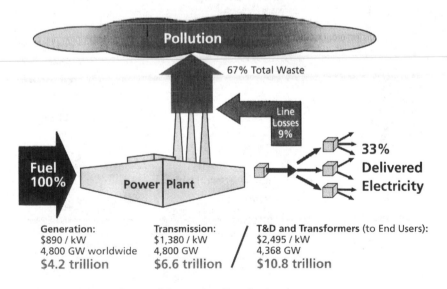

Figure 9. Conventional central generation flowchart.

Mellon Electric Industry Center suggests building only 78 percent of the 4,800 gigawatts as DG would provide equal or better reliability.[10] However, in developing economies, reliability may not be the driver. To be conservative, we have ignored the potential reduction in generation due to increased reliability inherent in larger numbers of smaller plants in the DG case. However, we did reduce required generation for the DG case to 4,368 GW, since there are no net line losses.

Figure 10 depicts the process of meeting expected world load growth with distributed generation. We estimated average capital costs for decentralized generation of $1,200 per kW, $310 more capital cost than a kilowatt of new central generation. Even with 9 percent less DG capacity, the capital costs for generation increase to $5.2 trillion, $1.0 trillion more than building central plants. Looking only at generation costs, DG is not competitive. However, the full decentralized generation case requires only 430 GW of new T&D, costing $0.6 trillion, a $6 trillion savings on T&D. *End users receive 4,638 GW in both cases, but society invests $5 trillion less for the DG case.*

Everyone knows that "you get what you pay for." What does the world give up by selecting a $5 trillion cheaper approach to meet projected electric growth? We extrapolated US analysis to the IEA Reference Case and found the world would give up the following by adopting the cheaper DG case:

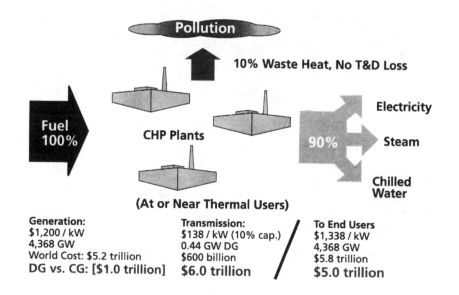

Figure 10. Combined heat and power flowchart.

- Consume 122 billion fewer barrels of oil equivalent (half of known Saudi oil reserves)
- Lost fossil fuel sales of $2.8 trillion
- Lost medical revenues from air pollution-related illnesses
- Potentially lost savings if governments opt to supply electric services to entire population instead of leaving 1.4 billion people without electric access
- Less global warming due to 50 percent less CO_2 emissions.

RECOMMENDED ACTIONS

If this analysis survives critical review, then what policy reforms will steer the power industry toward optimal decisions, given available technology? We offer two potential approaches, hoping to start the policy debate.

COMPREHENSIVE REFORM

Governments guide the electric industry with many rules, mandates, and limitations that collectively block competition and innovation, thus causing excessive costs and fuel usage. Small regulatory changes may nudge the power industry to slight course corrections, but are unlikely to break the central generation paradigm and optimize generation.

Immediately eliminating all current barriers to efficiency would cause the electric power industry to make better decisions. Each government could examine every rule that affects power generation and delivery and ask whether the social purpose behind that rule still exists. Then each state or country could enact comprehensive legislation that we term the Energy Regulatory Reform and Tax Act (ERRATA), to correct all of the mistakes in current law. ERRATA would deregulate all electric generation and sales, modernize environmental regulations to induce efficiency, and change taxation to reward efficiency.[11] Sadly, ERRATA legislation probably will not pass except in response to deepening environmental and economic pain.

ACTIONABLE REFORM, NATIONAL FOSSIL FUEL EFFICIENCY STANDARDS

A second possible approach simply rewards all fossil efficient power and penalizes fossil inefficient power. Each government could enact a Fossil Fuel

Efficiency Standard covering all locally used electricity, regardless of origin. This standard does not favor fuels, technologies, or participants. Here are the essential elements:

- Give all delivered megawatt-hours an equal allowance of incremental fossil fuel, regardless of age of plant, technology or ownership. Start with the national average fossil fuel per MWh for the prior year.
- Spread allowances over all generation of each owner, allowing owners to comply by increasing efficiency of existing plants, deploying new highly efficient plants, or purchasing fossil allowances from others.
- Reward plants requiring little or no fossil fuel, such as solar, wind, hydro, nuclear, and industrial waste energy recycling, by allowing them to sell fossil fuel credits.[12]
- Penalize inefficient fossil fuel plants by forcing them to purchase allowances for each MWh produced.
- Base allowances on delivered power to incorporate T&D losses from central generation.
- Credit displaced fuel to CHP plants that recycle heat.
- Force all generators to purchase adequate allowances or close their plants to ensure that the total allowance trading is economically neutral.
- Reduce the fossil fuel allowances per MWh each year according to a schedule.
- Adjust the schedule downward each year to correct for growth in total power delivered, guaranteeing that the total fossil fuel use will drop.

A Fossil Fuel Efficiency Standard would steer the power industry toward optimal choices. This will reduce power costs and emissions, which will improve local standard of living and improve the competitive position of local industry. Other states and nations will follow suit.

CONCLUSION

We have attempted to frame the consequences of meeting energy load growth with conventional central generation or deploying decentralized generation that recycles waste energy. The DG case saves the world $5 trillion in capital investment while reducing power costs by 40 percent and cutting greenhouse gas emissions in half. There are interesting implications for worldwide energy policy if this analysis stands up to critical review.

We hope readers and others will spell out concerns or suggest corrections so we can collectively improve the analysis of optimal future power generation. The needed policy changes are deep and fundamental and require a consensus about the best way to proceed. Together we might be able to change the way the world makes heat and power.[13]

NOTES

1. The IEA has issued an annual "World Energy Outlook" series since 1993. The publication projects many facets of the energy industry thirty years ahead. The projections are based on a "Reference Scenario that takes into account only those government policies and measures that had been adopted by mid-2002. A separate Alternative Scenario assesses the impact of a range of new energy and environmental policies that the OECD countries are considering."

2. *Energy Information Administration/Electric Power Monthly*, May 2004.

3. *Energy Information Administration/Monthly Energy Review*, June 2004.

4. Joseph Eto, of the Lawrence Berkeley National Laboratory, in a speech to NARUC, says outages cost the United States $80 billion per year. The EPRI Consortium for Electric Infrastructure to Support a Digital Society (CEIDS), *The Cost of Power Disturbances to Industrial & Digital Economy Companies,* June 2001, states power outages and other power quality disturbances are costing the US economy more than $119 million annually.

5. We assembled historical data for four central generating technologies—oil and gas-fired thermal plants (Rankine cycle), coal fired thermal plants, simple-cycle and combined-cycle gas turbines. Data for each technology and each year include capital costs per kW, load factor, and efficiency. We assumed a 25-year life to calculate annual capital amortization and the future wholesale price per MWh that would yield an 8 percent weighted average return on capital. Since new central generation requires new T&D, we converted estimates of $1260 per kW for T&D in 2000 and adjusted for inflation, then assumed a 35-year life for T&D to calculate required T&D charges. EIA did not keep line loss statistics prior to 1989, so we estimated prior years slightly below the current 9 percent losses. Summing produces the retail price needed for power from a central plant using a specific technology installed in that specific year. Finally, we converted everything to 2004 dollars.

6. Typical DG plants employ multiple generators with expected unplanned outages of 2 to 3 percent each. The probability of complete loss of power is found by multiplying expected unit unplanned outages by each other. Given the existing 10,286 generators operating in the US that are less than 20 megawatts of capacity, and the expectation, with barriers removed, of many DG plants inside every distribution network, spare grid capacity equal to 10 percent of installed DG should be more than adequate to cover unplanned outages.

7. Casten and Collins, *Recycled Energy: An Untapped Resource*, 2002; see http://www.primaryenergy.com.

8. Energy Information Administration, *Annual Energy Review* 2002, October 2003.

9. The "Optimizing Heat and Power" model has been adopted by the World Alliance for Decentralized Energy (WADE) and is being used by the European Union, Thailand, Nigeria, Canada, Ireland, and China to ask the best way to satisfy expected load growth. For model descriptions, contact Michael Brown, Director, at info@localpower.org.

10. Hisham Zerriffi. Personal communication. See *Distributed Resources and Micro-grids* by M. Granger Morgan of the Department of Engineering and Public Policy, Carnegie Mellon University, Sept. 25, 2003, for detailed analysis of how DG provides reliability with less spare capacity.

11. See Casten, Thomas R., *Turning Off The Heat* (Prometheus Books, 1998) chapter 10, for a more complete description of ERRATA.

12. Producers of electricity are given fossil fuel usage credits, meaning they are allowed to use a given amount of fossil fuels corresponding to efficiency, size of unit and other environmental parameters. Thus, the higher the efficiency of a company's unit, the less fossil fuel credits that company needs to use. The highly efficient plants and generation plants using a non-fossil fuel energy such as solar, wind, or hydro power would not need the full allowance and could sell the unused portion to less efficient fossil fueled plants. Such a system would provide added economic value to the efficient and non-fossil fueled plants and economic penalties to the inefficient fossil fueled plants.

13. The lead author adapted this article and chapter from his keynote address to the International Association for Energy Economics in Washington, D.C., July 10, 2004. A somewhat different version was published in the IAEE's journal *The Dialogue*.

AFTERWORD

How do things look now from the perspective of today's steeply higher energy prices? As we completed this analysis in mid 2004, both the fossil fuel and electric industries were about to undergo a fundamental sea change, which was not yet clear due to data reporting lags. Between 2001 and 2008 the prices of all fossil fuel increased by record percentage amounts. Coal has tripled, natural gas prices have been as much as seven times higher, and oil went from less than $20 per barrel to over $145 per barrel. These changes multiplied the value of power plants that recycle waste heat. In addition, the capital costs of electric generation and transmission equipment have rocketed upwards, reflecting the rising energy cost of producing the steel and copper for this equipment. Our estimates for new coal plant construction from 1990 through 2003 were largely theoretical because almost no new coal fired generation was built in the United States in that period. Since 2004 plans have been announced for 110 new central coal-fired generation plants with costs of $2,200 to $3,700 per megawatt of capacity—more than double the estimates in this article. We attribute the rising capital costs for large coal plants to tougher emissions requirements and rising equipment prices

What do these enormous fuel and capital cost increases imply for future US electric industry? The higher prices strengthen the 2004 conclusions. A major value of distributed or local generation is its capture and recycling of byproduct heat from electrical generation. The value of recycling normally wasted energy is a function of the amount and the cost of displaced boiler fuel. Local generation, by recycling waste energy continues to displace the same quantity of boiler fuel described in the article, but that fuel costs three to five times more than it did in 2001–2002. The avoidance of physical transmission and distribution equipment by local generation is as noted, but the cost of wires and transformers has shot up with the tripling of copper prices since 2000.

Finally, the US has seen another sea change—the rather sudden recognition of the need to reduce greenhouse gas emissions. In 2008 there have been voluntary prices for carbon emission avoidance of $4 to $10 per ton in the US and $40 per ton in Europe, where reductions are mandated. Any value for carbon dioxide reduction increases the value of local generation that recycles waste heat.

Bottom line, three years later? The need to embrace the most efficient generation has strengthened, and the value of changing the paradigm from central generation to local generation that recycles waste energy has tripled.

Part III

UNDERSTANDING PSEUDOSCIENCE, INVESTIGATING CLAIMS

THE PHILOSOPHY
BEHIND PSEUDOSCIENCE

Mario Bunge

José López-Rega was the evil gray eminence behind General Perón in his
dotage, as well as that of Perón's successor, his wife, Isabelita. He had
been a failed singer, a police officer, a bodyguard, and the author of best sellers
on business and the stars, love and the stars, and the like. He was a firm believer
in the occult and a practicing black magician. López-Rega believed not only
in astral influences and in the spirit world but also in his own ability to conjure up spirits and manipulate them. Once, he attempted to transfer the soul
of the dead Juan Perón into his dim-witted successor, Isabelita. However, this
task proved to be beyond his ability (see for example, Martínez, 1989).

López-Rega is not known to have dabbled in philosophy. However, like
everyone else, he did hold definite philosophical views. Among these were the
age-old myths of the immaterial soul, the possibility of paranormal cognition,
and the existence of supernatural beings. These beliefs underlie his conviction
that he was able to influence other people's behavior by sheer mental power, as
well as to get in touch with higher powers. In turn, these beliefs and practices
gave him the self-confidence, prestige, and authority he needed to perform his
sinister political maneuvers. Among other things, he organized a death squad
that carried out uncounted assassinations and torture sessions of political opponents during the 1973–76 period, when he was at the peak of his political power.

Thus, El Brujo ("the Wizard"), as the Minister of Public Welfare was popularly known, had the backing of millennia of philosophical myths.

Every intellectual endeavor, whether authentic or bogus, has an underlying philosophy and, in particular, an ontology (a theory of being and becoming) and an epistemology (a theory of knowledge). For example, the philosophy behind evolutionary biology is naturalism (or materialism) together with epistemological realism, the view that the world exists on its own and can be investigated. By contrast, the philosophy behind creationism (whether traditional or "scientific") is supernaturalism (the oldest variety of idealism) together with epistemological idealism (which involves the disregard for empirical tests).

To be sure, most scientists, as well as most pseudoscientists, are unaware that they uphold any philosophical views. Moreover, they dislike being told that they do. And the most popular among the respectable philosophies of science of the day, namely the logical positivists and Popper's followers, teach that science and philosophy are mutually disjointed rather than intersecting. However, this view is false. Indeed, nobody can help but employ a great number of philosophical concepts, such as those of reality, time, causation, chance, knowledge, and truth. And once in a while, everyone ponders philosophical problems, such as those of the nature of life, mind, mathematical objects, science, society, and what is good. Moreover, the neutrality view is dangerous, because it masks the philosophical traps into which bona fide scientists may fall, and it dissuades them from explicitly using philosophical tools in their research.

Since there is no consensus about the nature of science, let alone pseudoscience, I will inquire into the philosophies that lurk behind psychoanalysis and computationist psychology.

1. SCIENCE: AUTHENTIC AND BOGUS

We shall be concerned only with sciences and pseudosciences that claim to deal with facts, whether natural or social. Hence, we shall not deal with mathematics except as a tool for the exploration of the real world. Obviously, this world can be explored either scientifically or nonscientifically. In either case, such exploration, like any other deliberate human activity, involves a certain approach, that is, set of general assumptions, some background knowledge of the items to be explored, a goal, and a means or method of proceeding.

In a way, the general assumptions, the extant knowledge of the facts to be explored, and the goal dictate jointly the means or method to be employed. Thus, if what is to be explored is the mind, if the latter is conceived of as an immaterial entity and the goal is to understand mental processes in any old way, then the cheapest means is to engage in free speculation. Given such idealistic assumptions about the nature of the mind, it would be preposterous to try and catch it by exploring the brain. If, on the other hand, mental processes are assumed to be brain processes and if the aim is to understand the mechanisms underlying mental phenomena, then the scientific method, particularly in its experimental version, is mandatory. (This is the philosophical rationale of cognitive neuroscience.) That is, whether or not a scientist studies the brain in order to understand the mind depends critically upon her more or less tacit philosophy of mind.

In general, one starts research by picking a domain (D) of facts, then makes (or takes for granted) some general assumptions (G) about them, collects a body (B) of extant knowledge about the items in (D), decides on an aim (A), and in the light of the preceding, determines the proper method (M) to study (D). Hence, an arbitrary research project (p) may be sketched as the ordered quintuple p=(D, G, B, A, M). The function of this list is to keep track of the essentials in framing the following definitions.

A *scientific* investigation of a domain of facts (D) assumes that these are material, lawful, and scrutable, as opposed to immaterial (in particular supernatural), lawless, or inscrutable; and the investigation is based on a body of previous scientific findings (B), and it is done with the main aim of describing and explaining the facts in question (A) with the help of the scientific method (M). In turn, the latter may be described summarily as the sequence: choice of back ground knowledge; statement of problem(s); tentative solution (for example, hypothesis or experimental technique); run of empirical tests (observations, measurements, or experiments); evaluation of test results; eventual correction of any of the preceding steps, and new problems posed by the finding.

Contrary to widespread belief, the scientific method does not exclude speculation: it only disciplines imagination. For example, it is not enough to produce an ingenious mathematical model of some domain of facts the way mathematical economists do. Consistency, sophistication, and beauty are never enough in scientific research, the end product of which is expected to match reality—or at least to be true to some degree. Pseudoscientists are not to be blamed for exerting their imaginations but rather for letting them run loose. The place for unbridled speculation is art, not science.

The scientific method presupposes that everything can in principle be debated and that every scientific debate must be logically valid (even if no logical principles or rules are explicitly invoked). This method also involves two key semantic ideas: meaning and truth. Nonsense cannot be investigated, hence it cannot be pronounced false. (Think of calculating or measuring the time required to fly from one place to another using Heidegger's definition of time as "the maturation of temporality.") Furthermore, the scientific method cannot be practiced consistently in a moral vacuum. Indeed, it involves the ethos of basic science, which Robert K. Merton (1973) characterized as universalism, disinterestedness, organized skepticism, and epistemic communism (the sharing of methods and findings).

Finally, there are four more distinguishing features of any authentic science: changeability, compatibility with the bulk of the antecedent knowledge, partial intersection with at least one other science, and control by the scientific community. The first condition flows from the fact that there is no "live" science without research, and research is likely to enrich or correct the fund of knowledge. In sum, science is eminently changeable. By contrast, the pseudosciences and their background ideologies are either stagnant (like parapsychology) or they change under pressure from power groups or as a result of disputes among factions (as has been the case with psychoanalysis).

The second condition can be restated thus: to be worthy of the attention of a scientific community, an idea must be neither obvious nor so outlandish that it clashes with the bulk (though not the totality) of the antecedent knowledge. Compatibility with the latter is necessary, not only to weed out groundless speculation but also to understand the new idea as well as to evaluate it. Indeed, the worth of a hypothesis or of an experimental design is gauged partly by the extent to which it fits in with reasonably well-established bits of knowledge. (For example, telekinesis is called into question by the fact that it violates the principle of conservation of energy.) Typically, the principles of a pseudoscience can be learned in a few days, whereas those of a genuine science may occupy an entire lifetime, if only because of the bulky body of background knowledge they are based upon.

The third condition, that of either using or feeding other research fields, follows from the fact that the classification of the factual sciences is somewhat artificial. For example, where does the study of memory fall: in psychology, neuroscience, or both? And which discipline investigates the distribution of wealth: sociology, economics, or both? Because of such partial overlaps and interactions, the set of all the sciences constitutes a system. By contrast, the pseudosciences are typically solitary.

The fourth condition, summarized as control by the scientific community, can be spelled out this way. Investigators do not work in a social vacuum but experience the stimuli and inhibitions of fellow workers (mostly personally unknown to them). They borrow problems and findings and ask for criticisms; and, if they have anything interesting to say, they get both solicited and unsolicited opinions. Such interplay of cooperation with competition is a mechanism for the generation of problems and the control and diffusion of results; it makes scientific research a self-doubting, self-correcting, and self-perpetuating enterprise. This makes the actual attainment of truth less peculiar to science than the ability and willingness to detect error and correct it. (After all, everyday knowledge is full of well-attested trivialities that have not resulted from scientific research.)

So much for the distinguishing features of genuine factual science, whether natural, social, or biosocial. (More can be found in Gardner, 1983; Wolpert, 1992; Bunge, 1998a; and Kurtz, 2001.) By contrast, a *pseudoscientific* treatment of a domain of facts violates at least one of the above conditions, while at the same time calling itself scientific. It may be inconsistent, or it may involve unclear ideas. It may assume the reality of imaginary facts, such as alien abduction or telekinesis, self-replicating and selfish genes, or innate ideas. It may postulate that the facts in question are immaterial, inscrutable, or both. It may fail to be based on previous scientific findings. It may perform deeply flawed empirical operations, such as ink-blot tests, or it may fail to include control groups. It may fake test results, or it may dispense with empirical tests altogether.

Besides, the pseudosciences do not evolve or, if they do, their changes do not result from research. They are isolated from other disciplines, although occasionally, they interbreed with sister bogus sciences, as witnessed by psychoanalytic astrology. And, far from welcoming criticism, they attempt to fix belief. Their aim is not to search for truth but to persuade: they posit arrivals without departures and without journeys. Whereas science is full of problems, and every one of its findings poses further problems, pseudoscience is characterized by certainty. In other words, whereas science begets more science, pseudoscience is barren, because it generates no new problems. In sum, the main trouble with pseudoscience is that its research is either deeply flawed, or nonexistent. This is why, contrary to scientific research, pseudoscientific speculation has not delivered a single law of nature or of society.

So much for a sketchy characterization of both authentic and bogus science. Let us now apply our analysis to a couple of interesting recent cases: physical chemistry and neuropsychology.

2. TWO CASES: SELF-ORGANIZATION AND THE UNCONSCIOUS

Our first example is the treatment of self-organizing systems: complex wholes that get self-assembled in the absence of external forces. Self-organization, in particular, biological morphogenesis, is a wondrous but poorly understood process. No wonder that it has been the object of much pseudoscientific speculation peppered with high-sounding but empty expressions, such as "constructive force," "entelechy," "élan vital," "morphogenetic field," "autopoiesis," and the like. All such factors have often been regarded as being immaterial, hence beyond the reach of physics and chemistry. And they are neither described in any detail nor manipulated in the laboratory. Hence, talk of such factors is just hand waving, when not magic-wand waving.

By contrast, the scientific approach to self-organization is down-to-earth yet imaginative. Let us peek at a recent instance of this approach: the work of Adams, Dogic, Keller, and Fraden (1998). Colloids consisting of tiny rods and spheres were randomly suspended in a buffer sealed in glass capillaries, then left to their own devices and observed under a microscope. The rods were viruses, and the spheres plastic balls; the former were negatively charged, and the latter positively charged. After some time, the mixture separated spontaneously into two or more homogeneous phases. Depending on the experimental conditions, a phase may consist of layers of rods alternating with layers of spheres, or the spheres may assemble into columns.

Paradoxically, these various types of demixing are explained in terms of repulsions between the charged particles—which, intuitively, should preclude the crowding of particles with the same charge. And the equally paradoxical decrease in entropy (order increase) is explained by noting that the clumping of some of the colloids is accompanied by a raise in the translational entropy of the medium. In any event, the whole process is accounted for in strictly naturalistic terms. At the same time, the authors warn that their results are at variance with the pertinent theory—though of course not with any general physical systems. Such incompleteness is typical of factual science, by contrast to pseudoscience, where everything is cut and dried from the start.

Our second example is the study of the unconscious. Much has been written about it, most of it in a speculative vein, ever since Socrates claimed that, by clever questioning, he was able to ferret out tacit mathematical knowledge from an illiterate slave boy. Thanks to Eduard von Hartmann's best seller, *Die Philosophie des Unbewussten* (1870), the subject was already popular in 1900, when Freud first proposed his wild fantasies. Among other things, Freud reified the unconscious and attributed to it causal powers that allegedly accounted for a number of unexplained phenomena, such as slips of the tongue and the mythical Oedipus complex. But of course, it never occurred to him or to any of his followers to approach this subject in an experimental manner.

The scientific study of unconscious mental processes began a couple of decades ago with observations on split-brain and blindsight patients. Since then, the various brain-imaging techniques, such as PET scanning and functional MRI, have made it possible to ascertain whether someone feels or knows something even though he or she does not know that he feels or knows it. Moreover, these techniques make it possible to localize such mental processes in a noninvasive way. An example is the paper by Morris, Öhman, and Dolan (1998)—which, unsurprisingly, does not cite any psychoanalytic studies. Let us look at it.

The amygdala is the tiny brain organ that feels such basic and ancient emotions as fear and anger. If damaged, a person's emotional and social life will be severely stunted. The activity of the amygdala can be monitored by a PET scanner; this device allows the experimenter to detect a subject's emotions, and even to locate them in either side of the amygdala. However, such neural activity may not reach the conscious level. In this case, only a brain scanner can help.

For example, if a normal human subject is briefly shown an angry face as a target stimulus, and immediately thereafter an expressionless mask, he will report seeing the latter but not the former. Yet, the scanner tells a different story. It tells us that, if the angry face has been associated with an aversive stimulus, such as a burst of high-intensity white noise, the amygdala is activated by the target, even though the subject does not recall having seen it. In short, the amygdala "knows" something that the organ of consciousness (whichever and wherever it is) does not. Psychoanalysts could use this very method to measure the intensity of a male's hatred for his father. But they don't, because they don't believe in the brain: their psychology is idealistic, hence brainless. More on this in section 4.

The number of examples of pseudoscience can be multiplied at will. Astrology, alchemy, parapsychology, characterology, graphology, creation "science," "intelligent design," Christian "Science," dowsing, homeopathy, and memetics are generally regarded as pseudoscientific (see, for example, Kurtz, 1985; Randi, 1982; and *The Skeptical Inquirer*). On the other hand, it is less widely accepted that psychoanalysis and computationist psychology are also bogus sciences. This is why we shall examine them in section 3. But first, we must take a brief explicit look at philosophy, for some of it is bogus.

3. PHILOSOPHY: PROSCIENTIFIC AND ANTISCIENTIFIC

The above characterization of scientific research involves philosophical ideas of six kinds: logical, semantical, ontological, epistemological (in particular, methodological), ethical, and sociological. More specifically, it involves the notions of logical consequence and logical consistency; the semantic notions of meaning and truth; the ontological concepts of real fact and law (objective pattern); the epistemological concepts of knowledge and test; the principles of intellectual honesty; and the notion of a scientific community.

Why is this so? Because scientific research is, in a nutshell, the *honest* search for *true knowledge* about the *real world*, particularly its *laws*, with the help of both *theoretical* and *empirical* means—in particular the *scientific method*—and because every body of *scientific knowledge* is expected to be *logically consistent* and the subject of *rational discussion* in the bosom of a community of investigators. All the expressions in italics occur in (metascientific) discourses about any factual (empirical) science. And the discipline in charge of elucidating and systematizing the corresponding concepts is philosophy. Indeed, philosophy is the study of the most fundamental and transdisciplinary concepts and principles. Hence, philosophers are expected to be generalists rather than specialists. And some of us often assume the ungrateful task of passing judgment on the credentials of some pseudoscientific or ideological beliefs.

Now, different philosophical schools treat the above philosophical components of science differently or not at all. Recall briefly only four contemporary examples: existentialism, logical positivism, Popperianism, and Marxism.

Existentialism rejects logic and, in general, rationality; it espouses an extremely sketchy, nearly unintelligible, and even ridiculous ontology; and it has no use for semantics, epistemology, or ethics. No wonder that it has had no impact on science—except indirectly, and negatively, through its debasement of reason and support of Nazism. No wonder, too, that it has not produced an intelligible philosophy of science, let alone a stimulating one.

By contrast, logical positivism defends logic and the scientific method; but it has no defensible semantics; it has no ontology beyond phenomenalism ("there are only appearances"); its epistemology overrates induction and misunderstands or underrates scientific theory, which it regards as a mere data abstract; and it has no ethics beyond Hume's emotivism. Unsurprisingly, logical positivism misinterprets relativistic and quantum physics in terms of laboratory operations instead of as representing objectively existing physical entities that exist in the absence of observers (see for example, Bunge, 1973). Still, logical positivism is *scientistic*, and therefore far superior to the antiscience characteristic of postmodernism.

Popperianism praises logic but rejects the very attempt to do semantics; it possesses no ontology beyond individualism (or atomism, or nominalism); it values theory to the point of regarding experiment as only a way of testing hypotheses; it overrates criticism, underrates induction, and has no use for positive evidence; and it has no ethics beyond the Buddha's and Epicurus's and Hippocrates's injunction to do no harm. However, Popperianism has the merits of having defended a realist interpretation of physical theories and of having deflated inductivism. But Popper first underrated, and later accepted but misinterpreted, evolutionary biology as consisting exclusively of culling misfits; he opposed the psychoneural monism inherent in biological psychology; he rejected the materialist conception of history adopted by the most advanced historiographic school—that of the *Annales*; and he defended neoclassical microeconomics, which—as I will argue below—is pseudoscientific in being conceptually fuzzy and immune to empirical falsification.

As for Marxism, it has introduced some revolutionary ideas in social science, particularly the materialist conception of history and the centrality of social conflict. However, Marxian materialism is narrowly economicist: it underrates the roles of politics and culture (in particular, ideology). Moreover, Marxism, following Hegel, confuses logic with ontology. Hence, it is diffident of formal logic; its materialist ontology is marred by the romantic obscurities of dialectics, such as the principle of the unity of opposites; its epistemology is naïve realism (the "reflection theory of knowledge"), which makes no room for the symbolic nature of pure mathematics and theoretical physics; it glorifies social wholes at the expense of individuals and their legitimate aspirations; it exaggerates the impact of society on cognition; and it adopts the ethics of utilitarianism, which has no use for disinterested inquiry, let alone altruism. No wonder that, when in power, dialectical materialist philosophers have opposed some of the most revolutionary scientific developments of their time: mathematical logic, relativity theory, quantum mechanics, genetics, the synthetic theory of evolution, and post-Pavlovian neuropsychology.

In short, none of these four schools matches the philosophy inherent in science. I submit that any philosophy capable of understanding and promoting scientific research has the following characteristics (Bunge, 1974–1989):

Logical: Internal consistency and abidance by the rules of deductive inference; acceptance of analogy and induction as heuristic means, but no claim to *a priori* validations of analogical or inductive arguments.

Semantical: A realist theory of meaning as intended reference (denotation)—and as different from extension—together with sense or connotation. And a realist view of factual truth as the matching of a proposition with the facts it refers to.

Ontological: Materialism (naturalism)—all real things are material (possess energy), and they all fit some laws (causal, probabilistic, or mixed). Mental processes are brain processes, and ideas in themselves, however true or useful, are fictions. *Dynamicism*—all material things are in flux. *Systemism*—every thing is either a system or a (potential or actual) component of a system. *Emergentism*—every system has (systemic or emergent) properties that its components lack.

Epistemological: Scientific realism—it is possible to get to know reality, at least partially and gradually, and scientific theories are expected to represent, however imperfectly, parts or features of the real world. *Moderate skepticism*—scientific knowledge is both fallible and perfectible. However, some findings—for example, that there are atoms and fields, that there are no disembodied ideas, and that science pays—are firm acquisitions. *Moderate empiricism*—all factual hypotheses must be empirically testable, and both positive and negative evidence are valuable indicators of truth value. *Moderate rationalism*—knowledge advances through educated guessing and reasoning combined with experience. *Scientism*—whatever is knowable and worth knowing is best known scientifically.

Ethical: Secular humanism—the supreme moral norm is "Pursue the welfare (biological, mental, and social) of oneself and others." This maxim directs that scientific research should satisfy either curiosity or need and abstain from doing unjustifiable harm.

Sociological: Epistemic socialism—scientific work, however artisanal, is social, in that it is now stimulated, now inhibited, by fellow workers and by the ruling social order, and the (provisional) umpire is not some institutional authority but the community of experts. Every such community prospers with the achievements of its members, and it facilitates the detection and correction of error. (Warning: this is a far cry from both the Marxist claim that ideas are exuded and killed by society and the constructivist-relativist view that "scientific facts" are local social constructions, that is, mere community-bound or tribal conventions.)

I submit that the above philosophical principles are tacitly met by the mature or "hard" sciences (physics, chemistry, and biology); that the immature or "soft" sciences (psychology and the social sciences) satisfy some of them; and that the pseudosciences violate most of them. In short, I submit that scientificity is coextensive with sound philosophy.

Moreover, the reason the pseudosciences are akin to religion, to the point that some of them serve as surrogates for it, is that they share a philosophy, namely philosophical idealism—not to be mistaken for moral idealism. Indeed, pseudoscience and religion postulate immaterial entities, paranormal cognitive abilities, and a heteronomous ethics. I will spell this out.

Every religion has a philosophical kernel, and the philosophies inherent in the various religions share the following idealist principles: *Idealist ontology*—there are autonomous spiritual entities, such as souls and deities, and they satisfy no scientific laws. *Idealist epistemology*—some people possess cognitive abilities that fall outside the purview of experimental psychology: divine inspiration, inborn insight, or the capacity to sense spiritual beings or prophesy events without the help of science. *Heteronomous ethics*—all people are subject to inscrutable and unbendable superhuman powers, and they are not obliged to justify their beliefs by means of scientific experiment.

All three philosophical components common to both religion and pseudoscience are at variance with the philosophy inherent in science. Hence, the theses that science is one more ideology and that science cannot conflict with religion because they address different problems in different but mutually compatible ways are false. (More on religion in science in Mahner and Bunge, 1996.)

4. THE CASES OF PSYCHOANALYSIS AND COMPUTATIONIST PSYCHOLOGY

Do psychoanalysis and computationist psychology share the philosophical features that, according to section 3, characterize the mature sciences?

Psychoanalysis violates the ontology and the methodology of all genuine science. Indeed, it holds that the soul ("mind" in the standard English translation of Freud's works) is immaterial, yet can act upon the body, as shown by psychosomatic effects. However, psychoanalysis does not assume any mechanisms whereby an immaterial entity can alter the state of a material one: it just states that this is the case. Moreover, this statement is dogmatic, since psychoanalysts, unlike psychologists, do not perform any empirical tests. In particular, no laboratory has ever been set up by any psychoanalysts. Freud himself had emphatically dissociated psychoanalysis from both experimental psychology and neuroscience.

To mark the first centenary of the publication of Freud's *Interpretation of Dreams*, the *International Journal of Psychoanalysis* published a paper by six New York analysts (Vaughan et al., 2000) who purported to report on the first experimental test ever of psychoanalysis in the course of one century. Actually, this was no experiment at all, since it involved no control group. Hence, those authors had no right to conclude that the observed improvements were due to the treatment; they could just as well have been spontaneous. Thus, psychoanalysts make no use of the scientific method, because they do not know what this is. After all, they were not trained as scientists but only, at best, as medical practitioners.

The French psychoanalyst Jacques Lacan—a hero of postmodernism—admitted this and held that psychoanalysis, far from being a science, is a purely rhetorical practice: *"l'art du bavardage."* Finally, since psychoanalysts claim that their views are both true and effective without having submitted them to either experimental tests or rigorous clinical trials, they can hardly be said to proceed with the intellectual honesty that scientists are expected to abide by (even if they occasionally lapse). In sum, psychoanalysis does not qualify as a science. Contrary to widespread belief, it is not even a failed science, if only because it makes no use of the scientific method and ignores counterexamples. It is just quack psychology.

Computationist psychology claims that the mind is a set of computer programs that can in principle be implemented in either brains or machines—or perhaps even ghosts. That is, this popular school adopts the functionalist view that matter does not matter—that only function does. This view is encouraged by the idealist ontologies, whereas science investigates only concrete things on various levels: physical, chemical, living—thinking and nonthinking—or social. Moreover, the computationists beg the question whether certain mental processes are computations. They have no evidence that all mental processes are computational; they just assert this thesis.

But this thesis is false, since neither emotional nor creative processes are algorithmic, and only a fraction of cognitive processes are. For example, there can be no algorithms for acting spontaneously, asking original problems, formulating original hypotheses, forming fruitful analogies, or designing original artifacts, such as radically new algorithms, machines, or social organizations. Indeed, every algorithm is a procedure for performing operations of a specified kind, such as sorting, adding, and computing values of a mathematical function. By contrast, original scientific findings are not specifiable in advance—this being why research is necessary.

In sum, computationist psychology is nonscientific because it ignores negative evidence and it disregards the matter of mind—the brain that does the minding. Consequently, it isolates itself from neuroscience and social science—and disciplinary isolation is a reliable indicator of non-scientificity. The secret of its popularity lies not in its findings but in the computer's popularity, in that it does not demand any knowledge of neuroscience, and in the illusion that sentences of the form "X computes Y" explain, while in fact they only conceal our ignorance of the neural mechanisms. (Remember that there is no genuine explanation without mechanism, and that all mechanisms are material; see Bunge, 2006.)

So much for a sample of pseudoscience. The subject of its underlying philosophy is intriguing and vast, yet largely unexplored (see however, Flew, 1987). Just think of the many pockets of pseudoscience ensconced in the sciences, such as the anthropic principle, the attempt to craft a theory of everything, information talk in biochemistry, the "it's-all-in-the-genes" dogma in biology, human sociobiology, West Coast (purely speculative) evolutionary psychology, and game-theoretic models in economics and political science. Analyze an egregious error in science, and you are likely to find a philosophical bug.

5. BORDERLINE CASES: PROTO- AND SEMI-

Every attempt at classifying any collection of items outside mathematics is likely to meet borderline cases. The main reasons for such vagueness are either that the classification criteria themselves are imprecise or that the item in question possesses only some of the features necessary to place it in the box in question. Remember the case of the platypus, the egg-laying mammal.

Whatever the reason, in the case of science we find plenty of disciplines, theories, or procedures that, far from falling clearly either in the range of the scientific or outside it, may be characterized as proto-scientific, semi-scientific, or as failed science. Let us take a quick look at these cases.

A *proto-science*, or emerging science, is obviously a science *in statu nascendi*. If it survives at all, such a field may eventually develop either into a mature science, a semi-science, or a pseudoscience. In other words, when a discipline is said to be a proto-science, it is too early to pronounce it scientific or non-scientific. Examples: physics before Galileo and Huygens, chemistry before Lavoisier, and medicine before Virchov and Bernard. All of these disciplines matured quickly to become fully scientific. (Medicine and engineering can be called scientific even though they are technologies rather than sciences.)

A *semi-science* is a discipline that started out as a science, is usually called a science, yet does not fully qualify as such. I submit that cosmology, psychology, and economics are semi-sciences. Indeed, cosmology is still rife with speculations that contradict solid principles of physics. There are still psychologists who deny that the mind is what the brain does, or who write about neural systems "subserving" or "mediating" mental functions. And of course, many of the so-called Nobel Prizes in economics (which are actually prizes of the Bank of Sweden) are awarded to inventors of mathematical models that have no resemblance to economic reality—if only because they ignore production and politics—or to designers of economic policies that harm the poor. The game-theoretic models proposed by Thomas C. Schelling, which won a Nobel Prize in 2005, are a case in point. One of them designed the strategic bombing of the Vietnamese civilian population. The same game theorist also discovered that African Americans segregate themselves: they "feel more comfortable among their own color" (Schelling, 1978, 138–39).

In some cases, it is hard to know whether something is scientific, semi-scientific, or pseudoscientific. For instance, the vast majority of nineteenth-century physicists regarded atomism as a pseudoscience, because it had produced only indirect evidence for the atomic hypothesis. Worse, since there was no detailed theory of individual atoms, atomism was only weakly testable, namely through the predictions of statistical mechanics. But the theory became scientifically respectable almost overnight as a consequence of Einstein's theory of Brownian motion and Perrin's experimental confirmation of it. Only die-hard positivists, like Ernst Mach, opposed atomism to the last.

Another example: quantum theory is undoubtedly a paradigm of successful high-level science. But the Copenhagen interpretation of this theory is pseudoscientific, because it places the observer at the center of the universe, since it assumes that every physical event results from a laboratory procedure. That this thesis is blatantly false is shown by the facts that the theory holds for stars, which are of course uninhabitable, and that it contains no postulates describing any observers. (More on this in Bunge, 1973; Mahner, 2001.)

String theory is a suspicious character. It looks scientific because it tackles an open problem that is both important and difficult, that of constructing a quantum theory of gravitation. For this reason, and because it has stimulated mathematics, it is attracting some of the brightest young brains. But the theory postulates that physical space has six or seven dimensions rather than three, just to secure mathematical consistency. Since these extra dimensions are unobservable, and since the theory has resisted experimental confirmation for more than three decades, it looks like science fiction, or at least, failed science.

The case of phrenology, the "science of skull bumps," is instructive. It proposed a testable, materialistic hypothesis, namely that all mental functions are precisely localizable brain functions. But instead of putting this exciting hypothesis to the experimental test, the phrenologists sold it successfully at fairs and other places of entertainment: They went around palpating people's skulls and claiming to locate alleged centers of altruism, philoprogenitivity, imagination, and so on. The emergence of modern neuroscience finished phrenology.

The discrediting of phrenology cast doubt not only on radical localizationism but also on the scientific attempts to map the mind onto the brain. In particular, the brain-imaging devices invented over the past three decades were first greeted with skepticism because the very attempt to localize mental processes sounded like phrenology. But these new tools have proved very fruitful and, far from confirming the phrenological hypothesis (one module per function), it has given rise to many new insights, among them, the view that all the subsystems of the brain are interconnected. If a tool or a theory leads to important findings, it cannot be pseudoscientific, because one of the marks of pseudoscience is that it is built around an old superstition.

Finally, a word of caution. Most of us are suspicious of radically new theories or tools, and this is the case for either of two reasons: because of intellectual inertia or because it is necessary to grill every newcomer to make sure that it is not an impostor. But one must try to avoid confusing the two reasons. Inquisitive types like novelty, but only as long as it does not threaten to dismantle the entire system of knowledge.

6. PSEUDOSCIENCE AND POLITICS

Pseudoscience is always dangerous, because it pollutes culture and, when it concerns health, the economy, or the polity, it puts life, liberty, or peace at risk. But of course, pseudoscience is supremely dangerous when it enjoys the support of a government, an organized religion, or large corporations. A handful of examples should suffice to make this point.

Eugenics, once promoted by many bona fide scientists and progressive public intellectuals, was invoked by American legislators to introduce and pass bills that restricted the immigration of people of "inferior races" and led to the institutionalization of thousands of children regarded as mentally feeble. The racial policies of the Nazis were justified by the same "science" and led to the murder or enslavement of millions of Jews, Slavs, and Gypsies.

The replacement of genetics with the loopy ideas of the agronomist Trophim Lysenko, who enjoyed Stalin's protection, was responsible for the spectacular backwardness of Soviet agriculture, which in turn, led to severe food shortages. The same dictatorship replaced sociology with Marxism-Leninism, whose faithful rightly indicted the social flaws of the capitalist societies but neglected the study of equally acute social issues in the Soviet empire. The consequence was that these issues got worse, and no Soviet social analyst foresaw the sudden collapse of the empire.

More recent cases of the pseudoscience-politics connection are the issues of climate change, stem cell research, "intelligent design," and wildlife protection on the part of the present US government. Such interferences are bound to have negative impacts on science, medicine, and the environment. The latest case of government support of pseudoscience is the decision of the French health minister to remove from the official Web site a report that cognitive-behavioral therapy is more effective than psychoanalysis (French psychoflap, 2005).

CONCLUSION

Pseudoscience is just as philosophically loaded as science. Only, the philosophy inherent in either is perpendicular to that ensconced in the other. In particular, the ontology of science is naturalistic (or materialist), whereas that of pseudoscience is idealistic. The epistemology of science is realist, whereas that of pseudoscience is not. And the ethics of science is so demanding that it does not tolerate the self-deceptions and frauds that plague pseudoscience. In sum, science is compatible with the proscience philosophy sketched in section 2, whereas pseudoscience is not.

"So what?" the reader may ask. What is the point of the above exercise in border patrolling? Answer: it may help as a warning that a research project inspired by a wrong philosophy is likely to fail. After all, this is all we can do when evaluating a research proposal before the data are in: to check whether the project is trivial or worse, namely, contrary to the "spirit" of science, so that it may deserve the infamous Ig Nobel Prize (Bunge, 2004). Much the same holds, *a fortiori*, for the evaluation of ongoing research. For example, present-day particle physics is brimming with mathematically sophisticated theories that postulate the existence of weird entities that do not interact appreciably, or at all, with ordinary matter, as a consequence of which they are safely undetectable. (Some of these theories even postulate that space-time has ten or eleven dimensions instead of the four real ones.) Since these theories are at variance with the bulk of physics, and violate the requirement of empirical testability, they may qualify as pseudoscientific even if they have been around for a quarter of a century and get published in the most reputable physics journals.

Second example: all economics and management students are required to study neoclassical microeconomics. However, they are unlikely to use this theory in tackling any real-life economic problems. The reason for such uselessness is that some of the theory's postulates are wildly unrealistic and others excessively fuzzy, hence hardly testable. Indeed, the theory assumes that all the actors in a market are free, mutually independent, perfectly well-informed, equally powerful, immune to politics, and fully "rational"—or capable of choosing the options likely to maximize their expected utilities. But real markets are peopled by individuals and firms who have imperfect information and, far from being totally free, belong to social networks or even monopolies. Moreover, the expected utility in question is ill-defined, being the product of two quantities that are estimated subjectively rather than on the strength of hard data, namely the probability of the event in question and the corresponding utility to the agent. (Most of the time, the precise form of the utility function is not specified. And, when specified, the choice is not justified empirically.) Milton Friedman (1991) boasted that, in its present form, this theory is just "old wine in new bottles." In my view, the fact that the theory has remained essentially untouched for over a century despite significant progress in other branches of social science is a clear indicator that it is pseudoscientific. (More in Bunge, 1998b.)

The moral: before jumping headlong into a research project, check it for unsound philosophical presuppositions, such as the beliefs that mathematical sophistication suffices in factual science, that playing with undefined symbols can make up for conceptual fuzziness or lack of empirical support, or that there can be smiles (or thoughts) without heads.

In short, tell me what philosophy you use (not just profess), and I'll tell you what your science is worth. And tell me what science you use (not just pay lip service to), and I'll tell you what your philosophy is worth.

REFERENCES

Adams, Marie, Zvonimir Dogic, Sarah L. Keller, and Seth Fraden. "Entropically Driven Microphase Transitions in Mixtures of Colloidal Rods and Spheres." *Nature* 393 (1998): 249–351.

Bunge, Mario. *Chasing Reality: Strife over Realism.* Toronto: University of Toronto Press, 2006.

————. *Philosophy of Physics.* Dordrecht, Holland: Reidel, 1973.

————. *Philosophy of Science,* 2 vols. New Brunswick, N.J: Transaction Publishers, 1998a.

————. *Social Science under Debate.* Toronto: University of Toronto Press, 1998b.

————. "The Pseudoscience Concept, Dispensable in Professional Practice, is Required to Evaluate Research Projects." *Scientific Review of Mental Health Practice* 2 (2004): 111–14.

————. *Treatise on Basic Philosophy,* 8 vols. Dordrecht, Holland/Boston: Reidel-Kluwer, 1974–1989

Flew, Antony (ed.). *Readings in the Philosophical Problems of Parapsychology.* Amherst, NY: Prometheus Books, 1987.

French psychoflap. *Science* 307(2005): 1197.

Friedman, Milton. "Old Wine in New Bottles." *Economic Journal* 101(1991): 33–40.

Gardner, Martin. *Science: Good, Bad, and Bogus.* Oxford: Oxford University Press, 1983.

Kurtz, Paul. *Skeptical Odysseys.* Amherst, NY: Prometheus Books, 2001.

Kurtz, Paul (ed.). *A Skeptic's Handbook of Parapsychology.* Amherst, NY: Prometheus Books, 1985.

Mahner, Martin (Ed.). *Scientific Realism: Selected Essays of Mario Bunge.* Amherst, NY: Prometheus Books, 2001.

Mahner, Martin, and Mario Bunge. "Is Religion Education Compatible with Science Education?" *Science & Education* 5 (1996): 101–23.

Martínez, Tomás Eloy. *La Novela de Perón.* Madrid: Alianza Editorial, 1989.

Merton, Robert K. *The Sociology of Science.* Chicago: University of Chicago Press, 1973.

Morris, J. S., A. Öhman, and R. J. Dolan. "Conscious and Unconscious Emotional Learning in the Human Amygdala." *Nature* 393 (1998): 467–70.

Randi, James. *Flim-Flam!* Amherst, NY: Prometheus Books, 1982.

Schelling, Thomas C. *Micromotives and Macrobehavior.* New York: W. W. Norton, 1978.

Vaughan, Susan C., et al. "Can we Do Psychoanalytic Outcome Research? A Feasibility Study." *International Journal of Psychoanalysis* 81 (2000): 513–27.

Wolpert, Lewis. *The Unnatural Nature of Science.* London: Faber & Faber, 1992.

THE COLUMBIA UNIVERSITY "MIRACLE" STUDY
Flawed and Fraud

Bruce Flamm

On September 11, 2001, the United States of America was rocked by perhaps the most horrific event in its history. In the horrible and uncertain days following the destruction of the World Trade Center (and other attacks) by Islamic zealots, many Americans turned to prayer. Millions prayed in their homes and churches as their senators and congressmen prayed on the steps of the Capitol building and their president prayed in the White House. Bumper stickers, signs, and banners flooded the nation proclaiming, "God Bless America" and "Pray for America." Millions of faithful Americans prayed for a miracle or perhaps a sign from God. Three weeks later such a miracle occurred. The timing could not have been better.

On October 2, 2001, the *New York Times* reported that researchers at prestigious Columbia University Medical Center in New York had discovered something quite extraordinary (1). Using virtually foolproof scientific methods the researchers had demonstrated that infertile women who were prayed for by Christian prayer groups became pregnant twice as often as those who did not have people praying for them. The study was published in the *Journal of Reproductive Medicine* (2). Even the researchers were shocked. The study's results could only be described as miraculous. This was welcome and wonderful news for a shaken nation.

Columbia University issued a news release claiming that the remarkable study had several safeguards in place to eliminate bias and that the study itself was carefully designed to eliminate bias (3). This was no hoax. Media attention immediately focused on the miraculous study, and articles touting its spectacular results quickly appeared in newspapers around the world. Rogerio Lobo, chairman of the Department of Obstetrics and Gynecology at Columbia and the study's lead author, told Reuters Health that, "Essentially, there was a *doubling* of the pregnancy rate in the group that was prayed for" (4). Dr. Timothy Johnson, ABC News medical editor and *Good Morning America* commentator, stated, "A new study on the power of prayer over pregnancy reports surprising results; but many physicians remain skeptical" (5).

The facts I will relate here about the Columbia University "miracle" study confirm that those physicians who doubted the study's astounding results had extremely good reasons to be skeptical. It remains to be seen whether ABC's Dr. Johnson, a medical doctor who also serves as a minister at the evangelical Community Covenant Church in West Peabody, Massachusetts, will report or ignore the following shocking information that has since been revealed about the alleged study and its authors.

THE "MIRACLE" STUDY

In vitro fertilization (IVF) is the most advanced form of infertility treatment currently available and represents the last hope for women with severe infertility. Therefore, any technique that could increase the efficacy of IVF by even a few percent would be a medical breakthrough. Yet the Columbia University study claimed to have demonstrated, in a carefully designed randomized controlled trial, that distant prayer by anonymous prayer groups increased the success rate of IVF by an astounding 100 percent (2). The Cha/Wirth/Lobo study involved 219 infertility patients in Seoul, South Korea, who required in vitro fertilization. Twenty patients were excluded due to incomplete data, leaving 199 study subjects. After randomization, 100 patients were assigned to the study group to receive IVF plus prayer from Christian prayer groups in the United States, Canada, and Australia. The control group of ninety-nine patients received IVF but did not receive any prayers from these prayer groups. In vitro fertilization was performed in the usual fashion in both groups. The 100 patients in the study group were not informed that the groups were praying for them. Furthermore, none of the patients were even informed that they were being used as study subjects. The prayer groups, which were thousands of miles away from the study subjects, prayed over photographs that had been faxed to them from Korea. Remarkably, the pregnancy rate in the prayed-for group (50 percent) was almost twice as high as the pregnancy rate in the nonprayed-for group (26 percent, p= .0013). The highly significant results seem to indicate that something spectacular had occurred.

However, even a cursory review of the report reveals many potential flaws. For one thing, the study protocol was convoluted and confusing, involving at least three levels of overlapping and intertwining prayer groups. Tiers 1 and 2 each consisted of four blocks of prayer participants. Prayer participants in tier 1, block A, received a single sheet of paper with five IVF patient's pictures (a treatment "unit") and prayed in a directed manner with a specific intent to "increase the pregnancy rate" for these patients about whom they apparently had no information whatsoever. Prayer participants in tier 2, block A, apparently performed two different types of prayer. First, they prayed for their fellow prayer participants in tier 1, block A, with the intent to "increase the efficacy of prayer intervention." In other words, they were apparently praying to increase the effectiveness of their colleagues' prayers, whatever those prayers might be. Next they prayed in a nondirected manner for the study patients with the "intent that God's will or desire be fulfilled in the life of the patient." Similar prayers apparently took place in all of the other blocks. Finally, in addition to all of the above groups, tiers, blocks, and units, a separate group of three individuals prayed in a general nonspecific manner with the intent that "God's will or desire be fulfilled for the prayer participants in tiers 1 and 2." In other words, these final three prayer participants were praying to increase the efficacy of the second tier of prayer participants who were in turn praying to increase the efficacy of the first tier of prayer participants who were in turn praying for increased pregnancy rates in the study patients.

As can be seen from this brief description, the study protocol was so convoluted and confusing that it cannot be taken seriously. Of course, a simple protocol could have been used to determine if prayer was efficacious in increasing the success rate of IVF. One might simply instruct a few believers to pray for successful IVF in the study group while no one prayed for patients in the control group. With distant prayer the patients would not know if they were being prayed for, or not, so there would be no intention-to-heal or placebo bias. Contrast this with the study design described above and draw your own conclusions. This article is too brief to describe all of the study's flaws but readers who want more information are referred to two critiques I have published in *The Scientific Review of Alternative Medicine* (6, 7).

Briefly, here are a few problems I pointed out. Choosing a complex study design rather than a simple one requires explanation, however the authors give no reason for selecting a bewildering study design. Including prayers asking that "God's will or desire be fulfilled" introduced a vague and obfuscating concept that cannot be measured as an endpoint: no one knows what God's will is, hence any outcome could be viewed as a success. The authors made no attempt to discover how much prayer was being conducted outside the study protocol, perhaps to other gods, since only one-third of Koreans are Christians. Occam's razor (the principle that a simple explanation rather than a convoluted one is more often correct) demands that highly unlikely results be viewed with suspicion. Is it more likely that this study is flawed or fraudulent or that the authors have demonstrated the existence of a supernatural phenomena and thus have made perhaps the most important discovery in history?

THE COLUMBIA UNIVERSITY CONNECTION

The study's three authors were Kwang Cha, Rogerio Lobo, and Daniel Wirth. Kwang Yul Cha, M.D., was the director of the Cha Columbia Infertility Medical Center at the time of the "miracle" study but apparently severed his relationship with Columbia soon after the study was published. A page on Columbia's Web site, which has since been removed, described Cha as an "internationally renowned clinician and researcher." Cha is a graduate of the Yonsei Medical School in Seoul, South Korea. Professor Rogerio A. Lobo, M.D., recently stepped down as chairman of the department of obstetrics and gynecology at Columbia University. When the study results were announced, Dr. Lobo told the *New York Times* that the idea for the study came from Dr. Cha; however, the Columbia news release claimed that Dr. Lobo led the study. For two years both Dr. Cha and Dr. Lobo have refused to return my phone calls and e-mails asking questions about the study. The study's third author, Daniel Wirth, who will be described below, has no known connection with Columbia University other than his participation in the study.

In December 2001, the Department of Health and Human Services (DHHS), after being alerted by media coverage, launched an investigation into the lack of informed consent in the Columbia study.

Columbia University subsequently acknowledged noncompliance with its Multiple Project Assurance (MPA) and its own policies and procedures (8). Specifically, Dr. Lobo never presented the above research to the Institutional Review Board (IRB) of Columbia-Presbyterian Medical Center (CCPM). In response to the DHHS investigation Columbia University agreed to have its IRB perform an educational in-service for Lobo's department.

In addition, in December 2001, Columbia University Vice President Thomas Q. Morris, a physician, informed the DHHS *that Dr. Lobo first learned of the study from Dr. Cha six to twelve months after the study was completed* and that Lobo primarily provided editorial review and assistance with publication (8). This seems inconsistent with Lobo being listed as one of the study's authors. This also conflicts with the fact that Lobo was identified by both *The New York Times* and ABC News as the report's *lead* author. Lobo was also identified as the report's lead author in a news release posted for two years on the Columbia University Web site. Interestingly, the press release has recently been removed from the Columbia site. If the report's lead author did not conduct the international prayer study, who did?

THE MYSTERIOUS DANIEL WIRTH

The remaining author is a mysterious individual known as Daniel P. Wirth. In October 2002, one year after the Cha/Wirth/Lobo study was published, Mr. Wirth, along with his former research associate Joseph Horvath, also known as Joseph Hessler, was indicted by a federal grand jury (9). Both men were charged with bilking the troubled cable television provider Adelphia Communications Corporation out of $2.1 million by infiltrating the company, then having it pay for unauthorized consulting work. Police investigators discovered that Wirth is also known as John Wayne Truelove. FBI investigators revealed that Wirth first used the name of Truelove, a New York child who died at age five in 1959, to obtain a passport in the mid-1980s. Wirth and his accomplice were charged with thirteen counts of mail fraud, twelve counts of interstate transportation of stolen money, making false statements on loan applications, and five other counts of fraud. A federal grand jury concluded that the relationship between Wirth and Horvath extended back more than twenty years and involved more than $3.4 million in income and property obtained by using false identities. In addition to the Adelphia scheme, Wirth apparently found a way to defraud the federal government by collecting Social Security benefits totaling approximately $103,178 from 1994 to 2003 in the name of Julius Wirth. This man, possibly Daniel Wirth's father, died in 1994 but his benefits continued to be paid after his death via electronic funds transferred to the Republic National Bank.

Incredibly, at the time of the indictment, Horvath, Wirth's partner, was already in jail, charged with arson for burning down his Pennsylvania house to collect insurance money (10). The FBI investigation revealed that Horvath had previously gone to prison in a 1990 embezzlement and false identity case in California. Interestingly, the investigation also revealed that he had also once been arrested for posing as a doctor in California. It appears that the "doctor" who performed biopsies on human research subjects in Wirth's famous healing studies may have actually been Horvath impersonating a doctor. Horvath was a co-author on another of Wirth's studies in which salamander limbs were amputated and found to grow back more quickly when "healers" waved their hands over the wounds.

Both Wirth and Horvath initially pled not guilty to the felony charges, and over the next eighteen months their trial was delayed six times. However, on May 18, 2004, just as the criminal trial of the *United States v. Wirth & Horvath* was finally about to begin, both men pled guilty to conspiracy to commit mail fraud and conspiracy to commit bank fraud (11). Each man faced a maximum of five years in federal prison and agreed to forfeit assets of more than $1 million obtained through fraudulent schemes. Horvath, however, was found dead in his jail cell on July 13, 2004, an apparent suicide.

DANIEL WIRTH'S PRIOR RESEARCH

Wirth, identified as Doctor Daniel Wirth on several of his publications, has no medical degree. He holds a master's degree in parapsychology and a law degree. Wirth has a long history of publishing studies on mysterious supernatural or paranormal phenomena, mainly dealing with alternative and spiritual healing. Most of these studies originated from an entity called "Healing Sciences Research International," an organization that Mr. Wirth supposedly headed. This entity, which sounds like a medical center or impressive research facility, could only be contacted through a post office box in Orinda, California. Between 1992 and 1997 approximately eighteen research papers authored by D. P. Wirth were published, mostly in obscure paranormal journals (12-29).

Wirth has stated that his experiments "represent a seminal research effort within the field of complementary healing," and many faith healing advocates fully agree with his statement. Due to the apparently meticulous design and conduct of Wirth's randomized, double-blind controlled studies he has become the virtual poster boy of alternative healing methods, particularly Noncontact Therapeutic Touch (NCTT). In NCTT the "healer" does not actually touch the patient but supposedly alters undetectable "human energy fields" surrounding the patient. According to Wirth, NCTT apparently achieves its healing effect by an interaction of "energy fields" between the practitioner and the subject. The method requires the healer to 1) "center" his/herself both physically and psychologically, 2) "attune" to the "energy field" of the subject by "scanning" with the hands two to six inches from the body in order to detect imbalances within or blocks within the energy field, and 3) consciously redirect and "rebalance" the energy in those areas of blockage (24). The existence of these imagined human energy fields has never been proven. Even if such fields did exist, it is not clear how a healer could possibly detect or modify them. In fact, in a recent study twenty-one experienced NCTT practitioners were unable to detect any "human energy fields" under blinded conditions. The study concluded that failure to substantiate TT's most fundamental claim is unrefuted evidence that the claims of NCTT are completely groundless (30).

In addition to his extensive work on NCTT, Wirth has previously conducted several studies involving Christian faith healing. For example, he evaluated and reported on forty-eight patients treated by Greg Schelkun, a spiritual healer trained in the Philippines in the "Espiritista System" of faith healing (17). This system includes "psychic surgery," laying on of hands, and distant prayer healing. It has a Christian foundation in which the practitioner supposedly cultivates divine healing by entering a trace-like state and opening themselves to the healing power of the Holy Spirit. Schelkun asserts that he acts only as a channel for the "universal energies" of God and that any "miraculous cures" that occur are due solely to the Grace of God. Wirth evaluated patients treated by Schelkun for conditions ranging from ovarian cysts to AIDS and even cancer. Wirth found that 90 percent of patients *believed* that their condition was improved by the treatment.

In October 2001 narcotics officers raided the Santa Monica, California, office of Dr. William Eidelman, co-author of many of Daniel Wirth's papers. Eidelman is a believer in paranormal healing and an outspoken proponent of the medical use of marijuana. Officers presented a search warrant charging that Eidelman provided undercover narcotic agents with medical marijuana recommendations without valid medical grounds. On May 28, 2002, Eidelman's license to practice medicine was suspended.

JOURNAL OF REPRODUCTIVE MEDICINE

Perhaps the most fascinating aspect of this entire sordid saga can be summed up in one question: How did a bizarre study claiming extraordinarily unlikely and apparently supernatural results end up in a peer-reviewed medical journal? We may never know. For two years the editors of the *Journal of Reproductive Medicine (JRM)* refused to answer my calls or respond to letters about this study. The fact that study co-author Lobo serves on the Editorial Advisory Board of the *JRM* may or may not be relevant. It is known that Columbia University Vice President Thomas Q. Morris informed DHHS investigators that

Dr. Lobo first learned of the study from Dr. Cha *six to twelve months after the study was completed* and that Dr. Lobo primarily provided editorial review and *assistance with publication* (8).

On May 30, 2004, the *London Observer* made many of these events public for the first time in an article titled "Exposed: Conman's Role in Prayer-power IVF 'Miracle'"(31). The *Observer* article noted that the study was still posted on the *JRM* Web site and that phone calls from the *Observer* to the *JRM* were not returned. Three days after the scandal had been made public and linked to the journal, perhaps in response to an avalanche of inquiries, *JRM* co-editor-in-chief Dr. Lawrence Devoe finally stated that the *Journal of Reproductive Medicine* would remove the flawed Columbia study from its Web site and publish an editorial clarifying their author requirements. Both the *Observer* article and a June 7, 2004, article in *The New York Sun* stated that the authors did not respond to their requests for comment.

It must be emphasized that, in the entire history of modern science, no claim of any type of supernatural phenomena has ever been replicated under strictly controlled conditions. The importance of this fact cannot be overstated. One would think that all medical journal editors would be keenly aware of this fact and therefore be highly skeptical of paranormal or supernatural claims. One must therefore wonder if the Columbia researchers and the *JRM* editors were blinded by religious beliefs. Everything else being equal, if the claimed supernatural intervention had been Ms. Cleo manipulating Tarot cards rather than Christians praying, would the reviewers and editors have taken this study seriously? In any case, the damage has been done. The fact that a "miracle cure" study was deemed to be suitable for publication in a scientific journal automatically enhanced the study's credibility. Not surprisingly, the news media quickly disseminated the "miraculous" results.

DAMAGE CONTROL

Clearly, *JRM*'s belated decision to remove the Columbia study from its Web site will not correct the errors it made in publishing an absurd article and then persistently ignoring warnings about the mistake in doing so. Serious damage has been done. The editors were informed of several of the study's flaws within weeks of its publication and yet allowed the entire study to remain on their Web site for two years. During that time the public was never given any reason to doubt the study's validity or its miraculous claims. As a result of *JRM*'s inaction the Cha/Wirth/Lobo study has been cited in many other "healing" publications and on other Web sites as strong scientific evidence for the validity of faith healing. A Google search performed on June 4, 2004, for the terms, "Wirth, Columbia, prayer" found 686 sites; many of these links led to articles touting the miraculous results of the Cha/Wirth/Lobo study.

Worse yet, the Columbia study is now being cited by faith healers as a shining example of "healing" research of the highest scientific quality. For example, I recently wrote a letter to the editor of *Southern California Physician* critical of its article "Prescription for Prayer" and the appalling claim by noted faith healer Dr. Larry Dossey that some 1,600 studies have revealed "something positive" about intercessory prayer. I commented that if there were, in fact, something positive it certainly wouldn't take 1,600 studies to find it! Dr. Dossey's published response to my letter included the following convincing argument, "Controlled clinical trials and the peer-review process continue to serve us well. The most recent example of this process in action in the area of intercessory prayer is from Columbia Medical School—a positive, controlled clinical trial published in the respected, peer-reviewed *Journal of Reproductive Medicine*" (32). Yes, Dossey had used the hopelessly flawed Columbia "miracle" study to demonstrate the scientific validity of faith healing.

In the February 2004 edition of her nationally distributed newsletter, faith healing advocate Dr. Susan Lark cites the Cha/Wirth/Lobo study as strong evidence for the power of prayer (33). She notes that critics of faith healing have argued that most prayer studies have not been credible due to weak methodologies. However, she points out that "those researchers who believe in prayer are answering this critique quickly—and effectively. The fact is, the medical journals are rapidly filling with studies that are proving the power of prayer." She then presents the proof by describing the Columbia "miracle" study.

In a published critique of phony healing methods, a noted physician and chairman of the Dutch Union Against Quackery, Dr. Cees Renckens has this to say about the Cha/Wirth/Lobo study: "Very recently a seemingly impeccable paper proving absurd claims was published in a serious and (hitherto?) respected journal in the field of reproductive medicine" (34). Dr. Renckens also states, "Fraud is difficult to extract from an apparently impeccable paper, but everyone is invited to draw one's own conclusions about the trustworthiness of the authors. We do not believe anything of the story and are very much opposed to publishing this kind of absurdity in serious journals."

For both Columbia University and the *JRM*, the only honorable solution to this scandal is to fully and publicly disclose their mistakes and apologize for the attempted cover-up. Columbia erroneously submitted a profoundly flawed and absurd article and *JRM* erroneously published it. Simply claiming that they were duped by Wirth and attempting to blame him for their own mistakes would be unethical—and almost certainly false. It would also be a setback for science.

CONCLUSION

In summary, one of the authors of the Columbia Cha/Wirth/Lobo study has left the University and refuses to comment, another now claims he did not even know about the study until six months to a year after its completion and also refuses to comment. The remaining author is on his way to federal prison for fraud and conspiracy. *Fraud* is the operative word here. In reality, the Columbia University prayer study was based on a bewildering study design and included many sources of error. But worse than flaws, in light of all of the shocking information presented above, one must consider the sad possibility that the Columbia prayer study may never have been conducted at all.

Finally, Daniel Wirth's history of criminal fraudulent activity casts a dark shadow over many of the supposedly seminal publications in the field of alternative and faith healing. In light of these facts, *all* of his frequently cited publications must now be viewed with suspicion. While faith healers have performed rituals and cast out demons for millennia, they are now attempting to validate their claims with scientific methods and publish their results in peer-reviewed medical journals. It is one thing to tell an audience at a tent revival that prayers yield miracle cures but quite another thing to make the same claim in a scientific journal. By doing so, faith healers cross the line into the domain of science, a domain where superstitious and supernatural claims are not taken seriously.

LESSONS FROM THE 'MIRACLE' STUDY SCANDAL

• The real scandal here lies not in Wirth's actions but in those of Columbia University and the *Journal of Reproductive Medicine*. The scientific method is designed to detect and correct errors and misconduct. In this case the system failed in many places. In fact, if Wirth had not been arrested, the Cha/Wirth/Lobo study might have never been retracted.

• Faith healing advocates like Drs. Dossey and Lark will no doubt try to put a positive spin on this scandal by claiming that it has successfully weeded out a few bad apples from an otherwise pristine bunch. Nothing could be further from the truth.

• Extraordinary claims demand extraordinary evidence. Unless replicated under strictly controlled conditions, studies claiming to have demonstrated "miracle" cures belong in religious and paranormal magazines, not in scientific journals. This is true regardless of whether the claimed "miracle" involves supposed actions of deities, ghosts, psychic powers, or other "mysterious" phenomena.

• It is often claimed that faith healing may not work but at least does no harm. In fact, reliance on faith healing can cause serious harm and even death (35).

• In the entire history of modern science, no claim of any type of supernatural phenomena has ever been replicated under strictly controlled conditions. All scientists and editors of scientific and medical journals should be fully aware of this obvious fact.

• The "faith" in faith healing refers to an irrational belief, unsupported by evidence, that mysterious supernatural powers can eradicate disease. Science deals with evidence, not faith.

• Publication of absurd studies and pseudoscience in medical and scientific journals does serious damage to the public's perception of medical science and science in general.

REFERENCES

1. Nagourney, E. Study Links Prayer and Pregnancy. *New York Times.* 2001; October.
2. Cha, K. Y., D. P. Wirth, R. A. Lobo. "Does Prayer Influence the Success of In Vitro Fertilization-embryo Transfer?" *Journal of Reproductive Medicine* 46(2001): 781–87.
3. Eisner, R. "Prayer may Influence In Vitro Fertilization Success." *Columbia News.* This document remained on the Public Affairs News page of Columbia University Internet site for more than two years after the publication of the Cha/Wirth/Lobo study (http://www.columbia.edu/cu/news).
4. Schorr, M. "Prayer may Boost In-vitro Success, Study Suggests." *Reuters News Service* 2001; October.
5. Johnson, T. "Praying for Pregnancy: Study Says Prayer Helps Women Get Pregnant." ABC television's *Good Morning America* 2001; October 4.
6. Flamm, B. L. "Faith Healing by Prayer: Review of Cha, K. Y., Wirth, D. P., Lobo, R. A. Does Prayer Influence the Success of In Vitro Fertilization-embryo Transfer?" *Scientific Review Of Alternative Medicine* 6, no. 1 (2002): 47–50.
7. Flamm, B, L. "Faith Healing Confronts Modern Medicine." *Scientific Review Of Alternative Medicine* 8, no. 1 (2004): 9–14.
8. Carmone, M. A. *Letter to Thomas Q. Morris, MD, Vice President for Health Sciences Division,* regarding possible noncompliance with DHHS regulations for protection of human subjects in the conduct of the Cha et. al. Study.
9. "Pair Charged with Scheming Adelphia out of $2.1 Million." *Associated Press.* October 16, 2002.
10. Dale, M. "Arson Added to Charges Pending against Ex-Adelphia Manager." *Associated Press. Contra Costa Times.* February 5, 2003.
11. McDermott, J. "Mystery Man Admits to Conspiracy." *Morning Call Newspaper.* May 18, 2004. This article can be viewed in the news archives at http://www.mcall.com.

12. Wirth, D. P., J. R. Cram, R. J. Chang. "Multisite Surface Electromyography and Complementary Healing Intervention: A Comparative Analysis." *Journal of Alternative And Complementary Medicine* 3, no. 4 (Winter 1997): 355–64.

13. Wirth, D. P., J. R. Cram. "Multisite Electromyographic Analysis of Therapeutic Touch and Qigong Therapy." *Journal of Alternative And Complementary Medicine* 3, no. 2 (Summer 1997): 109–18.

14. Wirth, D. P., J. T. Richardson, W. S. Eidelman. "Wound Healing and Complementary Therapies: A Review." *Journal of Alternative And Complementary Medicine* 2, no. 4 (Winter 1996): 493–502.

15. Wirth, D. P., R. J. Chang, W. S. Eidelman, J. B. Paxton. "Hematological Indicators of Complementary Healing Intervention." *Complementary Therapies in Medicine* (January 1996): 14–20.

16. Wirth, D. P., J. T. Richardson, R. D. Martinez, W. S. Eidelman, M. E. Lopez. "Noncontact Therapeutic Touch Intervention and Full-thickness Cutaneous Wounds: A Replication." *Complementary Therapies in Medicine* (October 1996): 237–40.

17. Wirth, D. P. "The Significance of Belief and Expectancy within the Spiritual Healing Encounter." *Social Science & Medicine* 41, no. 2(1995): 249–60.

18. Wirth, D. P. "Complementary Healing Intervention and Dermal Wound Reepithelialization: An Overview." *International Journal of Psychosomatics* 42(1995): 48–53.

19. Wirth, D. P., J. R. Cram. "The Psychophysiology of Nontraditional Prayer." *International Journal of Psychosomatics* 41, no. 1–4 (1994): 68–75.

20. Wirth, D. P., M. J. Barrett. "Complementary Healing Therapies." *International Journal of Psychosomatics* 41, no. 1–4 (1994): 61–67.

21. Wirth, D. P., B. J. Mitchell. "Complementary Healing Therapy for Patients with Type 1 Diabetes Mellitus." *Journal of Scientific Exploration* 8, no. 3(1994): 367–77.

22. Wirth, D. P., M. J. Barrett, W. S. Eidelman. "Non-contact Therapeutic Touch and Wound Reepithelialization: An Extension of Previous Research." *Complementary Therapies in Medicine* 2 (October 1994): 187–92.

23. Wirth, D. P., D. R. Brenlan, R. J. Levine, C. M. Rodriguez. "The Effect of Complementary Healing Therapy on Postoperative Pain After Surgical Removal of Impacted Third Molar Teeth." *Complementary Therapies in Medicine* (July 1993): 133–38.

24. Wirth, D. P., J. R. Cram. "Multi-site Electromyographic Analysis of Non-contact Therapeutic Touch." *International Journal of Psychosomatics* 40, no. 1–4(1993): 47–55.

25. Wirth, D. P., et al. "Full Thickness Dermal Wounds Treated with Non-contact Therapeutic Touch; A Replication and Extension." *Complementary Therapies in Medicine* (July 1993): 127–32.

26. Wirth, D. P. "Implementing Spiritual Healing in Modern Medical Practice: Advances." *Journal of Mind-Body Health* 9(1993): 69–81.

27. Wirth, D. P., C. A. Johnson, J. S. Horvath, J. D. MacGregor. "The Effect of Alternative Healing Therapy on the Regeneration Rate of Salamander Forelimbs." *Journal of Scientific Exploration* 6 (1992): 375–91.

28. Wirth, D. P. "The Effect of Non-contact Therapeutic Touch on the Healing Rate of Full Thickness Dermal Wounds." *Nurse Healers Professional Associates* 13, no. 39(1992): 4–8

29. Wirth, D. P. "The Effect of Non-contact Therapeutic Touch on the Healing Rate of Full Thickness Dermal Wounds." *Subtle Energies* 1(1990): 1–20.
30. Rosa, L., E. Rosa, L. Sarner, S. Barrett. "A Close Look at Therapeutic Touch." *JAMA* Apr 1, 279, no. 13(Apr 1998): 1005–10.
31. Harris, D. "Exposed: Conman's Role in Prayer-power IVF 'Miracle'." *The Observer* May 30, 2004.
32. Dossey, L. "Response to Letter to the Editor." *Southern California Physician* (December 2001): 46.
33. Lark, S. "The Power of Prayer." *The Lark Letter: A Women's Guide to Optimal Health and Balance.* (February 2004): 1–3.
34. Renckens, C. N. M. "Alternative Treatments in Reproductive Medicine: Much Ado about Nothing." *Human Reproduction* 17, no. 3(2002): 528–33.
35. Flamm, B. L. "The Inherent Dangers of Faith Healing." *Scientific Review of Alternative Medicine.* In Press.

THIRD STRIKE FOR COLUMBIA UNIVERSITY PRAYER STUDY

BRUCE FLAMM

The preceding chapter describes in detail my exposé of one of the most outrageous scandals in the history of medical research (Flamm, 2004). To summarize, in 2001, *The New York Times* reported that researchers at prestigious Columbia University Medical Center in New York had made an astounding discovery. Anonymous prayers from people in the United States, Canada, and Australia had fully doubled the success rate of complex IVF infertility treatments performed on patients in Korea. The study's three authors were Kwang Cha, the director of the Cha-Columbia Infertility Medical Center, Rogerio A. Lobo, the chairman of the Department of Obstetrics and Gynecology at Columbia University, and Daniel Wirth, a mysterious man with no medical or scientific training.

To many readers, the bizarre study stunk to high heaven, and *The Journal of Reproductive Medicine* (JRM) was immediately bombarded with calls and letters from skeptical scientists and physicians. But things quickly went from bad to worse when the editors would not respond to calls and would not publish any letters critical of the supposedly miraculous study.

STRIKE ONE: FELONY FRAUD

The *JRM* editors' head-in-the-sand tactics successfully suppressed criticism of the study for two and a half years. It appeared that the "pray-for-pregnancy" study would never be removed from the legitimate scientific literature. Then, something quite unexpected happened. In June 2004, newspapers and magazines worldwide announced that prayer-study coauthor Daniel Wirth had been arrested by the FBI and had pled guilty to a twenty-year history of criminal fraud (Quackwatch, 2004). Mr. Wirth, the man who allegedly designed and conducted the prayer study, was sentenced to five years in a federal prison (Flamm, 2005). One down, two to go.

STRIKE TWO: DECEPTION

Staggered by the revelations about Wirth, the *JRM* editors finally broke their silence and announced that they had temporarily removed the controversial Cha/Wirth/Lobo study from their Internet site. However, they stopped short of retracting the prayer publication. After all, there were still two credible authors who could verify the study's miraculous results. Or could they? Beating a hasty retreat, Columbia University quickly announced that Professor Rogerio Lobo had only provided editorial assistance and actually had nothing to do with the alleged research and thus could not support the results or even verify that any data were ever collected. Shockingly, the study's lead author, the professor who had touted the study's supernatural results on the nationally televised Good Morning America show, now admitted he knew nothing about the research! Gerald Fischbach, dean of the Faculty of Medicine at Columbia University, reprimanded the disgraced Dr. Lobo and instructed him to remove his name from the absurd paper. And so he did (Carey, 2004). Two down, one to go.

STRIKE THREE: PLAGIARISM

The Cha/Wirth/Lobo paper suddenly morphed into the Cha/Wirth paper. Of the two remaining authors, one now resided in a federal penitentiary and the other slipped quietly away from Columbia University to run the Cha Fertility Center in Los Angeles. Undaunted, Lawrence Devoe, editor in chief of the *JRM*, decided that things were quieting down and it was time to replace the prayer study on his journal's Web site. And so he did. For the next two years, Devoe steadfastly refused to respond to critics and declined to retract the nonsensical study. Once again, it seemed that the flawed and almost certainly fraudulent study would never be excised from the legitimate scientific literature.

But fact is sometimes stranger than fiction. Yes, this story is about to take yet another almost unbelievable twist. On February 18, 2007, the *Los Angeles Times* reported that Kwang Cha, the only remaining unincarcerated author of the Cha/Wirth paper, had committed plagiarism (Ornstein, 2007). After a long investigation, Alan DeCherney, editor in chief of *Fertility and Sterility*, announced that a 2005 article by Kwang Cha et al. was a word-for-word, chart-for-chart copy of a paper previously published by a different author in a Korean medical journal.

In addition to plagiarism, Cha may have some other problems. According to the *Los Angeles Times*, Cha appeared to be violating state law by using M.D. after his name on Web sites and in news releases in California, in spite of the fact that he is not licensed to practice medicine in the state. Cha still claims to be a professor of obstetrics and gynecology at Columbia University, although the medical school's dean states that this was never the case (Cha, 2007). Cha's Web site also indicates that his astrological sign is Sagittarius and that his Chinese zodiac year is that of the dragon. The doctor's astrological credentials apparently remain uncontested. A credible author? Strike three.

GAME OVER?

On February 20, 2007, *The Chronicle of Higher Education* published an article about Kwang Cha's problems. When the *Chronicle* asked DeCherney at *Fertility and Sterility* what he thought about the validity of the 2001 Cha/Wirth prayer study, he revealed a remarkable fact. His journal had seen the manuscript before it was sent to the *JRM* and rejected it. "It's baloney," he said. "That's not in question."

Not a single credible author remains to support the outlandish study, and there is absolutely no reason to believe the absurd results. Yet it remains unretracted, stinking up the scientific medical literature. Sadly, the outrageous and unprecedented behavior of the editor of the *JRM* has damaged evidence-based medicine and the credibility of the scientific literature. How can anyone decide what articles can be relied upon for correct information if articles known to be fraudulent are not retracted?

Clearly, the Cha/Wirth/Lobo study has been demolished. Yet, the defenders of this preposterous publication just sit there, out of excuses, dazed and confused, waiting, perhaps, for divine intervention.

This story might be humorous if not for the fact that barbarians are clawing at the gates of the universities and medical schools that scientists and critical thinkers have valiantly struggled to build. Science is under attack by people who want schools to teach that the earth is 6,000 years old, that evolution is a foolish theory, and that faith-healing is more efficacious than modern medical care. During the past ten years, courses stressing the potential health benefits of spiritual beliefs have been integrated into the curriculum of the majority of US medical schools (Fortin and Barnett, 2004; Puchalski and Larson, 1998). The intractable resolve of those who have defiantly fought to protect the Cha/Wirth/Lobo prayer study should be a wake-up call for all who support science and evidence-based medicine. If mystics can win this battle, they should have no problem defending future faith-based "miracle" studies.

REFERENCES

Carey, Benedict. Researcher Pulls his Name from Paper on Prayer and Fertility. *New York Times* (December 4 2004.) Available at: http://www.nytimes.com/2004/12/04/science/04prayer.html?ex=1259902800&en=c2a3e6baeb4f5ebe&ei=5088&partner=rssnyt.

Cha, Kwang Yul. User Profile. *Blogger* Web site. January 2007. Available at: http://www2.blogger.com/profile/05830505851878978098.

Flamm, Bruce. "The Columbia University 'miracle' study: Flawed and Fraud." *Skeptical Inquirer* 28 no. 5 (September/October 2004.): 25–31. Available at: http://www.csicop.org/si/2004-09/miracle-study.html.

——. "The Bizarre Columbia University 'miracle' Saga Continues." *Skeptical Inquirer* 29, no. 2 (March/April 2005): 52–53.

Fortin, Auguste H. VI, and Katherine Gergen Barnett. "Medical School Curricula in Spirituality and Medicine." *Journal of the American Medical Association* 291, no. 23 (June 16 2004): 2883. Available at: http://jama.amaassn.org/cgi/reprint/291/23/2883.pdf.

Ornstein, Charles. "Credit for U.S. Journal Article at Issue." *Los Angeles Times* (February 18 2007). Available at: http://www.latimes.com/news/printedition/california/la-me-research18feb18,1,2222823.story?coll=la-headlines-pe-california.

Puchalski, Christina M., and David B. Larson. "Developing Curricula in Spirituality and Medicine." *Academic Medicine* 73, no. 9 (September 1998): 970–74.

Quackwatch "Indictment of Josepf Horvath and Daniel Wirth." *Quackwatch* Web site. June 1 2004. Available at: http://www.quackwatch.org/11Ind/wirthindictment.html.

ANOMALOUS COGNITION:
A Meeting of the Minds?

Amir Raz

"What exactly is anomalous cognition?" As a cognitive scientist, I wondered about this question as I was peering over an intriguing invitation to attend an exclusive Meeting of Minds (MoM) conference on this very topic.[1] I had been counting the days before the MoM, until finally in July 2007 about sixty researchers got together at the University of British Columbia in Vancouver, Canada. To avoid media coverage, the organizers targeted a select group of speakers, and attendance was by invitation only. I was surprised when the conference turned out to be a series of presentations, including reports of what are arguably the best accounts in favor of the possibility of things such as parapsychology and psychic influence, also known as psi. The meeting brought together behavioral scientists and experimental psychologists—most of the audience for the talks—a few skeptics, and a group of self-labeled psi researchers, most of the presenters. As a special treat, a handful of renowned panelists—two Nobel laureates and two distinguished professors of psychology—offered pithy summaries of their impressions following the presentations. It did not take long to realize that anomalous cognition is a new euphemism for the time-honored claims of psi, including extrasensory perception (ESP) and telekinesis.

Initially, I was not sure whether I was invited as a scientist, a skeptic, a magician, or as a friend of one of the organizers. Although I am not a parapsychologist, I am genuinely interested in what I refer to as *atypical cognition* and rarely shy away from investigating areas within my purview, even those considered as fringe by most of my colleagues. For example, I have been studying the brain computations that occur during planes of altered consciousness, including the cognitive neuroscience of phenomena such as sleep-deprivation, hypnosis, and meditation. At the same time, I consider myself a skeptic—of the deferentially inquisitive rather than gravely unyielding variety—who thrives on converging independent replications of rigorous empirical evidence, not on doctrinaire viewpoints. Finally, it was nice to see among the MoM guests a few fellow conjurors who are, foremost, scientists. Their presence was reassuring, if only to avoid thinking about my answer to the phrase "Are you the best magician among scientists or the best scientist among magicians?" which I have heard one too many times. In that crowd, I was neither.

Having spoken to one of the organizers a few weeks before the meeting, it was my understanding that the conference's leadership envisaged it as an opportunity to present some of the most compelling data sets in support of anomalous cognition and to urge "mainstream" scientists to foster sufficient open-mindedness to consider a more programmatic investigation into these fields based on these findings. That approach seemed fair and appropriate. Although it was unclear to me at that time what exactly anomalous cognition is, I thought then—as I do now—that it is certainly legitimate to advocate for the possibility of anomalous cognition, including psi. The agenda at the meeting, however, went beyond asking that "mainstream" scientists consider the possibility of psi: it intimated that scientific evidence for psi was solid and replicable. Furthermore, it went on to propose that a major goal of the MoM was to consider why scientific and lay communities do not appreciate the existence of psi.

Interestingly, a number of presenters who argued for the possibility of psi were mainstream researchers, at least in the sense that they had trained and worked in some of the world's most prestigious institutions of higher learning. While several speakers judiciously implied the possibility of psi, a few explicitly claimed that, based on rigorous data, several anomalous phenomena were veridical. It is perilous, however, to overlook the tenuous boundary between suggesting the possibility of certain phenomena and insinuating—not to mention explicitly submitting—that such anomalies actually exist. During the MoM several speakers blurred this boundary, some in letter and some in spirit, and a few unflinchingly crossed it.

As the conference unfolded, serious issues began to surface concerning the role of scientific evidence, replicability of findings, and philosophy of science. In addition, another question gradually emerged, one that scientists seldom ponder: when is it rational to end the pursuit of a hard-to-pin-down goal? In other words, when should one stop looking for evidence in support of an elusive effect?

As a matter of good practice, members of the scientific community tend to be skeptical. Science thrives on a skeptical approach, and scientists are typically conservative in what they consider a "generally accepted view." Two types of errors, however, stand in the way of any gatekeeper of science. One pertains to how nonexistent phenomena may pass as real or generally accepted; the other pertains to how real phenomena, which should be generally accepted, may pass as nonexistent. Scientists typically pay more attention to the former trap, and some consequently tend to be overzealous or dogmatically skeptical; members of this staunch group can be skeptical of their own belly buttons. The second trap, however, is usually less explored. If psi effects are real, then the scientific establishment needs to be careful not to deny a phenomenon that may later become a generally accepted view.

Carl Sagan popularized Marcelo Truzzi's dictum that extraordinary claims require extraordinary proof. Although Truzzi used the word "evidence" rather than Sagan's "proof," the former, too, had paraphrased earlier statements by great skeptics such as David Hume and Pierre-Simon Laplace. Most scientists still uphold the "extraordinary" motto; however, many of them might not realize that later in his life Truzzi recanted his own maxim. While we can speculate why he did, it remains unclear what constitutes an extraordinary claim. Does claiming to possess X-ray vision or that the sun will not shine tomorrow count as extraordinary? Deciding on what constitutes an extraordinary claim is probably related to our working knowledge—the proverbial *a priori* Bayesian probabilities with which we navigate the world. We typically use the inductive process to decide whether claims are extraordinary. It would be easier to accept X-ray vision, for example, if we suddenly discovered special receptors for that wavelength in the human body. The presence of such receptors is unlikely—if only because they have eluded us heretofore—but not impossible. That the sun will not shine tomorrow is perhaps a more extraordinary claim because our inductive experience, not to mention our knowledge of physics, suggests otherwise. In addition, while it may be difficult to agree on what would lend extraordinary support to a claim, scientists usually agree on what constitutes unimpressive evidence. Thus, for example, experimental results that do not replicate, effects that are very small and tenuous, flaws of design and methodology, insufficient sample size, inadequate statistical analyses, and lack of a theoretical basis may all contribute to weak evidence.

Conducting parapsychology experiments is an unprotected legal act: anyone can do it without a special license. At the conference, a few talks featured nonpsychologists, including physicists, engineers, and other professionals with little or no training in behavioral science, who nonetheless reported data from studies they conducted in experimental psychology. While at least some of these studies were markedly inadequate and contained glaring shortcomings, others consisted of more careful efforts, sometimes with intriguing results. Physicists with little training in behavioral science, however, are probably not the best professionals to conduct complex psychological experiments in the same way that experimental psychologists with little background in theoretical physics are likely suboptimal candidates to carry out empirical research in quantum mechanics. Of course, individuals who combine psychology with relevant interdisciplinary knowledge, including that from the exact, life, social, and engineering sciences, may have relative merits. In this regard, magicians—those performers who are well-versed in the art of human deception and trickery—may have especially good insights to offer. Whereas I have been an active magician and spent considerable time following claims of the paranormal, I am now a professional academic scientist, at least in the sense that a reputable university supports my research and salary. These credentials make me neither omniscient nor an authority on truth. But they do suggest at least some experience with and perhaps proficiency in assessing psi claims.

Science provides an evanescent form of truth. We never get there, but we can judge how close we are. One test that we can perform requires the convergence of evidence over multiple researchers, methods, labs, and periods. We should probably apply the same time-honored, scientific principle to the study of psi. The psi phenomena reported in the conference, however, tended to comprise very small, elusive effects that were difficult to replicate. In the few cases seemingly supported by replication or meta-analysis (a statistical method that can provide a more complete picture than individual small studies can), multiple caveats cast long shadows over the raw data and the inclusion/exclusion criteria of specific studies. Statistical analysis, however rigorous, is independent of the quality of the unprocessed information: it crunches both meaningful and less meaningful data indiscriminately. Thus, independent of the statistical methods, interpretation of the results is inconclusive at best.

It became clear that proponents of the existence of psi, who typically claim that evidence for psi is bona fide and replicable, largely base their claims on the results of several meta-analyses. It is precarious, however, to rely almost exclusively on the outcomes of meta-analyses for support. Meta-analytical studies are retrospective, not prospective, and confound exploratory with confirmatory investigation. In addition, in the known cases where more than one team of investigators have conducted a meta-analysis of the same research domain within psi, the conclusions have been strikingly different (for example, a psi proponent reported a meta-analysis of Ganzfeld studies with an average effect size that significantly differed from zero with odds of more than a trillion to one while another meta-analysis of the Ganzfeld data concluded that the average effect size was consistent with zero). This lack of robustness is difficult to reconcile.

Scientists, including the better and smarter of them, are fallible beings prone to the entire spectrum of human behaviors and blunders. People, including scientists, often ask unscientific questions: do you believe that hypnosis can reduce pain? Do you suppose that Prozac can help depression? Pristine scientists, however, do not *believe* or *suppose*. Instead, they look at the data and ask whether the evidence supports the hypothesis. At least in theory, researchers' beliefs should be immaterial to the results of their experiments, because science is about empirical evidence. In reality, however, the experimenter's beliefs may introduce a substantive bias to the interpretation of data and sometimes even to more nuanced aspects. For example, beliefs and attitudes may bias participant recruitment and influence their expectations, affect feedback, and may even subtly permeate data collection and analysis. At the MoM, it quickly became evident that people had strong beliefs. "What kind of data would make you change your mind?" I asked many a colleague. While several associates danced around the answer with grace and elegance, most coy responses amounted to one troublesome sentiment: "none."

In a short, informal gathering following the main MoM event, a few participants suggested that perhaps psi effects are not amenable to standard scientific scrutiny because the alleged effects, when they do occur, typically disappear soon after the initial experiment, thereby preventing replication. This "decline effect"—the tendency of psi phenomena to wane over time, sometimes reaching chance levels—is most peculiar. Another commonly reported outcome is the "experimenter effect": a difference in participants' performance as a function of the individual who is administering the experiment. It may be interesting to further pursue the latter, as it may also elucidate the therapeutic alliance we so desperately seek with our health practitioners. Nonetheless, we should heed Karl Popper, an influential twentieth-century philosopher of science, who taught us that a proposition or theory is scientific if it permits the possibility of being shown false—the falsifiability criterion. The history of science shows that many theories were not initially falsifiable not because they were not sufficiently well-operationalized in terms of measurable variables—as was the case in Freudian theories, for example—but because they were not fully developed. Such theories, however, have often served a valuable purpose. Proponents of psi may feel that they operate in a similar climate: they might not yet be ready for "prime time" but may want to use the controversy surrounding psi to generate interest and perhaps even a large body of research from which new theories and empirical findings can evolve.

Theory is important, and the life of the scientific theoretician is anything but easy because experiments are inexorable evaluators of one's work. These unfriendly judges—the experiments—never say yes to a theory and in the great majority of cases assert a flat-out no. Even in the most favorable of situations, they suggest only a "perhaps." Historically, rather than anchor their observations in a theoretical framework, most proponents of psi have focused on a technicality: their pivotal criterion for the presence of psi hinged on obtaining a statistically significant departure from chance. It became gradually evident, however, that in this way it was difficult to specify what properties typified psi and what criteria determined its absence. Nowadays, theories of psi abound, with most loosely brushing against quantum theory and generating no specific, testable, and falsifiable predictions. Such theories, some rather grandiose, appear especially disjointed, as they are not grounded in supporting experimental data.

"A wise man . . . proportions his belief to the evidence," wrote Hume in his 1748 essay *Of Miracles*. Having attended all the talks at the meeting, the collective evidence that I have examined does not support the hypothesis that psi phenomena exist. Neither I nor anyone else, however, can reject this hypothesis and conclude that such phenomena do not exist. For example, based on insufficient evidence we cannot decisively conclude that the Tooth Fairy does not exist. But the burden of "proof" rests with those who make the extraordinary claim. On the one hand, when intriguing nascent evidence presents itself, further investigation should ensue. On the other hand, skeptics will probably continue to maintain that psi is unlikely, and proponents will almost certainly continue to look for new ways to demonstrate their claims.

The air was effervescent as each panelist offered an extemporaneous eight-minute summary. Peppering their comments with humor and panache, the psychologists were largely unimpressed by the evidence and pointed to a number of the above-mentioned weaknesses. The Nobel laureates, however—one in physics and one in chemistry—echoed a favorable and more accepting tenor. One mentioned atmospheric science as a metaphor for the science of psi, suggesting that psi phenomena may be difficult to predict and replicate consistently in the same way that weather forecasts are nebulous. The other described his experiences with personal acquaintances whom he considered to be genuine psychics.

These last statements left me rubbing my ears in disbelief. On the one hand, albeit far from perfect, weather forecasts have gotten better over the past few decades and are certainly more reliable than outcome predictions from psi research. On the other hand, befriending individuals who claim psychic abilities is hardly firm grounds for scientific exchange.

Individuals, including intelligent persons, are infamously irrational, and one personal "psi experience" is often more compelling than multiple converging scientific accounts. Social psychologists have coined this phenomenon the "vividness" effect. Being a scientist, a prestidigitator, and a skeptic who is keenly aware of his bellybutton, I'd be curious to see compelling scientific demonstrations of psi (like a string of multiple successful experiments by several independent investigators producing lawful and replicable outcomes). Alas, I have found none to date. But when do you conclude that the effect you are seeking is unlikely? When do you stop looking?

Data in support of psi have so far failed to meet the acceptable scientific standards of lawfulness, replicability, objectivity, falsifiability, and theoretical coherence. A group of dogmatically skeptical individuals seems to consistently reject psi research because of granitic prejudices, but navel-denying skepticism is incongruent with good science. While some scientists may indeed reject psi out of prejudice, they typically do not "discriminate" against psi; they show a similar "prejudice" against any claim that seemingly violates fundamental principles of current scientific theory. A healthy first reaction to any departure from existing frameworks is to look for defects in the supporting evidence. If such defects are not apparent, it is time to insist on obtaining independent replications. Until such evidence is forthcoming, it would be difficult for the scientific community to accept a claim for an anomaly.

Highly biased perceptions of reality may be at odds with the findings of science, and establishing the existence of paranormal phenomena might well comprise an intractable task. If compelling evidence were to materialize, however, scientists should be willing to change their minds. Members of the scientific community should be amenable, at least in some measure, to the possibility of novel phenomena. At the same time, proponents of new claims should provide compelling "proof," and everyone should be sufficiently critical to dismiss claims that already have been found specious. While some of us may have concluded that the Tooth Fairy seems unlikely, others may keep on looking for her. . . . Still others may be undecided.

NOTE

1. Meeting of Minds. Invitational Workshop on Anomalous Cognition. University of British Columbia, July 15–16, 2007. Supported by grants from the Fetzer Institute, the Samueli Institute, the Bial Foundation, University of British Columbia, and the Institute of Noetic Sciences.

ANOMALOUS COGNITION?
A Second Perspective

Ray Hyman

In the previous chapter, Amir Raz provides an interesting and insightful account of the Meeting of the Minds (MoM) conference. His report offers the viewpoint of a young neuroscientist who is newly encountering the world of parapsychology and its claims. I thought I might complement his description with a few comments from my perspective. I have been a critic—I hope a constructive one—of parapsychological claims for fifty years. Together, our two accounts can better convey some of the issues stemming from this meeting.

I was invited to speak as a representative of the skeptical community. As a presenter, I felt my responsibility was to directly address the issues in the statement of the meeting agenda and goals. These issues, as spelled out in advance by the organizers, were:

- to bring together a distinguished set of researchers to consider the state of the evidence for anomalous cognition
- to discuss the methodological and theoretical challenges presented by such phenomena
- to address sociological barriers that have constrained academic discussion of this topic
- to examine the process and potential impact of the meeting.

My attention was captured by the following two quotations from the agenda:

> . . . meta-analyses of several classes of experiments published in peer-reviewed journals suggest that some effects, while small in magnitude, are highly repeatable. . . .
>
> Because of these and other reasons once commonly used to dismiss contemplation of anomalous cognition are becoming increasingly debatable, we believe the time has come to examine the taboo that has constrained serious scientific consideration of this evidence.

The obvious subtext of this meeting statement can be summarized in three propositions:

1. Psi (ESP and Psychokinesis) is real.
2. The evidence for psi is consistent and independently replicable.
3. The time has come for the scientific community to seriously consider the claims for psi.

My presentation directly challenged each of these propositions. I did so by using data and arguments provided by leading figures in parapsychology. I began by considering parapsychological claims of having demonstrated an "anomaly." Since the beginnings of modern science, the scientific community has repeatedly been confronted with claims of anomalies. Some, such as meteorites, discrepancies in the orbit of Uranus, discrepancies in the advancement of Mercury's perihelion, X-rays, and continental drift, eventually were shown not to be the result of mistaken observations or flawed methodology. Furthermore, they were supported by evidence that was consistent and independently verifiable. Given these circumstances, the claims were accepted, and scientific theories were appropriately modified to accommodate them.

Other claims, such as those for Martian canals, N-Rays, polywater, mitogenetic radiation, the "discovery" of the planet Vulcan, and cold fusion, were rejected because the evidence was inconsistent and could not be independently replicated. Interestingly, some of the defenders of these claims argued that the inconsistencies and failures of replication were properties of the claimed phenomena. The parapsychologists, who now admit that their evidence cannot be replicated, also argue that this failure to replicate is one of the unusual properties of psi!

I first addressed the apparent inconsistencies in parapsychological claims about the status of the evidence. Some parapsychologists, such as Jessica Utts and Dean Radin, repeatedly declare that the evidence for anomalous cognition is compelling and meets the most rigorous scientific standards of acceptability. Others such as Dick Bierman, Walter Lucadou, J. E. Kennedy, and Robert Jahn, openly admit that the evidence for psi is inconsistent, irreproducible, and fails to meet acceptable scientific standards. I quoted Radin's statement in his 1997 book *The Conscious Universe* that "we are forced to conclude that when psi research is judged by the same standards as any other scientific discipline then the results are *as consistent* as those observed in the hardest of the hard sciences" (italics in the original) I also quoted from Jessica Utts' 1995 Stargate report that, "Using the standards applied to any other area of science, it is concluded that psychic functioning has been established." Both Radin and Utts were present during my presentation. Neither took this opportunity to retract these claims. I can only assume that they still stand behind these strong assertions.

I was hoping Radin and Utts would provide an explanation of how they can maintain such a position in the face of mounting evidence and arguments within the parapsychological community that the reality of psi cannot be justified according to accepted scientific standards. Dick Bierman, the Dutch parapsychologist, for example, carefully re-analyzed major meta-analyses of parapsychological research on mentally influencing the fall of dice, the Ganzfeld psi experiments, precognition with ESP cards, psychokinetic influence on RNGs, and mind over matter in biological systems (Bierman, 2000). He looked especially at the relationship between effect size and the date when the studies in each of these research areas was conducted.

Bierman fitted a regression line to the data in each area. In all cases, the regression line revealed a consistent trend for the effect sizes to decrease with time and to eventually reach zero.[1] In addition to these linear trends from the meta-analyses, Bierman and other parapsychologists point to dramatic failures of direct attempts to replicate major parapsychological findings. These particular failed replications cannot be dismissed as being due to low power, which is the excuse commonly offered by Utts, Radin, and a few others. Bierman concluded, "In spite of the fact that the evidence is very strong, these correlations are difficult to replicate."

Other major parapsychologists also agree with Bierman's conclusions. Lucadou put it this way, "The usual classical criteria for scientific evidence are effect oriented. Experimental results of parapsychology seem unable to fulfill these requirements. One gets the impression that an erosion of evidence rather than an accumulation of evidence is taking place in parapsychology" (Lucadou, 2001). Kennedy put it this way, "Many parapsychological writers have suggested that psi may be capricious or actively evasive. The evidence for this includes the unpredictable, significant reversal of direction for psi effects, the loss of intended psi effects while unintended secondary or internal effects occur, and the pervasive declines in effect for participants, experimenters, and lines of research. Also, attempts to apply psi typically result in a few very impressive cases among a much larger number of unsuccessful results. The term unsustainable is applicable because psi is sometime impressive and reliable, but then becomes actively evasive" (Kennedy, 2003).

As the preceding quotations indicate, many leading parapsychologists acknowledge that the existence of psi cannot be demonstrated with evidence that meets currently accepted scientific standards. Most critically, these standards include the essential ingredient that the evidence has to be capable of being reliably reproduced by independent investigators. Lacking this basic ingredient, a claim cannot be considered seriously by the scientific community. Above all, it is this basic standard that has made contemporary science the preeminent—and the only—method for gaining trustworthy knowledge. The parapsychologists who admit that the evidence for psi cannot achieve this standard, however, still believe that psi exists and, in most cases, want the scientific community to take their claim seriously. How can they justify such a position?

My presentation dealt directly with this issue. Again, I was hoping for some sort of explanation or justification of this demand for special treatment. It seems to me that the parapsychologists, especially the organizers of the MoM conference, were pleading for a special exemption from the standard scientific criteria. They want the scientific community to accept their claims without having to pass the usual tests. I pointed out that this was not going to happen. N-Rays, Martian canals, and other claims of anomaly that did not pass these tests occupy the junk heap of discarded science. Why should claims of psi be treated any differently?

Finally, I speculated about what might happen if the parapsychologists, especially the organizers of the conference, achieved their goal of getting the scientific community to take their claims seriously. For example, what if the National Academy of Sciences appointed a committee to examine in detail the current evidence for psi? The committee members would obviously find the same results that the parapsychological community has already uncovered. The effect size for psi, in every major research program in parapsychology, declines over time and reaches zero. Major attempts to directly replicate a key parapsychological finding, even when possessing adequate power, fail. The scientific community, when apprised of these findings, would dismiss the parapsychological claims with even more force than they now do. Rather than earning the respect that it seeks, parapsychology's reputation as a serious research program would suffer greatly.

I ended with a quotation from Martin Johnson, a respected parapsychologist of a previous generation:

> I must confess that I have some difficulties in understanding the logic of some parapsychologists when they proclaim the standpoint, that findings within our field have wide-ranging consequences for science in general, and especially for our world picture. It is often implied that the research findings within our field constitute a death blow to materialism. I am puzzled by this claim, since I thought that few people were really so unsophisticated as to mistake our concepts for reality. . . . I believe that we should not make extravagant and, as I see it, unwarranted claims about the wide-ranging consequences of our scattered, undigested, indeed rather 'soft' facts, if we can speak at all about facts within our field. I firmly believe that wide-ranging interpretations based on such scanty data tend to give us, and with some justification, a bad reputation among our colleagues within the more established fields of science.

Johnson wrote those words over thirty years ago. However, they apply with even more force today. In spite of the many new directions that parapsychology has taken since 1976, the only consistent feature of parapsychological evidence is its inconsistency.

As I indicated, I took the organizer's agenda seriously. My presentation dealt directly with each of the issues raised by the organizers: the claim that psi was real and supported by replicable evidence; the implication that the scientific community was unfairly refusing to accept parapsychological claims; and the consequences of having the scientific community seriously consider such claims. I pointed to the apparent contradictions in the claims of the organizers that the evidence for psi was convincing and scientifically warranted and the admission by many contemporary parapsychologists that the evidence for psi does not and cannot meet scientific criteria. I suggested that if, indeed, the organizers succeed in their quest to gain the attention of the scientific community, the result would be a serious blow to the status of parapsychology.

I was expecting serious consideration of my specific challenges to the assumptions of the agenda. I also was expecting that the other presenters and discussants at the conference would deal directly with these issues. I wish I could relay to you how the presenters and the conference attendees responded. Unfortunately, there was a disconnect between the stated agenda and goals and what actually took place. As far as I could tell, I was the only presenter to directly address the issues spelled out in the meeting statement. The majority of presentations were irrelevant to the conference goals. Indeed, several appeared to actively distract from the goals.

The conference organizers and the parapsychologists in attendance failed to respond or even discuss my challenges to the claims in the meeting statement. No one seemed bothered by the contradictions inherent in the claim that the evidence for psi is rock solid or the admissions within the parapsychological community that the evidence for psi is capricious and irreproducible. I have no idea why the conference failed to follow its stated agenda. Perhaps I misunderstood. Maybe the agenda was advanced not as something for discussion but rather as a set of "truths" that were to be presupposed by the participants. I do not know.

What I do know is that although I attempted to put the stated goals of the agenda on the table for debate and discussion, no one seemed eager to do so. What I also know is that the conference agenda implies two requests that the parapsychological community is putting to the scientific community. Both of them are radical and unrealistic. The first request is that the scientific community accept the claim that psi is real. The second is that they do so by exempting parapsychologists from the requirement that they provide evidence according to acceptable scientific standards. Both requests amount to changing science as we know it. Obviously, the scientific community will not, and should not, acquiesce to these requests.

NOTE

In two cases, Bierman suggests that after reaching zero, the effect size shows signs of increasing again. However, this is questionable and appears to be an artifact of fitting a polynomial to data where the zero effect size has existed for a while. Under such circumstances, a second-degree polynomial will better fit the data than will a linear regression line.

REFERENCES

Bierman, Dick J. "On the Nature of Anomalous Phenomena: Another Reality between the World of Subjective Consciousness and the Objective World of Physics?" Edited by P. Van Loocke, in *The Physical Nature of Consciousness*. New York: Benjamins Publishing, 2000, pp. 269–92.

Johnson, Martin. "Parapsychology and Education." Edited by B. Shapin and L. Colby, in *Education in Parapsychology*. New York: Parapsychology Foundation, 1976, pp. 130–15.

Kennedy, J. E. "The Capricious, Actively Evasive, Unsustainable Nature of Psi: A Summary and Hypotheses." *Journal of Parapsychology* 67 (2003): 53–74.

Lucadou, Walter. "Hans in Luck: The Currency of Evidence in Parapsychology." *Journal of Parapsychology* 65 (2001): 3.

THE PEAR PROPOSITION
Fact or Fallacy?

Stanley Jeffers

For twenty-five years a remarkable group at Princeton University, the Princeton Engineering Anomalies Research (PEAR) group, has been pursuing a research program in what many would characterize as parapsychology. A recent article by this group, "The PEAR Proposition" (Jahn and Dunne, 2005) summarizes this quarter-century effort. The bulk of the research has been to show that human intent can remotely affect mechanical and electronic devices in a manner consistent with their intention. They have also reported experiments in remote perception. However, in this article I will take a critical look only at the first group of experiments. I first came upon the work of PEAR while on sabbatical leave in 1992. While browsing in the library, I came upon an article in the journal of the Institute of Electrical and Electronics Engineers (IEEE) with the eye-catching title "The Persistent Paradox of Psychic Phenomena: An Engineering Perspective" by R. Jahn (1982). A number of things struck me about this article: it appeared in a reputable, peer-reviewed, credible scientific journal; its author was clearly a colleague with credible scientific credentials, being formerly a Dean of Engineering; and the work was conducted at an institution with impeccable standards, particularly in the sciences. Jahn's article reviewed historical claims for parapsychology and justifiably rejected many of them for a host of obvious reasons, among them poor methodology, poor statistics, outright fraud, and so on. The last section discussed experiments conducted at Princeton whereby an electronic device was designed to produce a random series of pulses. These were counted in a pre-set time interval and it was established that the accumulated counts scattered around a mean and standard deviation, which conforms to a high degree to Gaussian statistics—that is, the statistics of random numbers.

Some of the first results obtained using this device are shown in figure 1. The distribution of the numbers does seem to conform to the expected random distribution about a mean of 100 and with a standard deviation of

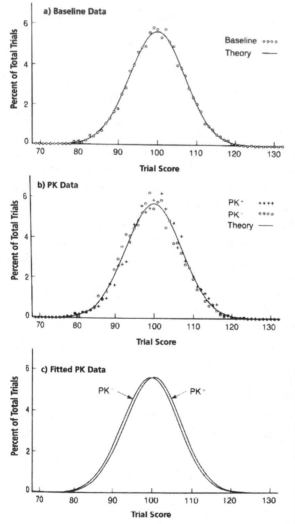

Figure 1. First formal results of one operator's intentions on REG output count distributions, superimposed on theoretical chance expectation: a) baseline data; b) high and low-intention data; c) best binomial fits to high and low data.

around seven. A fit derived from the Gaussian distribution with these parameters is shown as the solid line and labeled theory. The word theory is used in the sense of modeled behavior—here the assumption being that in the absence of any extraneous effects the device behaves in a random manner. Theory is not used in the sense of being derivative of a set of physical principles. In their many publications the PEAR group use theory to imply fitting of experimental data to statistical expectations.

The data, as published, appear to show a small offset between the data derived when someone is ostensibly attempting to bias the mean to be higher than it would be in their absence (or, more precisely, in the absence of their mental efforts to produce a bias) and conversely a small offset in the opposite sense when someone attempts to bias the output to be smaller than it would be otherwise.

In attempting to underpin the claims for statistical significance with a theoretical basis, the PEAR group make frequent appeals to quantum mechanics and quote approvingly of Schrödinger, Wigner, etc. They argue, in a metaphorical sense, that the parameters of quantum systems can be mirrored by psychological correlates. Some of the claims advanced by the PEAR group are post-dictions, for example, the claims for gender bias, baseline bind, etc. None of these are actually predicted by any of the many interpretations of quantum mechanics.

METHODOLOGICAL ISSUES

I have conducted several experiments in collaboration with others in this field (Jeffers, 2003). One characteristic of the methodology employed in experiments in which I have been involved is that for every experiment conducted in which a human has consciously tried to bias the outcome, another experiment has been conducted immediately following the first when the human participant is instructed to ignore the apparatus. Our criterion for significance is thus derived by comparing the two sets of experiments. This is not the methodology of the PEAR group, which chooses to only occasionally run a calibration test of the degree of randomness of their apparatus. We contend, although Dobyns (2000) has disputed our claim, that our methodology is scientifically more sound.

If the claims are credible, it should be possible for other groups to replicate them. To their credit, the PEAR group did enlist two other groups, both based at German universities (Jahn et al., 2000) to engage in a triple effort at replication. These attempts failed to reproduce the claimed effects. Even the PEAR group was unable to reproduce a credible effect.

BASELINE BIND OR BASELINE BIAS?

One favored way of analyzing and displaying data from experiments of this type is to calculate the accumulated deviation from normal expectations. If humans had successfully biased the data such that it now has a higher mean than in the absence of their efforts, then by forming successive differences in

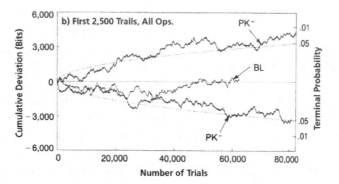

Figure 2. REG Grand cumulative deviations: All operators.

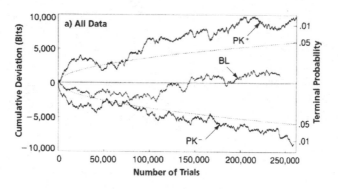

Figure 3. Cumulative deviations of all mean shifts achieved by the same operator as figure 2 over entire database of 125,000 trials per intention.

the data so biased with a data set in which there is presumably no bias, then one would get an ever-increasing sum. Similarly, if the subject had succeeded in biasing the data such that the data now had a lower mean, the accumulated sum would increment in the negative direction. No bias would result in a cumulative sum, which would hover around zero. Gaussian statistics allows one to assign a likely probability to the cumulative sum if no bias were present. Typically, if this probability, p value, is less than .05, then one concludes that a real bias is present. Inspection of the data in figure 2 appears to bear out the claims made. Curves labeled P+ (corresponding to efforts to bias the data to a higher mean value) do indeed accumulate to a running sum unlikely by chance at a p<.01, and curves labeled P– (corresponding to efforts to bias the data such that it has a lower mean) do accumulate in the negative sense with a p<.01. The baseline plot, BL, obtained from data in which no effort is made to bias the equipment does indeed, as per expectations, hover around zero.

In Jahn and Dunne's book *Margins of Reality*, a short chapter is devoted to "Baseline Bind." It is reported that "namely, of the seventy-six baseline series performed, seven or eight of the means would be expected to exceed the 0.05 terminal probability criterion, in one direction or the other, simply by chance. In fact not one of them does." In other words, the baseline data are too good. The means of the baseline data conform to the means of the calibration data, but the variance of the baseline data is less than that of the calibration data. This baseline bind effect is attributed to "the conscious or unconscious motivation on the part of the operators to achieve a 'good baseline.'" It is instructive to compare the baseline behavior of the data shown in figure 3 with that in figure 4. The data presented in figure 4 show an accumulated deviation which actually achieves significance according to PEAR criteria, as the terminal probability lies just outside the p=.05 envelope. The data in figure 4 represent all the data accumulated in PEAR's experiments.

When the data shown in figure 4 were first published, surprisingly there was no discussion about the behavior of the baseline data given the previous

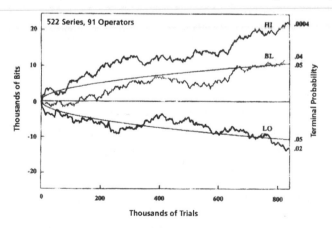

Figure 4. Cumulative deviations of all mean-shift results achieved by all 91 operators comprising a database of some 2.5 million trials.

claims regarding "baseline bind." The baseline data in figure 4 violate PEAR's own criteria for significance (for example, p<.05 terminal probability), and consequently—according to PEAR's own standards-must be regarded as evidence for nonrandom behavior in the baseline data. This has to call into question the claimed statistical significance of the data labeled HI and LO in the same plot.

CONCLUSIONS

In their book *Margins of Reality* Jahn and Dunne raise this question: "Is modern science, in the name of rigor and objectivity, arbitrarily excluding essential factors from its purview?" Although the question is couched in general terms, the intent is to raise the issue as to whether the claims of the parapsychological community are dismissed out of hand by mainstream science unjustifiably. This paper argues that in the light of the difficulties in replication (even by the PEAR group itself), the lack of anything approaching a theoretical basis for the claims made, and, perhaps most damaging, the published behavior of the baseline data of the PEAR group which by their own criteria indicate nonrandom behavior of the device that they claim is random, then the answer to the question raised has to be no. There are reasonable and rational grounds for questioning these claims. Despite the best efforts of the PEAR group over a twenty-five-year period, their impact on mainstream science has been negligible. The PEAR group might argue that this is due to the biased and blinkered mentality of mainstream scientists. I would argue that it is due to the lack of compelling evidence.

ACKNOWLEDGEMENTS

I thank all members of the PEAR team for all their encouragement and support, and Professor J. Alcock of York University for supportive interest and critical insights. Thanks also go to Professor Morris Freeman of the University of Toronto for his interest in the subject matter.

REFERENCES

Dobyns, Y. "Overview of Several Theoretical Models on PEAR Data." *Journal of Scientific Exploration* 14, no. 2 (2000): 163–94.

Ibison, M., and S. Jeffers. "A Double Slit Experiment to Investigate Claims of Conscious-ness-related Anomalies." *Journal of Scientific Exploration* 12, no. 4(1998): 543–50.

Jahn, R. G., and B. Dunne. *Margins of Reality: The Role of Consciousness in the Physical World* (New York: Harcourt Brace Jovanovich, 1987).

Jahn, R. G. "The Persistent Paradox of Psychic Phenomena: An Engineering Perspective." *Proceedings of the IEEE* 70, no. 2 (1982): 136–70.

Jahn, R. G., and B. Dunne. "The PEAR Proposition." *Journal of Scientific Exploration* 19, no. 2 (2005): 195–246.

Jahn, R., et al. "Mind/Machine Interaction Consortium: PortREG Replication Experiments." *Journal of Scientific Exploration* 14, no. 4 (2000): 499–555.

Jeffers, S. "Physics and Claims for Anomalous Effects Due to Consciousness." Edited by J. Alcock, J. Burns, and A. Freeman, in *Psi Wars: Getting to Grips with the Paranormal* (Exeter, UK: Imprint Academic, 2003), p. 135

PEAR LAB CLOSES, ENDING DECADES OF PSYCHIC RESEARCH

STANLEY JEFFERS

The Princeton Engineering Anomalies Research (PEAR) group is shutting down after some twenty-eight years of searching for proof of the paranormal. On February 10, 2007, PEAR issued a press release that stated, in part: "The PEAR program was established at Princeton University in 1979 by Robert G. Jahn, then Dean of the School of Engineering and Applied Science, to pursue rigorous scientific study of the interaction of human consciousness with sensitive physical devices, systems, and processes common to contemporary engineering practice. Over the next twenty-eight years, an interdisciplinary staff of engineers, physicists, psychologists, and humanists has conducted a comprehensive agenda of experiments and developing complementary theoretical models to enable better understanding of the role of consciousness in the establishment of physical reality."

If it has been the long-term goal of the PEAR group to be featured in the mainstream literature, then they have finally achieved their goal. The imminent closure of the PEAR laboratory was commented upon in both *The New York Times* (Carey, 2007) and *Nature* (Ball, 2007). The Canadian Broadcasting Company (CBC) featured a long and sympathetic interview with the director of PEAR, Professor Robert Jahn, on February 27. There are no plans by Princeton to continue to support work in this area. The university administration has maintained a discreet silence about the PEAR group and its remarkable research.

The startling claims of the PEAR group fall into the broad category of parapsychology, specifically psychokinesis (moving objects with the mind) and remote viewing (extrasensory perception). However, the PEAR team avoids terms such as *psychokinesis* and *telekinesis* in favor of less provocative terms such as *anomalous transfer of information* and *anomalous injection of information into the data stream*. The PEAR group has recently published a summary of the first twenty-five years of their work (Jahn and Dunne, 2005). A critical analysis of some of the PEAR claims has recently appeared (Jeffers, 2006, and reprinted in this volume preceding this note).

Much of the work of the PEAR group has employed "random event generators" (REGs), which are essentially electronic random number generators whose "operators" are invited, by dint of their own intentionality, to bias in such a way that the mean of the random number distribution would be either higher or lower than it would be in the absence of their intentional efforts. The claim is that some "operators" can achieve a bias consistent with their intentions at a level that, although minute, is statistically very unlikely to have arisen by chance.

In his CBC interview, Professor Jahn stood fervently by his claims and said that he would repeat this long effort "in a heartbeat." He remains convinced that his work reveals something profound about the nature of mind and matter. However, it is somewhat telling that, despite this long record of experimentation, very few in the academy have been convinced of the validity of the claims. Most of the work has been reported in the *Journal of Scientific Exploration*, a periodical specializing in claims for all kinds of physical, biological, and parapsychological anomalous effects. Two papers have appeared in more mainstream journals, the *IEEE* (back in 1992) and *Foundations of Physics*. The attitude of most of the academy has either been immediate rejection without a close examination of the evidence or simple indifference. One notable exception is the support offered to the PEAR group by one Nobel Laureate in physics, Brian Josephson. One waggish editor did offer to publish a PEAR paper "if it could be transmitted telepathically."

The work of the PEAR group does raise larger issues for the academy concerning academic freedom. Princeton, to its credit, has recognized Jahn's freedom to pursue a controversial area despite the obvious discomfort of some of the faculty, particularly in the physics department.

As with any other claim, the veracity of PEAR's claims will finally be settled by time-honored methods of science—and demand reproducibility. Here PEAR has a significant problem. To its credit, PEAR did engage two other groups of researchers at two different German universities in a three-way attempt at validating the claims. However, none of these groups—including PEAR itself—was able to reproduce the claimed effects.

Furthermore, as has been previously pointed out, there are some problems with the calibration data of PEAR's REG. As far back as 1987 (Jahn and

Dunne) the PEAR team claimed that the performance of their REGs when no one was invited to influence them showed a distribution that was better than Gaussian. This effect was dubbed "baseline bind." It was attributed to the *unconscious* actions on the part of "operators" to please the experimenters (how one can test for *unconscious* intentionality is unclear). However, the baseline data reported later over a long period exhibits a trend that is unlikely by chance at the $p=.05$ level. This was the level of statistical significance previously employed to claim a significant effect. I have argued that the later data exhibit baseline bias and hence the REG over the long term is not generating random numbers as claimed. This has to call the basic claims into question.

Although the PEAR lab will be no more, work in this area is expected to continue under the auspices of the International Consciousness Research Laboratories, a not-for-profit public foundation. One suspects that, without the cachet that attaches to the Princeton name, this group will have an even more difficult time convincing the skeptical community.

REFERENCES

Ball, P. When Research Goes PEAR Shaped. 2007. Available at http://www.nature.com.

Carey, B. "After 28 Years, Princeton Loses ESP Lab to the Relief of Some." *New York Times*, February 10 2007.

Jahn, R. G., and B. Dunne. *Margins of Reality: The Role of Consciousness in the Physical World*. New York: Harcourt Brace Jovanovich, 1987.

———. The PEAR Proposition, *Journal of Scientific Exploration* 19, no. 2 (2005): 195–246.

GARY SCHWARTZ'S ENERGY HEALING EXPERIMENTS
The Emperor's New Clothes?

Harriet Hall

Gary Schwartz believes many things. He believes in psychics, mediums, and life after death, and he believes there is scientific evidence to support these beliefs. Schwartz is now focusing his powers of belief on a new field: energy medicine. In a new book, *The Energy Healing Experiments: Science Reveals Our Natural Power to Heal*, he explains that we all emit human energy fields, that we can sense each other's fields, and that healers can influence these fields to heal illnesses and injury. He believes these are not just theories but scientifically supported facts.

The book starts with three "gee-whiz" testimonials of supposed energy healing (which are frankly not very convincing and could be easily outdone by any self-respecting purveyor   of quack remedies). He goes on to describe experiments done in his own lab that he claims establish not only our ability to detect and alter human energy fields but our ability to detect the thoughts and intentions of others. In the final part of the book, he descends into blethering about quantum physics, the oneness of the universe, the connectedness of all things, and the possibility that energy awareness will solve all of mankind's problems.

He claims to have demonstrated many things. First, he claims to have shown that a subject can sense when a researcher's hand is being held over his or her own

290

hand and can sense when the researcher's hands are being held near his or her ears from behind. Other experiments supposedly show that people can tell when someone is looking at them or thinking about them. He goes on to describe purported measurements of subtle human energy emissions, Reiki influences on lab cultures of bacteria, and photography of biophoton emission from plants, among other phenomena of dubious reality or significance.

He makes a big deal of the fact that humans emit electromagnetic energy (as picked up by EKG, EEG, etc.), and he would like to think energy healers can pick up that energy and decode it in the same way your radio picks up Rush Limbaugh out of the atmosphere. And then he would like to think that energy healers can send something back into the patient's body to enable healing. He misses the crucial fact that there is information encoded in the electromagnetic waves your radio detects, but there is no reason to think there is any analogous information coming from the body, much less any way to change that information and send it back to produce healing. I only wish we *could* use "energy healing" on radio and TV waves to improve the quality of programming!

He makes a big deal of the fact that everything affects everything else. He seems to mean this in a holistic, metaphysical, New Age, "the universe is one and is conscious and we can create our own reality" sense. Science recognizes that small events can have far-reaching effects, but that doesn't mean one thing can predict or control another. The flap of a butterfly's wings may set up initial atmospheric conditions that will result in a tornado somewhere else, but that doesn't mean you can predict the tornado or deliberately use a butterfly to cause one. Theoretically, a change in the magnitude or position of your body mass will enter into the overall gravity equations of the universe, but that doesn't mean one thing can control or predict another. You could hardly expect to meaningfully influence someone out there beyond Alpha Centauri by losing ten pounds or moving to Antarctica. You can't expect to change the EEG of an astronaut in the Space Station by exercising to change your own EKG. We are talking about very small influences. If a gnat pushes an elephant, it's not likely to fall over; it's not likely to even notice. And then there are inconvenient complications like quantum theory and chaos theory.

The only thing of substance in the book is the experiments, which lose credibility because they were not accepted for publication in mainstream peer-reviewed journals. Schwartz claims this is because of politics. He says prestigious journals tend to reject positive-energy studies. He doesn't believe that his studies could have been rejected because they didn't meet the standards of good science. I feel sorry for him: he's a smart guy, he means well, he really believes he has found something wonderful, but he has a blind spot and just doesn't get it when others try to point out the flaws in his experimental methods and reasoning. (See Ray Hyman, "How Not to Test Mediums: Critiquing the Afterlife Experiments," *Skeptical Inquirer*, January/February 2003, and the follow-up exchange between Schwartz and Hyman, May/June 2003, plus the critical letters to the editor in that issue.)

To put the accusation of "politics" into perspective, consider the *Helicobacter* experiments. When researchers first suggested that ulcers might be caused by bacteria, they were laughed at. They published their results, peer review had a field day, other labs looked into the idea, more data came in, results from various lines of research coalesced, and within a mere ten years it became standard practice to treat ulcers with antibiotics. It didn't matter that the idea sounded crazy at first; science responded to good evidence. (See Kimball C. Atwood IV, "Bacteria, Ulcers, and Ostracism," *Skeptical Inquirer*, November/December 2004.) If Schwartz had evidence of equal quality, he would get an equal hearing by the scientific community.

Sure, Schwartz has some data that he finds convincing. So did the discoverers of N-rays, polywater, and cold fusion. Good science demands that we withhold judgment until data can be replicated in other labs and validated by other methods—especially when the data come from a researcher as clearly prejudiced as Schwartz. Even the best researchers can fall prey to errors of unconscious bias and unrecognized pitfalls in experimental design.

A good scientist considers *the entire body* of available evidence, not just the claims of one group of researchers. Schwartz only describes experiments that support his beliefs. Not until the end of the book does he even bring up the fact that other experiments have directly contradicted his findings. He finally gets around to mentioning Emily Rosa's landmark experiment, published in the *Journal of the American Medical Association* in 1998, which showed that therapeutic touch practitioners could not sense human energy fields as they claimed. She tested twenty-one experienced practitioners of therapeutic touch.[1] They all thought they could detect Rosa's human energy field and feel whether she was holding her hand over their right or left hand, but when they were prevented from seeing where her hand was, their performance was no better than chance.

Rosa was nine years old at the time, and the article grew out of her school science fair project. The experiment was beautiful in its simplicity. Adult true believers had published much research on the techniques and effects of therapeutic touch, but in the true spirit of childlike questioning, Rosa went back to basics and asked the crucial question: "Is the phenomenon itself real? Can they really feel something or is it possible they are fooling themselves?" Amazingly, no researcher had ever asked that question before. They had ignored one of the basic principles of the scientific method as explained by Karl Popper:

it's easy to find confirmation for any hypothesis, but every genuine test of a hypothesis is an attempt to falsify it.

Schwartz dismisses her experiment as having five "potential problems":

1. It was a science-fair project done by a young girl.
2. She was the only experimenter.
3. She randomized by flipping a coin, which he calls "an unreliable procedure."
4. One of the authors was the founder of Quackwatch.
5. The subjects did *worse* than chance.

These objections are just silly; they are either inaccurate or are *ad hominem* attacks:

(1.) It shouldn't make any difference whether Rosa was a young girl or an old man or a sentient purple octopus from an alien planet. It shouldn't matter whether she did the experiment for an elementary school project, a doctoral dissertation, a Coca Cola commercial, or a government grant. What matters is the quality of the evidence. In this case, her project was well designed and executed, had clearly significant findings, and was of high enough quality to be approved for publication in a prestigious peer reviewed medical journal.

(2.) She was not the only experimenter. Others were involved; the experiment was repeated under expert supervision on *Scientific American Frontiers*. This should preclude any accusations of deliberate cheating or inadvertent failure to follow the protocol properly. Rosa was the only one to carry out the trials, but what would multiple testers have added to the experiment? The results didn't depend on any special ability or quality of hers, but on the ability of the subjects who claimed they could sense anyone's energy fields. For the televised trials, they even got to "feel" the "energy" from each of Rosa's hands and choose which one they wanted her to use in the trials. About half chose her left hand and half her right. No one objected, "I can't feel energy from either hand."

(3.) Flipping a coin is not an "unreliable procedure"—unless the flipper is deliberately cheating. I hope Schwartz didn't intend to suggest that. The number of heads and tails was approximately equal, and the distribution appeared random. The editors of JAMA found the method acceptable. There are situations where coin-flipping could legitimately be criticized, for instance in psi experiments where researchers are looking for minuscule differences in large bodies of data and even their computerized random number generators have been criticized for not being "perfectly" random. But in this experiment, the results were clearly significant; it is hard to envision how a different method of randomization could have altered the results. The coin flip was only used to determine which of the subject's hands she would hold her hand over. The subjects claimed to be able to sense energy fields with either hand, so it shouldn't have made a bit of difference to their perception. Faulty randomization might have allowed the subjects to perceive a pattern and guess, which would have tended to give false positive results rather than the negative results Rosa got.

(4.) One of the authors, the founder of Quackwatch, was admittedly skeptical of therapeutic touch. Yes, someone with possible bias was indirectly involved in the experiment. If that is an objection, there is an even greater objection to Schwartz's own experiments: he and his colleagues are all strongly biased toward belief in energy phenomena and they were directly involved in their experiments.

(5.) It is simply not true that the subjects did "worse than chance." Their performance was consistent with chance. If they *had* done worse than chance (significantly worse) that would have tended to support Schwartz's claim that some kind of effect was present, even though it would have been the reverse of what he claimed to find.

In my opinion, none of these "problems" invalidates the conclusion that the therapeutic touch practitioners failed to do what they claimed they could do. And if he thinks these were valid problems, why didn't he simply repeat her experiment in his own lab with multiple experimenters and a more reliable method of randomization? He could have published a failed replication study, and the scientific community could have proceeded to evaluate both studies and sort out the truth. In reality, Rosa's experiment was a great example of a young child being able to see more clearly than prejudiced adults—a real "Emperor's New Clothes" story.

I see a lot of "potential problems" in Schwartz's research—not just *ad hominem* problems but flaws of experimental design. To start with his most basic experiment: his subjects were blindfolded, sat facing the experimenter with their hands on their laps, and tried to detect which hand the experimenter was holding his hand over. The experimenter held his hands together between trials to keep his hand temperature constant. The subjects often didn't think they could tell, but they were asked to guess, and their guesses were statistically significant.

The first problem is that blindfolds don't work. Rosa knew this. Instead, she had her subjects put their arms through holes in a screen and covered the gaps with a towel to preclude any possibility of conscious or unconscious visual cues. She also had subjects lay their arms on a table instead of on their laps, thus reducing the chance of their detecting subtle clues from the person sitting in front of them. Another problem is that when the researcher holds his hands together, that raises the skin temperature and raises the possibility that heat is being detected rather than any other type of energy. And if Schwartz's results are real, independent researchers should be able to replicate them using the same protocol. Apparently they have not been replicated elsewhere. In fact, Rosa's experiment amounts to an independent attempt to replicate Schwartz's basic experiment, only with better controls; and it failed to confirm his results.

If a rigorous scientist thought he had found evidence that people could detect "human energy fields," he would maintain a healthy skepticism; he would immediately try to prove himself wrong, and he would enlist his colleagues to help show him where he might have gone wrong. He would try to rule out all other possible explanations (the subject might be sensing heat, sound, motion, air currents, might be able to see under the blindfold, etc.). If the phenomenon proved robust, he would try to refine his understanding by doing things like varying the distance to see if it obeyed the inverse square law and interposing a sheet of cardboard or glass to see if the effect could be blocked. Then he would try to use instruments to measure what kind of energy was being sensed.

When a believer thinks he has found something to justify his belief, his approach tends to be less rigorous. Instead of subjecting his original experiment to outside scrutiny, he tends to do more new experiments to try to convince others that he is right. Schwartz goes off on a tangent doing other experiments that purportedly show that the subject is not sensing the energy field but is actually sensing the conscious intention of the experimenter. In one, he claims to show that persons can tell whether someone standing behind them is staring at their head or at their back! If he really believed energy medicine was some kind of psychic thought transmission, he would concentrate on that route of research, but instead he keeps trying to document the ability to detect measurable physical energy fields. His thinking is confused, and he's trying to eat his cake and have it too.

Schwartz's style of reasoning was revealed when an experiment to influence *E. coli* bacteria with Reiki didn't produce the desired results. Instead of accepting that it didn't work, he tried to find a way to make the experiment look like it worked. He did some inappropriate "data mining" and tried to show that before the trials where the Reiki practitioners apparently failed, they had been under more stress than before the trials where they apparently succeeded.

He finds a gifted individual who can detect whether a wooden box has a rock in it or not—his success rate is 95 percent for natural crystals, although barely chance for man-made crystals. Unfortunately, before this individual can be tested properly in an independent lab, he develops medical problems and loses his ability. (It's strange how often these inconvenient things happen when psychic claims are involved.) Schwartz is mystified by the work of John of God, the Brazilian spiritual healer who performs bloodless, painless surgery.

He doesn't recognize that this charlatan is merely using old gimmicks from the carnival sideshow repertoire to fool the gullible. Schwartz also believes science has established that the human mind can change the pH of water over long distances. He is far less skeptical about such claims than the average scientist.

Schwartz has tried to bolster his credibility by getting a former Surgeon General's endorsement. In Richard Carmona's foreword, he says he has seen things he can't easily explain and says we don't have all the answers. He helped establish the National Center for Complementary and Alternative Medicine (NCCAM, which he curiously refers to as the National Center for Alternative and Complementary Medicine). The purpose of the NCCAM was allegedly to test complementary and alternative medicine (CAM) and find out which treatments worked and reject those that didn't. But in its entire history, despite consistently negative results, it has never dared to reject anything. Carmona is currently CEO of Canyon Ranch Health, where Schwartz is the Director of Development of Energy Healing. Canyon Ranch offers integrative medical wellness services, including therapeutic touch. Carmona says, "Where the science supports these integrative concepts of energy medicine, let's use them. Where there is not enough science, let the studies begin and continue."

What about "if there is no convincing science or plausible mechanism to support them, let's stop wasting our time chasing moonbeams?" All of energy medicine hinges on one basic claim: that people can detect subtle human energy fields. If Schwartz is wrong about that, the rest of the claims for so-called "energy medicine" fizzle away.

Since 1996, the James Randi Educational Foundation (JREF) has offered a substantial reward (currently $1,000,000) to anyone who can demonstrate an ability to detect a "human energy field" under conditions similar to those of Rosa's study. Of the more than 80,000 American therapeutic touch practitioners who claim to have such ability, only one person attempted to demonstrate it. She failed. The JREF challenge is admittedly not a definitive scientific test, but prudence would seem to dictate that if no one can even meet this simple challenge, we shouldn't be wasting research money on what is probably a myth.

Others have attempted to establish the "science" of energy medicine and have failed.[2] Even the NCCAM, which is willing to consider almost any possibility in alternative medicine, is skeptical. It distinguishes between real energy (sound waves, electromagnetism, and other energies measurable by physicists) and the kind of "putative" energy Schwartz is trying to validate. It concludes that the "putative" energy approaches "are among the most controversial of CAM practices because neither the external energy fields nor their therapeutic effects have been demonstrated convincingly by any biophysical means."

Schwartz sounds like a scientist. He tries to talk the talk and walk the walk.

He even makes some skeptical noises to try to convince us he is objective. But there is also a lot of very unscientific language in his book.

For instance:

> Human rage and pain, especially generated by terrorism and war, create a global energetic climate whose negative effects can extend from the physical and environmental—potentially including climate—to the psychological and ultimately spiritual. ... [P]ollution is not simply chemical, it is ultimately energy based and therefore conscious as well.

Really? Conscious pollution? So maybe if we talk nice to pollution it will cooperate and go away? Or should we try doing Reiki to lower the atmospheric CO_2 levels? Does Al Gore know about this?

"Energy medicine" is an emperor whose new clothes still look awfully transparent to critical thinkers and to the scientific community no matter what glorious colors and fabrics Schwartz and his colleagues imagine they are seeing.

NOTES:

1. "Therapeutic touch" is a bit of a misnomer because these practitioners don't actually touch but simply massage the air a few inches from the patient's body. They are convinced that they are detecting and manipulating the energy field, balancing and smoothing it, and correcting any abnormalities, thus allowing the body to heal itself.

2. Hall, H. "A review of Energy Medicine: The Scientific Basis." *Skeptic* 11, no. 3 (2005): 89–93. Available at http://quackfiles.blogspot.com/2006/01/review of energy-medicine-scientific.html.

3. http://nccam.nih.gov/health/backgrounds/energymed.htm.

REFERENCES:

Rosa, L., E. Rosa, L. Sarner, and S. Barrett. "A Close Look at Therapeutic Touch." *Journal of the American Medical Association* 279 (1998): 1005–1010.

Schwartz, Gary E., with William L. Simon. *The Energy Healing Experiments: Science Reveals Our Natural Power to Heal.* New York: Atria Books, 2007.

TESTING NATASHA

Ray Hyman

Our assignment might seem straightforward. A seventeen-year-old Russian girl, Natasha Demkina, says she can look at people and "see" the status of their internal organs. The Discovery Channel asked Richard Wiseman, Andrew Skolnick, and me to test her claim for their television program, The Girl with X-ray Eyes. You might think that testing Natasha's claims would be routine. The test of a psychic claim, however, is rarely cut-and-dried. Most such claims do have much in common. Each also offers unique challenges. We had to conduct the test of Natasha's claim to fit the constraints of a television program. We had only a month to devise a protocol that would be acceptable to all parties. After everyone agreed to the procedure, we had less than a week to locate a testing site in New York City and to find seven willing and suitable test subjects.[1]

THE CLAIM AND ITS SUPPORT

Monica Garnsey, director and producer of the program, told us how Natasha operates and what she claimed to do. Many news sources and reports on the Internet described her accomplishments. (This information was consistent with what we observed when Natasha diagnosed volunteers at the Open Center in New York City the day before the test.) Garnsey e-mailed us the following information from Russia, where she was taping material for the television program:

> I double-checked a few things with her last night. Since the age of ten, a few days after having a religious dream, and also having had an operation to have her appendix removed that went wrong, swabs were left in her and she had to have another operation, Natasha has claimed to be able to see into people. . . . Natasha can see through clothing, but not see what someone is holding behind their back. She cannot see inside people if she shuts her eyes. Daylight is better. She does not need to talk to them to diagnose. She can also diagnose from a photograph. She usually scans people all over first, by making them stand up fully clothed and looking them up and down; delivers a general diagnosis; and then goes into more detail when the patients have discussed their concerns with her. She says she can *certainly* see ribs, heart, lungs, initially in general "like in an anatomy book," but can see right down to the cell level if she concentrates. She says that she can examine the whole body, but it can give her a bad headache if she does too much. The idea of restricting the test to the chest area appeals [to her], though her claims extend further than that.

Natasha's story is like thousands of other accounts. Alleged psychics and their supporters make claims that, if true, defy the physical limitations and laws of modern science. The proponents support the reality of these claims with testimonials of outstanding successes. They argue for the reality of the claim passionately and unreservedly. Although some proponents have had scientific training, none of the supporting evidence comes from well-controlled scientific studies.

In the long history of psychical research, not one of these claims has produced convincing scientific evidence for the existence of paranormal ability. A few researchers have claimed that they did have scientific proof for a paranormal claim. Scrutiny by other scientists, however, showed that the "scientific proof" had serious flaws. Furthermore, none of these claims could be independently replicated.

The evidence supporting Natasha's abilities comes from selected anecdotes of reactions to her readings. No matter how subjectively compelling, the context of such readings makes it impossible to separate how much of the apparent success is due to such possibilities as: guessing; external clues from the client's physical appearance and observable behavior; feedback from the client's spoken and bodily reactions; or actual paranormal powers. A meaningful test would allow Natasha to show her powers and, simultaneously, control for guessing and the use of normal sensory clues.

PROBLEMS WITH TESTIMONIAL SUPPORT OF NATASHA'S CLAIMS

The stories told by Natasha's proponents are consistent with her having X-ray vision. This does not show that she does have X-ray vision because the same stories are consistent with many other alternatives. Two possibilities are the following: 1) her statements have no connection with the client's condition but appear to do so because of luck, selective reporting, and/or other reasons that I will discuss; or 2) her statements accurately reflect the subject's condition, but this information comes through normal means such as the subject's appearance and behavior. Consider, first, the ways that her statements can falsely appear to describe the patient's condition.

SHE MIGHT HAVE NO KNOWLEDGE ABOUT THE CLIENT'S CONDITION BUT GET CREDIT ANYWAY

Natasha has been giving readings to a steady flow of clients for more than six years. By now the number of such readings is huge. Her supporters naturally emphasize the most striking examples of apparent hits. The number of diseases and internal parts that could be defective is limited. Some conditions, such as cancer and heart problems, are more common than others. We should expect that her supporters will find some examples of "correct" diagnoses. With so many diagnoses, a certain number will match the client's condition just by chance.

To evaluate a diagnostic procedure properly we need to clearly decide what is a "hit" and what is a "miss." Most important, we should set the criteria before we know the outcome. In Natasha's readings, no clear and objective standards were ever established. This allows for her generally vague utterances to be retrofitted to what the client or observer knows to be true. An example of such retrofitting occurred when Natasha was doing a reading in London. Dr. Chris Steele, described by *The Daily Mail* (January 29, 2004) as one of her champions, was observing. The newspaper quotes him as saying, "Natasha doesn't know any medical terms at the moment. With one person this week she was trying to describe a kidney stone, and her translator came up with the words, 'sand' and 'gravel' before I suggested stones. When kidney stones start off, they do look like sand." Dr. Steele gives her credit for correctly diagnosing kidney stones. Yet we have no idea what Natasha was "seeing" or what she had in mind. Dr. Steele made the medical diagnosis, not Natasha.

Other features of Natasha's readings foster the illusion of accuracy. When she tells clients something that agrees with previous medical diagnoses, they credit her with a hit.

Similarly, when she tells the clients something that disagrees with previous medical diagnoses, they still credit her with a hit; the clients and her supporters argue that she picked up on something that the medical professionals missed. We witnessed some examples of this when we watched her giving readings to volunteers at the Open Center in New York. She told one volunteer that she saw a problem with his right shoulder. After the reading, this volunteer told Monica that he had not previously realized something was wrong with his shoulder. Neither his previous medical examinations nor anything in his experience suggested something was wrong with his shoulder. I thought, as a result, he might be skeptical about Natasha's claim. Instead, he was impressed. He decided she had detected a problem that neither he nor his doctors had noticed.

POSSIBILITIES OF NATASHA PICKING UP CLUES BY NON-PARANORMAL MEANS

I have described just some ways that testimonials can appear to support Natasha's claim even if she is picking up no information about her clients. Those possibilities would suffice to make such testimonials useless as evidence for her ability. The testimonials become even more suspect when we realize how the circumstances of her readings allow her to pick up information about her client without having X-ray vision. Natasha is looking directly at her client when she does her diagnosis. This means that we cannot rule out the possibility that she is picking up clues from subtle (and not-so-subtle) client reactions. To make matters worse, the clients begin a session by asking Natasha questions about their concerns. This provides obvious clues about their condition. I watched one reading where the client began asking Natasha about her back. This narrows considerably the number of possibilities that Natasha needs to consider. Natasha can also gain considerable information from verbal exchanges with the client.

Another source of clues is how the clients react, both verbally and nonverbally, to her statements. Some of her clients say that they find it unsettling when Natasha is staring at them. This could enhance the tendency for individuals to react to her statements with subtle, unwitting bodily movements, breathing changes, pupil dilations, and other signs of emotional and cognitive states. Although psychological research has documented how humans frequently provide unconscious clues to their current thoughts and emotions, most people seem unaware of this possibility. The research also shows that subtle clues can influence us without our consciously realizing it.

One classic case involved the German horse Clever Hans. [2] In the early twentieth century, Hans became a celebrity in Germany and throughout the world. People could ask him questions about addition, the identity of musical pieces, about foreign words, spelling, and many other topics. Hans would answer by tapping his hoof or by nudging an alphabet board with his nose. He usually was correct. Prominent educators certified that he had the intelligence and competence of a thirteen- or fourteen-year-old German student. Oskar Pfungst, a German psychologist, investigated Hans with exemplary thoroughness. He eventually discovered that Hans was clever only in having "horse sense." Typically, a questioner would focus on the horse's right hoof, which Hans used to tap out the answer. When questioners focused on the hoof, they would almost imperceptibly lean forward and become tense as they watched the horse tap out the answer. This slight leaning and tensing were Hans's cues to begin tapping. When Hans had tapped the appropriate number of times, the questioner would unconsciously relax and move his or her head upwards very slightly. Often this movement was one millimeter or less. This was Hans's clue to stop tapping.

Pfungst then carried out experiments to confirm this finding. He played the role of Hans. He would invite people to stand beside him and think of a number. Pfungst would then begin tapping with his right hand. He would stop when he thought he detected a very slight bodily movement—usually a very slight displacement of the subject's head. These movements were extremely subtle, rarely more than a millimeter in extent. Pfungst amazed his volunteers, stopping his tapping at the number they had in mind.

Pfungst tried this experiment with twenty-five persons ranging in age from five years to adult. He succeeded in picking up cues from all but two of them. They insisted they were unaware of giving him any information. Pfungst used the same method to divine other kinds of thoughts the subjects had in mind. The subjects again denied that they had provided any clues about what they were thinking. Other psychological experiments have confirmed these results. Some skilled performers have made careers out of pretending to read minds when, in fact, they were relying upon subtle and unwitting clues provided by their volunteers.

Some reports supporting Natasha's claim describe outcomes consistent with the possibility that she is picking up such clues. For example, a Russian reporter says that he became a convert to Natasha's cause when she found the exact spot on his arm where he had fractured his wrist many years before. In another case, a reporter from a British tabloid validated Natasha's ability when Natasha succeeded in identifying the location of the fractures she had received in an accident. Both cases seem ideal for picking up the sorts of clues that Pfungst found that most people provide without realizing they are doing so.

What I have just written does not show that Natasha lacks X-ray vision. We do not know from the evidence offered by her proponents whether she docs or does not have a paranormal capacity to see into people's bodies. What we do know is that the accounts that seem to support Natasha's claim are consistent with both normal and paranormal possibilities. We also know that nonparanormal mechanisms can and do operate in the real world. We do not know that paranormal ability, such as that claimed for Natasha, exists. So far, no one has displayed such ability with scientific credibility. Given these two possible explanations for Natasha's apparent successes, rationality tells us to bet on the nonparanormal one. We should demand convincing evidence that is scientifically acceptable before we give credence to the paranormal claim.

THE TEST PROTOCOL

With input from Richard and me, Andrew wrote the test protocol, titled "Test Design and Procedures for Preliminary Study of Natasha Demkina." The goal was to make every aspect of the test explicit. The protocol stated how we would conduct the test and how we would interpret the results. We wanted all parties to be clear about what would and would not be considered a "successful" outcome. What makes a scientific experiment or a test meaningful is just such an explicit commitment to the interpretation of the outcome *before we observe the data*. This is a critical distinction between the post hoc interpretation of testimonial evidence and the prior commitment to specified outcomes of a meaningful test. Natasha's defenders apparently fail to grasp this essential point.

The written protocol protects the interests of all parties. Natasha and her supporters had the opportunity to study the document, to suggest modifications, and finally to agree or disagree with its provisions. The protocol also protects the investigators against a variety of false accusations about how we conducted the test.

We made sure to include in the protocol the statement that the "test is not in any way a definitive test. Deciding the truth of Natasha's claims with comfortable certainty is too simple and brief. It can only help to decide whether further studies of Natasha's claimed abilities are warranted." This statement is worth elaborating. Understanding what the test can and cannot do is essential. Even under ideal circumstances this test could not clearly decide if Natasha does or does not have X-ray vision. Any scientific hypothesis—especially a paranormal one—cannot be confirmed or disconfirmed by one test or one experiment. Scientific investigation requires a series of experiments. Each new experiment builds on the results of previous ones. The more we learn from the early experiments, the better we can understand what we need to control and what we can safely ignore. If the hypothesis is implausible and/or controversial—as Natasha's claim certainly is—then the original investigators must replicate their findings. In addition, independent investigators must also replicate the findings before they gain scientific credibility.

We knew that our test could not distinguish between two possibilities: (1) she can make correct matches using external clues; or (2) she can make correct matches using paranormal X-ray vision. The alternatives we could control or reduce were that she gets correct matches just by luck or that her correct matches are due to those factors that make vague statements seem like hits.

We were also aware that our test could only detect a large effect. Natasha's claim can be considered in several contexts. The testimonials imply that she is highly accurate. This has practical consequences. If clients are depending upon her for medical diagnoses, Natasha's readings should be reliable. Otherwise, she can do much harm. Of course, Natasha could possess paranormal powers, but they could be weak and erratic. Such unreliable and weak ability would be useless for medical diagnosis, but would still be of theoretical interest. We lacked the resources and time to try to detect such a weak effect. We used all our resources to obtain seven subjects. If we had been trying to test for a moderate or weak effect, we would have had to use many more subjects. Given the constraints of our task, this was impossible. Our test, then, was aimed at detecting a large effect. We reasoned that if she possessed the reliability of diagnosis that her proponents claimed, our test would reveal this. Such an effect would encourage us to investigate her abilities in more detail.

The outcome of the test could be from zero to seven correct matches. We set the criterion for success at five correct matches. We clearly stated this criterion in the test protocol and all parties agreed to this in advance. Although Natasha's mother says that her daughter never makes a mistake, we did not want to demand that Natasha perform perfectly. We wanted to give her some margin for error. Keep in mind that if she got five or more correct this would be consistent with her having the X-ray power that she claims. Yet it would also be consistent with the possibility that she was matching the target condition by normal means such as the appearance and behavior of the subjects.

THE TEST

Richard Wiseman, Andrew Skolnick, and I collaborated in designing the test. We arrived at a mutually satisfactory plan after exchanging several e-mails. The task of finding appropriate subjects, and coordinating the many details was left to Andrew. He had less than one week to accomplish all this. He had to do this from Amherst, more than 350 miles from New York City.

Austin Dacey, then executive director of the Center for Inquiry-MetroNY, obtained an excellent set of rooms for the test at the City College of New York and helped recruit several subjects. Dr. Barrie Cassileth, Chief of Integrative

Medicine Service at Memorial Sloan-Kettering Cancer Center, helped us with the daunting task of assembling seven appropriate and willing subjects. On the morning of the day of the test we learned that two of the subjects had withdrawn. Again, Andrew and Austin saved the day by finding two replacements at the last moment. (Andrew's separate article about certain aspects of the tests was published following mine in the May/June 2005 *Skeptical Inquirer.*)

During the test, we seated the seven subjects in a semicircle facing the chair where Natasha sat. Each volunteer had an internal condition that should be easy to detect if Natasha's claim is correct. The target conditions were as follows: One patient had metal surgical staples in his chest from open heart surgery; one had a section of her esophagus surgically removed; one had a large section of one lung removed; one had an artificial hip replacement; one had a missing appendix (we discovered afterwards that another subject also had a missing appendix, which he didn't mention when we recruited him. Natasha chose neither of these two as the one with the missing appendix); one had a large brain tumor removed and now has a large hole in his skull covered by a metal plate; and the final subject had none of these target conditions.

During the test, when Natasha was looking at the subjects, the subjects wore sunglasses whose lenses were covered with opaque tape. This prevented the subjects from knowing when Natasha was looking at them. This also prevented Natasha from picking up clues from their eye movements or pupillary dilations (which are a sign of emotional reaction). Before the test, I instructed and rehearsed the subjects on how to behave. They were to sit as still as possible when Natasha was in the room. If Natasha needed to observe them in a standing position, I would tell Natasha to turn her back while they stood up and when they sat again. We used similar precautions if Natasha needed to look at them in profile. These precautions reduced the possibility of reactions by the subjects from knowing which target condition Natasha was currently studying. We also wanted to reduce external movements (for example, the subject with a hip replacement might give herself away from her efforts to stand or to change the position of her body).[3]   The test room was large and had chairs for our seven subjects, for Natasha and two interpreters. One interpreter was Natasha's friend Sveta Skarbo. We allowed her in the test room to make Natasha feel comfortable. The other interpreter was supplied by the Discovery Channel. Ideally, only I, as the head proctor, Richard Wiseman as my co-investigator, Natasha and the two interpreters, and the seven subjects should have been present during the test. The realities of television production and the requests of Natasha's companions forced us to compromise here, and in some matters of protocol. The test room also included a television crew of three persons from the production company (Shine, Ltd.); Austin Dacey, who was videotaping the proceedings for CSICOP; Joe Nickell as an observer; a still photographer from the Discovery Channel; and Will Stewart, a British journalist living in Russia who was acting as a representative for Natasha. Except for the subjects (and Austin Dacey), everyone in the test room, including myself, was blind to the condition of each subject.

A small room, attached to rear of the test room, was used for briefing Natasha. We could retreat to this room when we wanted to discuss matters out of sight and hearing of the subjects. Because Andrew was in charge of recruiting the subjects and was not completely blind to their conditions, he stayed out of the testing room. He remained in the briefing room during the entire test (which lasted more than four hours). We used this room to brief Natasha before each of the six required matches (once she had made six matches, the seventh was determined by default). Before each trial Andrew gave her a clear description, along with images and diagrams, of the target condition that she was to match to a subject. We also discussed any of Natasha's questions or concerns in this room.

Andrew and I met with Natasha in this room before the test to review the procedure and to remind her about the details of the protocol. She had agreed to this protocol, which Monica had shown her five days previously. We reviewed each condition that we would ask her to detect. She expressed concerns about the removed appendix and the resected esophagus. She was worried that if the appendix had been removed long enough ago it might have grown back. Andrew assured her that appendices do not grow back. Her concern about the resected esophagus was that individuals might normally differ in the length of their esophagus and this could mislead her. Andrew told her that instead of the length she should look for the scar that completely encircled the place where the two ends of the resected esophagus had been surgically joined.

The test consisted of six trials. On each trial Andrew gave Natasha a test card that clearly described, in Russian and English, the condition she was to match to a subject. The card contained an illustration of the target organ or condition. Andrew also showed her relevant illustrations from an anatomy text. When she was satisfied, I accompanied Natasha to the test room, where she sat between the two interpreters and equidistant from each subject. After Natasha had studied the subjects for the given condition, she chose the subject she believed had the specified condition. She would circle the subject's number on the test card and both of us would sign the card. We then returned to the back room to prepare for the next condition and trial.

We wanted to make the test as comfortable and nonstressful for Natasha as possible. I made sure not to rush or pressure her in any way. I gave her all the time she wanted to make each match. She took one hour to make the first match—which was to find the subject who had a large section of the top of her left lung surgically removed. She required more than four hours to complete the matches of conditions to the seven subjects. Throughout this process I repeatedly asked her if she was comfortable and if we could do anything to make the process more agreeable to her. She could ask for a break in the proceedings whenever she wished. Her mother had decided to remain outside both the test and briefing rooms because she wanted to be with Natasha's younger sister. Midway through the proceedings, Natasha told us she would feel better if her mother could be in the briefing room. I immediately agreed to her request. [4]

THE OUTCOME

Natasha succeeded in correctly matching four target conditions out of a possible seven. Our protocol required that Natasha get five or more correct matches to "pass" our test.

Understandably, Natasha's supporters were disappointed. They expressed their misgivings about the test on the television documentary, in media interviews, on Web sites, and through e-mails. They accused the testers of bias and of deliberately manipulating the procedure to prevent Natasha from succeeding. Natasha has complained that if she had gotten five correct she would have been a success. Isn't four close enough?

Our answer is that five was the minimum score that everyone agreed upon. It was also the minimum score that would convince us of a possible ability to diagnose subjects with sufficient reliability to be useful. We designed our test to detect a large effect. We were looking for something that would distinguish Natasha's claims from many similar ones. We wanted a good reason to justify using the additional time and resources to investigate her ability further.

Although Natasha's score did not meet our criterion for "success," it is possible that she can pick up information about the subject's condition. Some of her choices might show some accuracy on her part, although of a low level. If this is true, her correct matches could be the result of three possibilities:

1. She gathered some information paranormally. That is, she can see into people's bodies, but imperfectly.
2. She gathered information by deliberately exploiting available clues such as outward appearances and behavior of the subjects.
3. She obtained information unconsciously from available clues. To me, this is the most likely explanation, other than chance or in addition to chance. Much recent work in psychology demonstrates implicit learning: how people unconsciously learn to exploit a variety of clues, often subtle ones.

Both inherent and unforeseen limitations of our test provided possible clues to the target conditions for some subjects. I already discussed the daunting task of finding seven appropriate subjects. We had to settle for a less than optimal set of subjects. These subjects differed sufficiently in outward appearance to provide possible clues about their conditions. Another problem occurred through two violations of the test protocol. Together these problems created the possibility for identifying the target conditions—by external, normal means—for the following four subjects:

(1). The "control" subject, the one who had no internal medical condition, was obviously the youngest of the group. He also looked in good physical condition and appeared much healthier. He was a good candidate for the person with no defects.

(2). The subject with the staples in his chest (because of major heart surgery) was male, the oldest of the group and looked the least healthy. He was an obvious choice for the person with the staples in his chest.

(3) A breach of protocol occurred on the first trial. Natasha posed a question and her interpreter translated it aloud in front of the subjects. The question, contrary to our protocol, allowed the subjects to know that Natasha was looking for the subject with part of her lung removed. Here it was possible that, knowing which condition Natasha was looking for, the subject with the missing lung might have given herself away through bodily reaction.

(4) After the test was over, I learned that Natasha and her companions, because of an apparent misunderstanding, had arrived at the test site before we had expected them. They waited outside the test building where they reportedly observed at least two of the test subjects climb the long flight of stairs and enter the test building. This breach of protocol may have provided them clues about which subjects did or did not have the artificial hip.

We do not know if Natasha took advantage of the clues I've described in the previous four paragraphs. However, it is suggestive that these were just the four subjects for whom Natasha achieved her correct matches. The probability that she was relying upon nonparanomal clues increases when we consider her misses. She wrongly picked the subject who was wearing a baseball cap as the one who had the metal plate in his head. Conceivably, she picked this subject because one might assume (falsely in this case) that the subject was trying to cover a scar on his head. We should also emphasize that her failure to correctly match the subject with the metal plate in his head further argues against any fledgling paranormal powers. If she truly can see into bodies, she should have easily detected the large area of missing skull along with the metal plate covering the hole

Our test included five subjects for whom external clues were available concerning their internal condition. The clues correctly pointed to the true target condition for four subjects. The external clue for the fifth subject falsely pointed to the hole in the skull. In each of these five cases Natasha made her choice consistent with how the external clue was pointing.

Because a single test, even one done under ideal conditions, cannot settle a paranormal claim, we conceived our test as the first stage of a potential series. The first stage would not necessarily rule out nonparanormal alternatives. If Natasha could pass the first stage, this would justify continuing onto the next stage. If she passed that stage, then we would continue studying her claim. On the other hand, if she failed at any of the early stages, this would end our interest in her claim.

Keep in mind that the burden of proof belongs to the parties making an extraordinary claim. Extraordinary claims require extraordinary proof. Our test had its limitations. None of these limitations, however, worked against Natasha's claim. If anything, they may have artificially enhanced her score. Our task was not to prove that Natasha does not have X-ray vision. Rather, Natasha and her supporters had the responsibility to show us that she could perform well enough to deserve further scientific investigation. This they failed to do.

(After the original version of this chapter first appeared in the *Skeptical Inquirer*, a number of comments and criticisms were received. Some commentators said, for example, the criterion for "passing" the test—five correct matches out of seven—was too high. But If Natasha has anything like the ability she claims, she should have easily matched each condition to the appropriate subject. Her claims imply a type of X-ray vision of extremely high resolution. Our test required X-ray vision of very low resolution. Consequently, if her claim is true, we should expect her to match all seven conditions correctly. We set the criterion at five, rather than at seven, because we wanted to give her some leeway. Getting only four correct matches is highly inconsistent with her claim.

(Some critics claimed that our test lacked sufficient power. By this they mean that even if Natasha's claim is true, our test had little chance to show it. In fact, the power of our test was adequate and much higher than the critics maintain.

(My detailed responses to these and other criticisms are posted on the CSICOP web site at http://www.csicop.org/specialarticles/natasha2.html.)

ACKNOWLEDGMENTS

I thank Richard Wiseman (University of Hertfordshire) and Andrew Skolnick (Commission for Scientific Medicine and Mental Health) for their many constructive criticisms to the earlier drafts of this paper. Richard convinced me to eliminate over half the material I had intended to include. This was a great improvement.

NOTES

1. We debated about how to refer to the seven volunteers who had conditions which Natasha had to detect. Each of the candidate terms such as *volunteer, participant, patient,* or *client* seemed ambiguous or not quite correct. Although not completely satisfactory, we decided to refer to these individuals as *subjects*.

2. Pfungst, O. *Clever Hans*. New York: Henry Holt & Co, 1911. Also see Vogt, E. Z., and Hyman, R. *Water Witching U.S.A.* Chicago: University of Chicago Press, 2000.

3. Here is another compromise we had to make in the test. Ideally, everyone in the test situation should be blind as to the true target condition for each subject. In our case, the subjects were not blind to their own conditions. Because the subjects had to be in the test room and Natasha had to study them visually, the test lacked this blindness. The use of the opaque sunglasses hopefully kept the subjects blind as to which target condition Natasha was looking for on a given trial, but this is not completely satisfactory.

4. At the start of the test some initial confusion existed as to who would be allowed into the test and briefing rooms. This was quickly corrected and Natasha's mother and Will Stewart were given the option of staying in one of these rooms.

"DR." BEARDEN'S VACUUM ENERGY

Martin Gardner

One of the strangest books ever written about modern physics was published in 2002, and reprinted two years later. Titled *Energy from the Vacuum* (Cheniere Press), this monstrosity is two inches thick and weighs three pounds. Its title page lists the author as "Lt. Col. Thomas E. Bearden, PhD (US Army retired)."

"Dr." Bearden is fond of putting PhD after his name. An Internet check revealed that his doctorate was given, in his own words, for "life experience and life accomplishment." It was purchased from a diploma mill called Trinity College and University—a British institution with no building, campus, faculty, or president, and run from a post office box in Sioux Falls, South Dakota. The institution's owner, one Albert Wainwright, calls himself the college "registrant."

Bearden's central message is clear and simple. He is persuaded that it is possible to extract unlimited free energy from the vacuum of space-time. Indeed, he believes the world is on the brink of its greatest technological revolution. Forget about nuclear reactors. Vacuum energy will rescue us from global warming, eliminate poverty, and provide boundless clean energy for humanity's glorious future. All that is needed now is for the scientific community to abandon its "ostrich position" and allow adequate funding to Bearden and his associates.

To almost all physicists this quest for what is called "zero-point energy" (ZPE) is as hopeless as past efforts to build perpetual motion machines. Such skepticism drives Bearden up a wall. Only monumental ignorance, he writes, could prompt such criticism.

The nation's number two drumbeater for ZPE is none other than Harold Puthoff, who runs a think tank in Austin, Texas, where efforts to tap ZPE have been underway for years. In December 1997, to its shame, *Scientific American* ran an article praising Puthoff for his efforts. Nowhere did this article mention his dreary past.

Puthoff began his career as a dedicated Scientologist. He had been declared a "clear"—a person free of malicious "engrams" recorded on his brain while he was an embryo. At Stanford Research International, Puthoff and his then-friend Russell Targ claimed to have validated "remote viewing" (a new name for distant clairvoyance), and also the great psi powers of Uri Geller. (See my chapter on Puthoff's search for ZPE in *Did Adam and Eve Have Navels?* Norton, 2000.)

Bearden sprinkles his massive volume with admirable quotations from top physicists, past and present, occasionally correcting mistakes made by Einstein and others. For example, Bearden believes that the graviton moves much faster than the speed of light. He praises the work of almost every counterculture physicist of recent decades. He admires David Bohm's "quantum potential" and Mendel Sach's unified field theory. Oliver Heaviside and Nikola Tesla are two of his heroes.

Bearden devotes several chapters to antigravity machines. Here is a sample of his views:

> In our approach to antigravity, one way to approach the problem is to have the mechanical apparatus also the source of an intense *negative energy* EM field, producing an intense flux of Dirac sea holes into and in the local surrounding space-time. The excess charge removed from the Dirac holes can in fact be used in the electrical powering of the physical system, as was demonstrated in the Sweet VTA antigravity test. Then movements of the mechanical parts could involve movement of strong negative energy fields, hence strong curves of local space-time that are local *strong negative gravity fields.* Or, better yet, movement of the charges themselves will also produce field-induced movement of the Dirac sea hole negative energy. This appears to be a practical method to manipulate the metric itself, along the lines proposed by Puthoff et al. [217]

The 217 superscript refers to a footnote about a 2002 paper by Puthoff and two friends on how to use the vacuum field to power spacecraft. Bearden's anti-gravity propulsion system is neatly diagrammed on page 319. "Negatively charged local space-time," says the diagram, "acts back upon source vehicle producing anti-gravity and unilateral thrust."

In the 1950s, numerous distinguished writers, artists, and even philosophers (like Paul Goodman, William Steig, and Paul Edwards) sat nude in Wilhelm Reich's "orgone accumulators" to absorb the healing rays of "orgone energy" coming from outer space. Bearden suspects (in footnote 78) that orgone energy "is really the transduction of the time-polarized photon energy into normal photon energy. We are assured by quantum field theory and the great negentropy solution to the source charge problem that the instantaneous scalar potential involves this process." I doubt if the Reichians, who are still around, will find this illuminating.

To my amazement Bearden has good things to say about the notorious "Dean drive"—a rotary motion device designed to propel spaceships by inertia. It was promoted by John Campbell when he edited *Astounding Science Fiction*, a magazine that unleashed L. Ron Hubbard's Dianetics on a gullible public and made Hubbard a millionaire. Only elementary physics is needed to show that no inertial drive can move a spaceship in frictionless space. On pages 448–453 Bearden lists eighty patents for inertial drives. They have one feature in common: none of them works.

Counterculture scientists tend to be bitter over the "establishment's" inability to recognize their genius. Was not Galileo, they like to repeat, persecuted for his great discoveries? This bitterness is sometimes accompanied by paranoid fears, not just of conspiracies to silence them, but also fears of being murdered. Bearden's pages 406–53 are devoted to just such delusions.

Several kinds of "shooters" are described that induce fatal heart attacks. He himself, Bearden writes, has been hit by such devices. An associate, Stan Meyer, died after a "possible" hit by a close-range shooter. Another ZPE researcher was killed by a bazooka-size shooter. Steve Marikov, still another researcher, was assaulted by a sophisticated shooter and his body thrown off a rooftop to make it appear a suicide. When his body was removed, the pavement "glowed."

One day at a Texas airport a person three feet from Bearden was killed with symptoms suggesting he was murdered by an ice-dart dipped in curare! "That was apparently just to teach me 'they' were serious." The colonel goes on to explain that "they" refers to a "High Cabal" who were offended by a friend's "successful transmutation of copper (and other things) into gold. . . . We have had numerous other assassination attempts, too numerous to reiterate. . . . Over the years probably as many as fifty or more overunity researchers and inventors have been assassinated . . . some have simply disappeared abruptly and never have been heard from since." *Overunity* is Bearden's term for machines with energy outputs that exceed energy inputs.

Any significant researcher should be wary of "meeting with a sudden sui-cide" on the way to the supermarket. Another thing to beware of, is a cali-brated auto accident where your car is rammed from the rear, and you are shaken up considerably. An ambulance just happens to be passing by moments later, and it will take you to the hospital. If still conscious, the researcher must not get in the ambulance unless accompanied by a watchful friend who understands the situation and the danger. Otherwise, he can easily get a syringe of air into his veins, which will effectively turn him into a human vegetable. If he goes to the hospital safely, he must be guarded by friends day and night, for the same reason, else he runs a high risk of the "air syringe" assassination during the night. Simply trying to do scientific work, I find it necessary to often carry (legally) a hidden weapon. Both my wife and I have gun permits, and we frequently and legally carry concealed weapons. As early as the 1930s, T. Henry Moray—who built a successful COP>1.0 power system outputting 50kW from a 55 lb power unit—had to ride in a bulletproof car in Salt Lake City, Utah. He was repeatedly fired at by snipers from the buildings or sidewalk, with the bullets sometimes sticking in the glass. He was also shot by a would-be assassin in his own laboratory, but over-powered his assassin and recovered.

Obviously, I'm not competent to wade through Bearden's almost a thou-sand pages to point out what physicists tell me are howlers. I leave that task to experts, though I suspect very few will consider it worth their time even to read the book. To me, a mere science journalist, the book's dense, pompous jargon sounds like hilarious technical double-talk. The book's annotated glos-sary runs to more than 120 pages. There are 305 footnotes, 754 endnotes, and a valuable seventy-three-page index.

The back cover calls the book "the definitive energy book of the twenty-first century." In my opinion it is destined to be the greatest work of outlandish science in both this and the previous century. It is much funnier, for instance, than Frank Tipler's best-seller of a few decades ago, *The Physics of Immortality*.

LESSONS OF THE "FAKE MOON FLIGHT" MYTH

James Oberg

Depending on the opinion polls, there's a core of Apollo moon flight disbelievers within the United States—perhaps 10 percent of the population, and up to twice as large in specific demographic groups. Overseas the results are similar, fanned by local attitudes toward the US in general and technology in particular. Some religious fundamentalists—Hare Krishna cultists and some extreme Islamic mullahs, for example—declare the theological impossibility of human trips to other worlds in space.

Resentment of American cultural and political dominance clearly fuels other "disbelievers," including those political groups who had been hoping for a different outcome to the Space Race—for example, many Cuban schools, both in Cuba and places where Cuban school teachers were loaned, such as Sandinista Nicaragua, taught their students that Apollo was a fraud.

Like a counterculture heresy, the "moon hoax" theme had been lingering beyond the fringes of mainstream society for decades. A self-published pamphlet here, or a "B-grade" science fiction movie there, or a radio talk show guest over there—for many years it all looked like a shriveling leftover of the original human inability to accept the reality of revolutionary changes.

But in the last ten years, an entirely new wave of hoax theories have appeared—on cable TV, on the Internet, via self-publishing, and through other "alternative" publication methods. These methods are the result of technological progress that Apollo symbolized, now ironically fueling the arguments against one of the greatest technological achievements in human history.

NASA's official reaction to these and other questions was both clumsy and often counter-productive. On the infamous Fox Television moon hoax program, which was broadcast several times in the first half of 2001, a NASA spokesman named Brian Welch appeared several times to counter the hoaxist arguments (Welch was a top-level official at the Public Affairs Office at NASA Headquarters, who died a few months later). The poor TV impression he gave (a know-it-all "rocket scientist" denouncing each argument as false but usually without providing supporting evidence) may have been due to deliberate editing by the producers to make the "NASA guy" look arrogant and contemptuous. But to a large degree it accurately reflected NASA's institutional attitude to the entire controversy. The disappointing results of participating seemed to strengthen the view within NASA that the best response was no response—to avoid anything that might dignify the charges.

Roger Launius, then the chief of the history office at headquarters, was an exception to NASA's overall unwillingness to engage the issue. As an amateur space historian and folklorist, I had been discussing with him for years the need for NASA to fulfill its educational outreach charter and to issue a series of modest *monographs* (a historian's term for a single-theme pamphlet-length publication) on many different widespread cultural myths about space activities. These ranged from allegations of UFO sightings (and videotapings) by astronauts, to the discovery of alien artifacts on the Moon and Mars and elsewhere, to miraculous and paranormal folklore associated with space activities, to the hoax accusations. Launius, nearing retirement in early 2002, decided it was time for a detailed response to the Apollo hoax accusations, and offered me a sole-source contract to write a monograph that analyzed why such stories seemed so attractive to so many people. Launius departed NASA soon thereafter, leaving the project in the care of a junior historian, Stephen Garber.

My requests for inputs from various NASA offices and public educational organizations soon reached the ears of news reporters, and some print stories appeared in late October. Although NASA officials were somewhat taken aback by the publicity, they were at first inclined to defend the project on educational grounds.

Then, on Monday, November 4, 2002, the eve of the national elections, ABC's *World News Tonight* anchor Peter Jennings chose the subject for his closing story: "Finally this evening, we're not quite sure what we think about this," he intoned. "But the space agency is going to spend a few thousand dollars trying to prove to some people that the United States did indeed land men on the moon."

Jennings described how "NASA had been so rattled" it "hired" somebody "to write a book refuting the conspiracy theorists." He closed with a misquotation: "A professor of astronomy in California said he thought it was beneath NASA's dignity to give these Twinkies the time of day. Now, that was his phrase, by the way. We simply wonder about NASA."

Jennings was referring to Philip Plait, an educator (not a professor) in California who runs the Bad Astronomy Web site that discusses many mythical aspects of outer space. What Plait actually had said was that he felt it *was* proper for NASA to respond, but that it did seem "beneath their dignity" to be forced to do it. Contrary to Jennings's account, Plait fully supported the monograph contract.

But that TV insult did it as far as NASA management was concerned. Their dignity called into question, and fearing angry telephone calls from congressmen returning to Washington after the election, they decided to revoke the contract. They paid for work done to date and washed their hands of the project.

Many educators contacted me in dismay. Like them, and unlike the NASA spokesmen, I had always felt that "there is no such thing as a stupid question." And to me the moon hoax controversy was not a bothersome distraction, but a unique opportunity.

This is the way I see it. If many people who are exposed to the hoaxist arguments find them credible, it is neither the fault of the hoaxists or of their believers—it's the fault of the educators and explainers (NASA among them) who were responsible for providing adequate knowledge and workable reasoning skills. And the localized success of the hoaxist arguments thus provides us with a detection system to identify just where these resources are inadequate.

I intend to complete the project, depending on successfully arranging new funding sources. The popularity of this particular myth is a heaven-sent (or actually, an "outer-space-sent") opportunity to address fundamental issues of public understanding of technological controversies.

MAGNET THERAPY:
A Billion-dollar Boondoggle

Bruce L. Flamm

About a year ago Leonard Finegold at Drexel University and I decided to look into the controversial field (no pun intended) of "magnet therapy." As a physics professor, Finegold knows a bit about magnets and magnetic fields. As a physician and former research chairman, I know a bit about therapy and medical research. Perhaps a physicist and a physician could shed some light on this interesting topic. We knew that magnets were touted as a treatment for many medical conditions and we knew that they were popular. But we were both quite surprised to learn just how popular they are. In the United States, their annual sales are estimated at $300 million (Brody, 2000), and globally more than a billion dollars (Weintraub, 1999). You can get a rough idea of the magnitude of the magnet healing industry by doing a Google search for *magnet healing*. A search in January 2006 yielded 459,000 Web pages, many of them claiming that magnets have almost miraculous healing power. Do they? Professor Finegold and I reviewed the literature on magnet therapy and found very little supporting evidence. An abbreviated version of our review was recently published in the *British Medical Journal* (Finegold and Flamm, 2006). What follows are a few comments on the magnet healing industry, a brief synopsis of our *BMJ* paper, and a look at magnet therapy from a theoretical point of view.

318

MAGNET THERAPY IS BIG BUSINESS

If you try the Internet searching experiment described above you will notice that in addition to almost half a million pages dealing with magnet therapy, Google automatically provides a list of "sponsored links." Your computer screen will fill with the names of companies that have paid to help you find their site. What do these sites offer? If you click on www.magnetsandhealth.com, you will learn that "magnets help to flush out toxins in our body" and that "our magnet products have both beauty and health benefits, they increase blood flow and they increase the oxygen level in the body." Really? They also point out that their magnets are small and mobile, which "allows you to heal the ailments of yourself and your family without having major interruptions in your life and routine. You also get all the benefits without having to go for expensive sessions with a magnetic therapist or having to take expensive courses of drugs which can also have harmful side effects." The message seems quite clear: Why bother with doctors and medicines when magnets are safe and effective?

Another of the scores of sponsored links is www.magnetictherapymagnets .com. This site is interesting because, in addition to selling dozens of magnetic healing devices for humans, it doesn't forget about Fido. For only $11.95 plus shipping they will send you an amazing pet collar that will "keep your cat or dog in excellent health and vitality with constant magnetic therapy." My wife and I are now kicking ourselves for spending thousands of dollars on veterinary care over the past several years. If we had only bought that collar!

I don't mean to pick on these two companies or imply that their claims are any more outrageous than any others. In fact, there are now hundreds of companies selling similar devices and making similar claims.

Among the companies touting magnet therapy I was surprised to find the Sharper Image, a seemingly reputable outfit. They offer a device called a "Dual-Head Personal Massager with Magnetic Therapy." It is somewhat phallus-shaped, small enough to fit in a purse, and claims to be a "discreet personal massager with two independent vibrating heads." That certainly seems enticing enough, but they insist that it does far more that your average vibrator. "A smaller pinpoint node enhances its massage with magnetic therapy for focused treatment." Hmm . . . magnet therapy for focused treatment.

Some companies actually claim that their magnets prevent, reverse, and cure cancer. For example, at one site purveyors of cancer-curing magnets will sell you, for only $2,595, the "Dr. Philpott Designed and Approved Polar Power Super Bed Grid." According to the site, "This is the strongest, deepest penetrating, permanent static magnet, biomagnetic therapy device available anywhere that we know of. It is used in many of Dr. Philpott's magnetic research protocols for prevention and reversal of cancer and other serious disease that requires a full systemic deep penetrating treatment of the whole body." Similarly absurd claims can be found at www.stopcancer.com/magnets.htm.

Do the legions of magnet therapists and magnet purveyors really believe the incredible claims that they make? Are they well-meaning but misguided individuals or con artists willing to say anything to make a buck? Both types are most likely involved.

STUDIES ON MAGNET THERAPY

The overall conclusion of our *BMJ* review was stated in the first sentence, "We believe there is a worldwide epidemic of useless magnet therapy" (Finegold and Flamm, 2006). As you can imagine, this statement was not well received in the magnet healing community. We found that many studies on "magnet therapy" were published in "alternative" journals as opposed to peer-reviewed medical journals. Many studies included too few patients to reach statistically significant conclusions. Others had problems with their placebo control groups. For example, study subjects realized that they were wearing a magnetic bracelet rather than a placebo bracelet when it attracted paper clips or other small metal objects. In light of the vast amounts of money spent each year on supposedly therapeutic magnets, surprisingly few legitimate randomized controlled trials have been conducted to evaluate their efficacy. An excellent critique of magnet therapy by Quackwatch founder Stephen Barrett, M.D., can be found at www.quackwatch.org.

IS MAGNET THERAPY EVEN THEORETICALLY POSSIBLE?

In reality, many people find anecdotal reports of healing, particularly from athletes or other trusted celebrities, to be more convincing than scientific studies.

There are certainly many people touting magnet therapy. But is there, even theoretically, any way that magnets could have any healing effect? In our *BMJ* review we restricted our comments to typical magnetic devices claimed to have therapeutic value: these use "static" magnets like those used to attach paper notes to a refrigerator door. In this context, *static* means nonmoving and has nothing to do with static electricity. Moving magnets or pulsed electromagnets can create electric fields and electromagnetic radiation that could have some effect on living tissue. In contrast, a typical nonmoving magnet produces only a magnetic field. Is there anything in the human body that is affected by magnetic fields? Surprisingly, the answer appears to be no. This seems counterintuitive since most people know that oxygen in our blood is carried by hemoglobin and that hemoglobin contains iron. This is why iron tablets are often recommended for the treatment of anemia. However, the iron in hemoglobin is not ferromagnetic (see www.badscience.net). If hemoglobin contained ferromagnetic iron it would be simple to separate red blood cells from other bloods cells with a magnet. Several studies have shown that static magnetic fields do not affect blood flow (see www.hfienberg.com/clips/magnet.htm and www.quackwatch.org). Perhaps more important, if hemoglobin contained ferromagnetic iron people might explode or be flung across the room when exposed to the extraordinarily powerful magnetic field of a MRI scan. For a fascinating look at things that can go wrong when ferromagnetic materials get too close to the powerful magnetic field of an MRI machine, visit http://mripractice.tripod.com/mrpractice/id69.htm and www.simplyphysics.com/flying_objects.html.

However, for the sake of argument, what if some effect of magnets on human tissue could be demonstrated? What is the likelihood that it would be a therapeutic or healing effect? Probably slim to nil. By analogy, consider chemical compounds. The number of known chemical compounds is on the order of ten million. However, only a handful have ever been shown to have any therapeutic effects. Yet millions are toxic. It would be most unwise to eat or drink anything found on the shelves of a typical chemistry lab. If a magnet had an effect on human tissue, there is no reason to believe that it would necessarily be a healing effect.

Moreover, even the rare chemical compound that has healing effects usually does so only in very specific dose ranges. Almost any prescription drug can harm or kill you if you ingest enough of it. If, theoretically, a magnet had some effect on human tissue and if, astoundingly, the effect was beneficial rather than toxic, would one not expect there to be an optimal dosage? Yet, advertised healing magnets vary widely in their field strength. Many magnet purveyors claim that the more powerful the magnet, the greater the healing effect. This sounds good but makes little sense. All known effective therapies—including medications, x-rays, and lasers—become toxic or damaging at high levels. Nevertheless, the "magnet therapist" who debated me on BBC radio immediately after our paper was published chided me for not understanding that some magnet healers fail because they don't use strong enough magnets. She was so convincing that I think she actually believes this. The BBC radio host made a point of stating that the magnet therapist was "certified." By whom, I wondered? The Intergalactic Association of Magnetic and

Crystal Healers? After the show my colleague Professor Finegold, who was raised in the United Kingdom, informed me that the word "certified" has a derogatory mental health connotation in the UK. Perhaps the BBC host was not flattering my opponent.

Some magnet advocates contend that no one has conclusively proven that magnets cannot heal. Of course, they have it backwards. When it comes to healing, the burden of proof is on the seller, not the buyer. One is supposed to prove that a therapy works before marketing it to the public. If this were not true, medical companies could save billions by selling all sorts of untested drugs and devices. In reality, the government insists that every medicine and therapeutic device be meticulously tested for both safety and efficacy. This protective system generally works and only rarely do unsafe or ineffective products slip through and reach the public. Sadly, it seems that no such protective laws exist for magnets, crystals, amulets, magic potions, or other claimed miracle cures.

Finally, in the firestorm of criticism that followed the publication of our *BMJ* article, a frequent complaint was that I don't have an "open" mind. It might be more fair to say that my mind is open—but not to nonsense. If properly conducted research demonstrates a genuine healing effect of static magnets, I will cheerfully incorporate magnet therapy into my clinical practice. Until that time, I hope that parents will take their sick children to evidence-based physicians rather than "certified" magnet healers.

NOTES

1. Brody, J. "Less Pain: Is it in the Magnets or in the Mind?," *New York Times*, November 28, 2000, p. F9.

2. Weintraub, M. "Magnetic Bio-stimulation in Painful Diabetic Peripheral Neuropathy: A novel Intervention—A Randomized, Double-placebo Crossover Study," *American Journal of Pain Management* 9 (1999): 8–17.

3. Finegold, L., and B. L. Flamm. "Magnet Therapy: Extraordinary Claims, But no Proved Benefits," *British Medical Journal* 332 (2006): 4.

OXYGEN IS GOOD—
EVEN WHEN IT'S NOT THERE

Harriet A. Hall

Oxygen is not just in the air; it's on the shelves. It has been discovered by alternative medicine and is being sold in various forms in the health supplement marketplace. Back when I was an intern, we used to joke that there were four basic rules of medicine:

1. Air goes in and out.
2. Blood goes round and round.
3. Oxygen is good.
4. Bleeding always stops.

Alternative medicine has latched onto rule number three and won't let go. The rationale, apparently, is that oxygen is required to support life; therefore more oxygen should make you more healthy. It's not clear how this relates to alternative medicine's advice on anti-oxidants, but that's irrelevant.

OXYGEN IS GOOD, so we should put it in our soft drinks and breathe it at oxygen bars. Take an oxygen tank home with you—you might feel better. The oxygen vendor might feel better, too. Dr. Andrew Weil, the renowned health guru, tells patients with chronic fatigue to ask their doctors to prescribe oxygen for a home trial. Sure, why not? The money it costs will literally vanish "into thin air," but who cares? OXYGEN IS GOOD.

Ignore the fact that you could find out whether you need oxygen by testing your blood oxygen saturation with that little-clip-thingy-they-stick-on-your-finger-in-the-ER (aka pulse oximeter). Who cares if your blood is fully saturated with oxygen already? OXYGEN IS GOOD. If your oxygen saturation is a little less than 100 percent, there is no evidence that raising it will help with anything. If it is a lot less, and you do need oxygen, any competent doctor should be able to figure that out. But try an oxygen tank anyway: OXYGEN IS GOOD. Is this starting to sound like a mantra? It should. This is religious belief I am talking about, not science.

OXYGEN THERAPIES

Just breathing oxygen is boring: any dumb animal can do that. Why not drink it? Alternative medicine found ways to use oxygen in a liquid form. No, not liquid oxygen; that's kind of chilly (around minus 298 degrees Fahrenheit). You could buy a $1,600 home water cooler to infuse oxygen into your drinking water and call that liquid oxygen. Or you could use chemicals that release oxygen. Ozone gives off oxygen. So do hydrogen peroxide and chlorite compounds. They're corrosive chemicals, but if you mix them with liquid and drink them, they will release friendly little oxygen bubbles in your stomach and dilution will counteract the corrosive effect. Corrosion may be bad, but OXYGEN IS GOOD.

Proponents of hydrogen peroxide found it worked wonders for everything from multiple sclerosis and cancer to hemorrhoids and the common cold. They tried bathing in it, drinking it, injecting it into their veins, and pumping it up their rectums. The proponents of hydrogen peroxide never reported any adverse effects. On the other hand, the medical literature reported that hydrogen peroxide caused deaths from gas embolism, gangrene, seizures, strokes, and other complications. How could that be? Didn't those scientists know that OXYGEN IS GOOD?

Could oxygen cure cancer and AIDS? They tried ozone by enema. They tried oxygen under pressure in a hyperbaric chamber. They even put cancer patients into a coma with overdoses of insulin in the hope that it would somehow regulate oxygen delivery to the cancer cells. Scientists insisted there was no proof that these therapies worked, but they couldn't prove to the true believers that they didn't work. Scientists insisted that these therapies could be harmful, but how could anything as natural as oxygen possibly be harmful? OXYGEN IS GOOD.

Invention proceeded apace. Alternative "science" found a way to put "electrically activated" oxygen in water. No one knows what "activated oxygen" means, but it sounds impressive. OXYGEN IS GOOD, and if it's activated it ought to be even better.

Once it has been "activated" in water, you can take oxygen safely and conveniently in the form of drops. You can put a few drops under your tongue, or drink them in a glass of water, and you will feel better. Of course, the amount of oxygen in a few drops of water is many orders of magnitude less than that contained in every breath. Of course, little or no oxygen is absorbed from the stomach. But fish get their oxygen from water, and if fish can do it, we ought to be able to do it too, gills or not! OXYGEN IS GOOD. They recommend that you drink it, spray it in your nose, gargle with it, spray it on cuts, spray some on your houseplants, and spray it on vegetables, chicken, seafood, and pork to decontaminate them. You can even use it as a natural alternative to antibiotics or give it to your pets to prevent doggy breath. OXYGEN IS GOOD—for just about everything.

SELLING OXYGEN—EVEN WITHOUT OXYGEN

Several brands of activated oxygen appeared on the market, each better than all the rest. Finally, one company came up with the best idea yet. They would sell plain water with a bit of salt and a trace of minerals, charge $10 an ounce for it, and *pretend* it contained activated oxygen.

The Rose Creek Company marketed oxygenated water without any oxygen in it. They said so themselves. Actually they claimed there was oxygen in it that "couldn't be detected" in the laboratory. Their excuse was that the precision equipment couldn't measure *over* 40 ppm of oxygen. I guess if they pour a whole pot of coffee into their coffee cup, the cup stays empty; and if they overinflate their tires they create a vacuum. This probably has something to do with homeopathy or an alternate reality. OXYGEN IS GOOD, even if the lab can't find it.

They called it "Vitamin O." They claimed that it would effectively prevent and treat pulmonary disease, headaches, infections, flu, colds, and even cancer. They claimed that it regulates metabolism, aids digestion, relaxes the nervous system, boosts energy, promotes sound sleep, and sharpens concentration and memory. They were selling 50,000 bottles a month until the Federal Trade Commission (FTC) spoiled the fun. A consent agreement required them to pay $375,000 for consumer redress and to stop making false statements. They were prohibited from "making any unsupported representation that the effectiveness of 'Vitamin O' is established by medical or scientific research or studies."

They changed the name of their company to R-Garden, Inc. and modified their ads, but they weren't happy. They knew that OXYGEN IS GOOD (it was definitely good for business), and they really wanted some scientific studies so they could thumb their noses at the FTC; so they hired the first scientist they could find who would stop laughing long enough to do some tests. They had to settle for an anthropologist. He reasoned that the oxygen that wasn't there should raise the blood oxygen level of people with anemia. So he decided to measure the partial pressure of oxygen and carbon dioxide in arterial blood (PaO_2 and $PaCO_2$), hypothesizing that both would rise with "Vitamin O," but not with placebo. This next bit is going to be a little technical, so you may skip the next two paragraphs if you wish.

If you are anemic, you have less hemoglobin in your blood to carry oxygen. But the blood gases, the PaO_2 and $PaCO_2$, have nothing to do with hemoglobin or anemia. They are measures of gas pressure, not of total oxygen carried by hemoglobin. So people with anemia should have a normal PaO_2. They might be able to raise their PaO_2 a little by hyperventilating, but this would lower their $PaCO_2$.

Anemic patients do not have low PaO_2 measurements. In this study, all of the anemic patients did have low baseline PaO_2. The researcher didn't notice that this was abnormal. When subjects took "Vitamin O" their PaO_2 went up, but mostly not up to normal. A couple of them developed a PaO_2 of over 100 mm Hg, which is not generally thought to be possible on room air without serious hyperventilation. He didn't notice anything unusual about this. He also discovered that the $PaCO_2$ went up right along with the PaO_2. Ordinarily this would indicate hypoventilation, possibly from severe lung disease, but he wasn't alarmed. He thought it would eliminate more nasty waste matter and would probably cause "greater youthfulness, improved mobility, better circulation, sharper mental clarity, enhanced heart and lung functions, and increased physical energy." (Huh?) Of course, in several patients the $PaCO_2$ went down as the PaO_2 went up, but this didn't bother him. Apparently he believes, with Ralph Waldo Emerson, that a foolish consistency is the hobgoblin of little minds.

He concluded that he had definitely proved that oxygen is present in "Vitamin O." In other words, if you find unexplained abnormal blood gas values and they change in a way that wouldn't be caused by increased oxygen intake, that proves that oxygen intake was increased, which proves that the increased oxygen intake had to come from the "Vitamin O." His logic didn't work any better than his blood gas analyzers.

His subjects were all Hutterites, members of an Anabaptist sect who live communally in groups of 60–150 on collective farms, mainly in the western US and Canada, and who remain aloof from outside society. He didn't explain how he persuaded them not to remain aloof from his experimental interventions. He noted that "no protocol or consent forms" were needed, since the head minister could make all the decisions for his flock and just order them to cooperate. It seems that Hutterites make even better experimental subjects than prisoners, because you have to get informed consent from prisoners. He didn't bother with any statistical analysis of his data. He didn't provide any references. He didn't expect that anyone would try to replicate his findings. As far as he was concerned, his definitive experiment had proved once and for all that there was oxygen in "Vitamin O." He broke just about every rule of scientific experimentation. After all, he wasn't searching for scientific truth, just for enough words on paper to meet the FTC's requirements.

In another study, the same researcher discovered an unbelievably high incidence of chronic fatigue syndrome among the Hutterites, and found that "Vitamin O" relieved their symptoms better than placebo. In his report of that study he gave an intriguing definition of his experimental method:

"The three key features of this approach in assessing the efficacy of natural man-made substance [*sic*] were employed here: randomization, blindedness, and measurement of predetermined outcomes."

I don't know what a "natural man-made substance" is; I agree that some kind of "blindedness" was at work; and if the outcomes were truly "predetermined," that would explain a lot.

In essence, the company funded studies that pretended to be scientific so they could pretend to have proven that the pretend oxygen in their product is really there (and that it really works). This kind of fuzzy thinking is typical of alternative medicine advocates. They ask you to maintain your health with preventive, natural treatments that have not been proven effective, and to mistrust conventional medical treatments that have been proven effective. They exploit fears of pollution, food additives, pesticides, and side effects from pharmaceuticals. They play the science game to appease their critics, but they don't really believe in science: they believe that truth can be accessed through intuition. They seldom disprove theories or report negative results; they rarely search for alternative explanations, suggest that further studies be done for confirmation, or seek peer review. They believe in the mystical ability of the body to heal itself through communication with the vital forces of the universe. How could science understand these complex, holistic phenomena? Scientists aren't even smart enough to sell water for $10 an ounce.

Science doesn't know everything. Intuition counts for something, doesn't it? There are more ways of knowing than scientists can imagine. Anyway, isn't scientific "truth" wishy-washy? Newton contradicted Copernicus, and Einstein contradicted Newton. Paradigms are shifty. Eventually scientists will have to give up their materialistic theories and will come to understand that the Universe is just one big mind interconnected by a continuum of ineffable cosmic quantum something-or-others.

Proponents of alternative oxygen therapy will never be convinced it doesn't work. As Lewis Carroll explained in *Alice in Wonderland*, practice makes perfect:

"One can't believe impossible things."

"I daresay you haven't had much practice," said the Queen. "When I was your age, I always did it for half-an-hour a day. Why, sometimes I've believed as many as six impossible things before breakfast."

"Vitamin O" is still being sold. Testimonials abound. You can't argue with true believers. OXYGEN IS GOOD.

FURTHER READING

Hall, Harriet. "A Failed Attempt to Prove That There Is Oxygen in 'Vitamin O.'" *The Scientific Review of Alternative Medicine* 7, no. 1 (2003): 29–33.

Marks, Stan. "Vitamin O?" Online at www.rgarden.net/forms/stanmarksinterview.htm.

Heinerman, John. "Proving the existence of elemental oxygen in a liquid nutritional product ("Vitamin O") through blood gas analyses of therapy/placebo-supplemented Hutterites." Online at www.rosecreekvitamino.com/original _research2002.html.

Barrett, Stephen. "FTC Attacks 'Stabilized Oxygen Claims'." Online at www.quackwatch.org/04ConsumerEducation/News/vitamino.html.

FTC Charges. Online at www.ftc.gov/os/1999/9903/rosecreekcmp.htm.

FTC Consent Agreement. Online at www.ftc.gov/os/2000/05/index.htm.

Heinerman, John. "Electrically-activated oxygen supplementation selectively improves energy efficiency in Hutterites demonstrating classic symptoms of chronic fatigue syndrome." Online at www.rosecreekvitamino.com/chronic_fatigue.html.

UPDATE: Apparently oxygen is still good. Amazon.com sells 27 different health supplements categorized as "Vitamin O." The R-Garden company received another warning letter from the FDA in 2005, but it is still in business and appears to be thriving.

UNDERCOVER AMONG THE SPIRITS:
Investigating Camp Chesterfield

Joe Nickell

C amp Chesterfield is a notorious spiritualist enclave of Chesterfield, Indiana. Dubbed "the Coney Island of spiritualism," it has been the target of many exposés, notably a book by a confessed fraudulent medium published in 1976. A quarter century later I decided to see if the old deceptions were still being practiced at the camp; naturally my visit was both unannounced and undercover.

THE BACKGROUND

Modern spiritualism began in 1848 with the schoolgirl pranks of Maggie and Katie Fox at Hydesville, New York. Although four decades later the sisters confessed that their "spirit" rappings had been bogus, in the meantime the craze of allegedly communicating with the dead had spread across America, Europe, and beyond. At séances held in darkened rooms and theaters, "mediums" (those who supposedly contacted spirits for others) produced such phenomena as slate writing, table tipping, and "materializations" of spirit entities.

As adherents grew in number, spiritualist camp meetings began to be common, and some groups established permanent spiritualist centers.

329

Perhaps none developed such an unsavory reputation as Camp Chesterfield, which opened in 1891. Even today, spiritualist friends of mine roll their eyes accusatorially whenever Chesterfield's name is mentioned, and they are quick to point out that the camp is not chartered by the National Spiritualist Association of Churches. The introduction to an official history of Chesterfield (*Chesterfield Lives* 1986, p. 6), admits it is surprising the camp has survived, given its troubled past:

> In fact, in its 100 years of recorded history, Camp Chesterfield has been "killed off" more than once! There have been cries of "fraud" and "fake" (and these were some of the nicer things we have been called!) and of course, the "exposés" came along with the regularity of a well-planned schedule. Oh yes! We have been damned and downed—but the fact remains that we must have been doing something right because: CHESTERFIELD LIVES!!

Be that as it may, the part about the exposés is certainly true.

A major exposé came in 1960 when two researchers—both sympathetic to spiritualism—arranged to film the supposed materialization of spirits. This was to occur under the mediumship of Edith Stillwell, who was noted for her multiple-figure spirit manifestations, and the séance was to be documented using see-in-the-dark technology. While the camera ran, luminous spectral figures took form and vanished near the medium's cabinet, but when the infrared film was processed the researchers saw that the ghosts were actually confederates dressed in luminous gauze, some of whom were recognizable as Chesterfield residents. They had not materialized and dematerialized but rather came and went through a secret door that led to an adjacent apartment (Keene, 1976, p. 40; Christopher, 1970, p. 174). One of the researchers, himself a devout spiritualist, was devastated by the evidence and railed against "the frauds, fakes and fantasies of the Chesterfield Spiritualist camp!" (O'Neill, 1960)

An even more devastating exposé came in 1976 with the book *The Psychic Mafia* written by former Chesterfield medium M. Lamar Keene. Saying that money was "the name of the game" at Chesterfield, Keene detailed the many tricks used by mediums there which he dubbed "the Coney Island of Spiritualism." He told how "apports" (said to be materialized gifts from spirits) were purchased and hidden in readiness for the séance; how chiffon became "ectoplasm" (an imagined mediumistic substance); how sitters' questions written on slips of paper called billets were secretly read and then answered; how trumpets were made to float in the air with discarnate voices speaking through them; and how other tricks were accomplished to bilk credulous sitters (Keene, 1976, pp. 95–114).

Keene also told how the billets were shrewdly retained from the various public clairvoyant message services held at Chesterfield. Kept in voluminous files beneath the Cathedral, the billets—along with a medium's own private files and those shared by fellow scam artists—provided excellent resources for future readings. There were other exposés of Camp Chesterfield. In 1985 a medium from there was making visits to Lexington, Kentucky, where he conducted dark-room materialization séances. He featured the production of apports, the floating-trumpet-with-spirit-voices feat, and something called "spirit precipitations on silk." To produce the latter, the sitters' "spirit guides" supposedly took ink from an open bottle and created their own small self-portraits on swatches of cloth the sitters held in their laps. I investigated when one sitter complained, suspecting fraud. Laboratory analyses by forensic analyst John F. Fischer revealed the presence of solvent stains (shown under argon laser light). A recipe for such productions given by Keene (1976, pp. 110–11)—utilizing a solvent to transfer pictures from newspapers or magazines—enabled me to create similar "precipitations" (Nickell with Fischer 1988). The prepared swatches had obviously been switched for the blank ones originally shown.

UNDERCOVER

I had long wanted to visit Camp Chesterfield, and in the summer of 2001, following a trip to Kentucky to see my elderly mother and other family members, I decided to head north to Indiana to check out the notorious site.

Now, skeptics have never been welcome at Chesterfield. The late Mable Riffle, a medium who ran the camp from 1909 until her death in 1961 (*Chesterfield Lives*, 1986) dealt with them summarily. When she heard one couple using the f-word—*fraud*—she snarled, reports Keene (1976, p. 48), "We do not have that kind of talk here. Now you get your goddam ass off these hallowed grounds and don't ever come back!"

Another skeptic, a reporter named Rosie who had written a series of exposés and was banned from the grounds, had the nerve to return. Wearing a "fright wig," she got into one of Riffle's séances and when the "spirits" began talking through the trumpet the reporter began to demean them. According to Keene (1976, pp. 48–49), Riffle recognized Rosie's voice immediately and went for her. "Grabbing the reporter by the back of the neck, she ushered her up a steep flight of stairs, kicking her in the rump on each step and cursing her with every profanity imaginable."

With these lessons in mind, I naturally did not want to be recognized at Chesterfield—not out of fear for my personal safety but so as to be able to observe unimpeded for as long as possible. When in my younger years I was a private investigator with an international detective agency, I generally used my own name and appearance and, for undercover jobs, I merely wore the attire that was appropriate for a forklift driver, steelworker, tavern waiter, or other "role" (Nickell, 2001).

The same is true for several previous undercover visits to paranormal sites and gatherings (including a private spiritualistic circle which included table-tipping and other séances that I infiltrated in 2000). Since I am often the token skeptic on television talk shows and documentaries on the paranormal, I have naturally feared I might be recognized, but I rarely made any effort to disguise myself and usually had no problem.

However, for my stint at Camp Chesterfield, I felt special measures were called for so I decided to alter my appearance, shaving off my mustache (for the first time in over thirty years!), and replacing my coat-and-tie look with a T-shirt, suspenders, straw hat, and cane. I also adopted a pseudonym, "James Collins," after the name of one of Houdini's assistants. From July 19 to 23, "Jim," who seemed bereft at what he said was the recent death of his mother, limped up and down the grounds and spent nights at one of the camp's two hotels (devoid of such amenities as TV and air conditioning). The results were eye-opening, involving a panoply of discredited spiritualist practices that seemed little changed from when they were revealed in *The Psychic Mafia*.

BILLET READING

I witnessed three versions of the old billet scam: one done across the table from me during a private reading, and two performed for church audiences, one of them accomplished with the medium blindfolded.

The first situation—with the medium working one-on-one with the client—involves getting a peek at the folded slip while the person is distracted. (Magicians call this misdirection.) For instance, while the medium directs the sitter's attention, say by pointing to some numerological scribblings (as were offered in my case), she can surreptitiously open the billet in her lap with a flick of the thumb of her other hand and quickly glimpse the contents. As expected, the alleged clairvoyant knew exactly what was penned on my slip—the names of four persons who had "passed into spirit" and two questions—but did not know that the people were fictitious.

One aspect of the reading, which was held in the séance room of her cottage, was particularly amusing. At times she would turn to her right—as if acknowledging the presence of an invisible entity—and say "Yes I will." This was a seeming acknowledgment of some message she supposedly received from a spirit, which she was to impart to me. I paid the medium thirty dollars cash and considered it a bargain—although not in the way the spiritualist would no doubt have hoped.

At both of the billet readings I attended that were conducted for audiences (one in a chapel, the other in the cathedral), a volunteer stood inside the doorway and handed each of us a slip of paper. Printed instructions at the top directed us to "Please address your billet to one or more loved ones in spirit, giving first and last names. Ask one or more questions and sign your full name."

On the first occasion I made a point of seeming uncertain about how to fold the paper and was told that it was to be simply doubled over and creased; if it was done otherwise, I was told strictly, the medium would not read it. I did not ask why, since I tried to seem as credulous as possible, but in fact I knew that there were two reasons. First, of course, the billets needed to be easy to open with a flick of the thumb, and second, it was essential that they all look alike. The reason for the latter condition lay in the method employed: After the slips were gathered in a collection plate and dumped atop the lectern (where they could not be seen from our vantage point) the medium would pick one up and hold it to his or her forehead while divining its contents. The trick involves secretly glancing down at an open billet. A sitter who had closed his slip in a distinctive way (such as by pleating it or folding it into a triangle) might notice that the billet being shown was not the one apparently being viewed clairvoyantly.

The insistence on how the paper must be folded indicated trickery. And that was confirmed for me at one session through my writing the names of non-existent loved ones and signing with my pseudonym. From near the back of the chapel I acknowledged the medium's announcement that he was "getting the Collins family." After revealing the bogus names I had written, he gave me an endearing message from my supposedly departed mother that answered a question I had addressed to her on the billet. However, my mother was actually among the living and, of course, not named Collins.

The other public billet reading I attended was part of a Gala Service held in the Cathedral. The medium placed adhesive strips over her eyes followed by a scarf tied in blindfold fashion. This is obviously supposed to prove that the previously described method of billet reading was not employed, but according to Keene (1976, p. 45), who performed the same feat, "The secret here was the old mentalist standby: the peek down the side of the nose." He adds: "No matter how securely the eyes are blindfolded, it's always possible to get enough of a gap to read material held close to the body." Unfortunately, at this reading my billet was not among those chosen, so there were no special communications from the non-existent persons whose names I had penned.

SPIRIT WRITING

Another feat practiced by at least three mediums at Chesterfield is called "spirit card writing." This descends from the old slate effects that were common during the heyday of spiritualism, whereby (in a typical effect) alleged otherworldly writing mysteriously appeared on the inner surfaces of a pair of slates that were bound together (Nickell, 2000). In the modern form (which exists in several variants), blank cards are placed in a basket along with an assortment of pens, colored pencils, etc. After a suitable invocation, each of the cards is seen typically to bear a sitter's name surrounded by the names of his "spirit guides" or other entities and possibly a drawing or other artwork. The sitter keeps the card as a tangible "proof" of spirit power.

At Chesterfield one afternoon I attempted to sign up for a private card-writing séance later that evening at the home of a prominent medium (who also advertises other feats including "pictures on silk"). When that session proved to be filled, I decided to try to "crash" the event and soon hit on a sub-terfuge. I placed the autograph of "Jim Collins" on the sign-in sheet for the following week, then showed up at the appointed time for the current séance a few hours later. I milled about with the prospective sitters, and then we were all ushered into the séance room in the medium's bungalow.

So far so good. Unfortunately when he read off the signees' names and I was unaccounted for, I had some explaining to do. I insisted I had signed the sheet and let him discover the "error" I had made. Then, suitably repentant and deeply disappointed, I implored him to allow me to stay, noting that there was more than one extra seat. Of course, if the affair were bogus, and the cards pre-pared in advance, I could not be permitted to participate. Not surprisingly I was not, being given the lame excuse (by another medium, a young woman, who was sitting in on the session) that the medium needed to prepare for the séance by "meditating" on each sitter's name. (I wondered which of the two types of mediums she was: one of the "shut-eyes," simple believers who fancy that they receive psychic impressions, or one of the "open mediums," who acknowledge their deceptions within the secret fraternity [Keene 1976, p. 23].) Even without my admission fee, I estimate the medium grossed approximately $450.

The next day I sought out one of the sitters who consoled me over my not having been accepted for the séance. She showed me her card which bore a scat-tering of names like "Gray Wolf" in various colors of felt-pen handprinting—all appearing to me on brief inspection to have been done by one person. The other side of the card bore a picture (somewhat resembling a Japanese art print) that she thought had also been produced by spirits, although I do not know exactly what was claimed by the medium. I did examine the picture with the small lens on my Swiss army knife which revealed the telltale pattern of dots from the halftone printing process. The woman seemed momentarily discomfited when I showed her this and indeed acknowledged that the whole thing seemed hard to believe, but she stated that she simply chose to believe. I nodded understandingly; I was not there to argue with her.

"DIRECT VOICE"

My most memorable—and unbelievable—experience at Camp Chesterfield involved a spirit materialization séance I attended at a medium's cottage on a Sunday morning. Such offerings are not scheduled in the camp's guidebook but are rather advertised via a sign-in book, and, perhaps an accompanying poster, on the medium's porch. As my previous experience showed, it behooved one to keep abreast of the various offerings around the village. So I was out early in the morning, hobbling with my cane up and down the narrow lanes. Soon, a small poster caught my eye: "Healing Séance with Apports." It being just after 6:00 a.m., and the streets silent, I quietly stepped onto the porch and signed up for the 10:00 a.m. session.

At the appointed time seven of us had gathered and the silver-haired medium ushered us into the séance room which she promptly secured against light leakage, placing a rolled-up throw rug at the bottom of the outer door and another rolled cloth to seal the top, and closing a curtain across an interior door. She collected twenty-five dollars from each attendee and then, after a brief prayer, launched into the healing service. This consisted of a "pep talk" (as she termed it) followed by a brief session with each participant in which she clasped the person's hands and imparted supposed healing energies.

It eventually came time for the séance. A pair of tin spirit trumpets standing on the floor by the medium's desk suggested we might experience "direct voice," by which spirits supposedly speak, the trumpets often being used to amplify the vocalizations. The medium began by turning off the lamps and informing us that "dark is light." Soon, in the utter blackness, the voices came, seeming to be speaking in turn through one of the trumpets. Keene (1976, pp. 104–108) details various means of producing "levitating" trumpets, complete with luminescent bands around them "so that the sitters could see them whirling around the room, hovering in space, or sometimes swinging back and forth in rhythm with a hymn." But here, however, we were left to our imaginations. Mine suggested to me that the medium was not even bothering to use the large trumpet, which might prove tiresome, but may have been utilizing a small tin megaphone—another trick described by Keene.

Some mediums were better at pretending direct voice than others; sometimes, according to one critic, "All the spirit voices sounded exactly like the medium . . ." (Keene 1976, p. 122). Such was the case at my séance. The first voice sounded just like the medium using exaggerated enunciation to simulate an "Ascended Master" (who urged the rejection of negativity); another sounded just like the medium adopting the craggy voice of "Black Elk" (with a message about having respect for the Earth); and still another sounded just like the medium using a perky little-girl voice to conjure up "Miss Poppy" (supposedly one of the medium's "joy guides").

At the end of the séance, after the lights were turned back on, one of the trumpets was lying on its side on the floor, as if dropped there by the spirits—or, as I thought, simply tipped over by the medium. Finally, we were invited up to get our "apports."

APPORTS

Supposedly materialized or teleported gifts from the spirits, apports appear at some séances under varying conditions—sometimes tumbling out of a spirit trumpet, for example. Keene (1976, p. 108) says those at Camp Chesterfield were typically "worthless trinkets" such as broaches or rings often "bought cheap in bulk." One medium specialized in "spirit jewels" (colored glass) while another apported arrowheads; special customers might receive something "more impressive." Camp Chesterfield instructed its apport mediums to ". . . *please ask your guides* to bring articles of equal worth to each sitter and not to bring only one of such articles as are usually in pairs (earrings or cufflinks, for instance)" (quoted from "The Medium's Handbook" by Keene 1976, p. 63).

At our séance the apports were specimens of hematite which (like many other stones) has a long tradition of alleged healing and other powers (Kunz, 1913). The shiny, steel-gray mineral had obviously been tumbled (mechanically polished), as indicated by surface characteristics shown by stereomicroscopic examination, and was indistinguishable from specimens purchased in shops that sell such New Age talismans.

The medium handed each of us one of the seven stones after picking it up with a tissue and noting with delight our reaction at discovering it was icy cold! This was a nice touch, I thought, imparting an element of unusualness as if somehow consistent with having been materialized from the Great Beyond—although probably only kept by the medium in her freezer until just before the séance when it was likely transferred to a thermos jar. We were told that each apport was attuned to that sitter's own energy "vibrations" and that no one else should ever be permitted to touch it. If someone did, we were warned, it would become "only a stone."

I left Camp Chesterfield on the morning of my fifth day there, after first taking photographs around the village. As I reflected on my experiences, things seemed to have changed little from the time Keene wrote about in *The Psychic Mafia*. Indeed the deceptions harkened back to the days of Houdini and beyond—actually, all the way back to 1848 when the Fox sisters launched the spiritualist craze with their schoolgirl tricks.

ACKNOWLEDGMENTS

I am grateful to Brant Abrahamson of Brookfield, Illinois, who was prompted by my investigations of spiritualism to send me some materials he obtained while stopping at Camp Chesterfield during a trip. The materials helped seal my resolve to visit the site.

REFERENCES

Christopher, Milbourne. *ESP, Seers & Psychics* (New York: Thomas Y. Crowell Co, 1970).

Hyde, Julia. *Letter to a Mrs. Peck*, written at Lily Dale, April 25; typescript text in file on the Hyde house, Genesee Country Village, Mumford, NY, 1896.

Keene, M. Lamar. *The Psychic Mafia*. Reprinted (Amherst, NY: Prometheus Books, 1997), p. 1976.

Kunz, George Frederick. *The Curious Lore of Precious Stones*, reprinted (New York: Dover, 1971). pp. 6, 80–81.

Nickell, Joe, with John F. Fischer. "Adventures of a Paranormal Investigator," edited by Paul Kurtz, in *Skeptical Odysseys*. (Amherst, NY: Prometheus Books), 2001. pp. 219–32.

———. *Entities*. (Amherst, NY: Prometheus Books, 1995), pp. 17–38.

———. *Secrets of the Supernatural*. (Buffalo, NY: Prometheus Books, 1988).

———. "Spirit painting, part II." *Skeptical Briefs* 10, no. 2 (June 2000): 9–11.

O'Neill, Tom. 1960. Quoted in Keene 1976, p. 40, and Christopher 1970, p. 175.

HOW TO "HAUNT" A HOUSE

Benjamin Radford

On Tuesday, November 11, 2003, I received a call at the Center for Inquiry in Amherst, New York, from a local woman asking for help. She believed her house was haunted by ghosts or evil spirits and didn't know where else to turn. In any other business office, this might have been seen as a prank, but at the Committee for the Scientific Investigation of Claims of the Paranormal (now the Committee for Skeptical Inquiry), we take these things seriously.

The case would turn out to be a fascinating one with many of the classic haunting phenomena, including "cold spots," poltergeist activity, animals acting strangely, and spooky footsteps and voices. There were even reports of a demonic face and a ghost attack! It sounded like just my kind of place.

I soon headed to a neighborhood south of Buffalo. Passing cookie-cutter houses, I soon found the place: a somewhat rundown but respectable two-story, pale-blue building. The houses were close together, separated by little more than short, concrete driveways leading to tiny backyards. As I gathered my tape recorder, camera, and notebook, a woman in her late twenties hesitantly poked her head out the door. I climbed onto the porch, where she introduced herself as Monica. Her husband, Tom, a stocky Hispanic man of about forty, shook my hand and led me through a dimly lit living room. We walked into the kitchen, where I took a seat at a small dinette table and interviewed the couple.

Monica and Tom, with their two-year-old daughter, had lived in the house for about three years. The couple slept in separate bedrooms, both upstairs. Tom suffered from apnea, a sleep disorder, and used a machine that forced air into his nose while he slept. Monica slept in an adjacent room with their daughter. Most of the ghostly events happened just as Tom was drifting off to sleep. Tom was almost always the main person affected, and he was the first to hear the sounds. Sometimes when they were both downstairs, they would hear steps on the house's creaky staircase.

The disturbances were sporadic but had recently increased. The wife said, "It got more active when he said to me [on October 27], 'You know, Monica, I hear this, it is pulling my machine,' and I said, 'I hear it too, or I feel it, too.' When we admitted it to each other, it got tremendously worse. It started the Monday before Halloween, and then it got really bad." Tom summoned a priest to bless the house on October 30, but the exorcism failed, and the disturbances got worse. Later that night, the family fled their home and were afraid to return. Both Monica and Tom were convinced that some sort of spirit was residing in their house. Monica said, "If somebody had come to me and said that there was a ghost in the house, I would have never believed them until I lived through what I lived through here. . . . We just want to move back into our home."

HAUNTING PHENOMENA

The family cited many unexplained phenomena that led them to believe they were being haunted:

1. Tom felt a hard tapping on his feet, near his ankle. This almost always happened at night in bed while he was going to sleep (or while he was asleep). "I get a tapping on my feet, not a repetitive tap, a trying-to-wake-you-up tap." This happened three or four times. Since then, Tom said, "I won't sleep in this room."

2. On at least one occasion, Tom said he was physically attacked by the spirit. Tom's bed shook as if from a ghostly kick. "If I don't pay attention to it [the tapping], it will kick the bed—it will hit the side of the bed. I feel my whole body move. . . . Then, if I go back to sleep, I start to get a sound sleep, that's when it kicks again."

3. The couple sometimes heard footsteps in the vacant hallway and stairs. The stairs were incredibly creaky, and the sound distinctive. "After a little while, you'll hear walking up and down the stairs. So I think it's my wife, getting up to go to the bathroom or something, and I don't see any lights on, so I get up to look, and I turn the lights on, and nothing." Monica, who said she

doesn't get up much at night, described "constant walking, up and down the stairs all through the night, from midnight until six in the morning."

4. At one point, Monica suggested that they take photos of the darkened house's hallways, rooms, and corners. "If you take pictures in the dark, you should be able to see things," Tom said, showing me a small stack of photos. While most of them were very ordinary, the couple pointed out three or four that seemed to show strange white orbs, pinpoints of light, and an eerie, inhuman face reflected in a tabletop. Tom showed the photos to a local radio psychic, who, he said, told him that his "house was full of ghosts."

5. While downstairs, the couple sometimes heard faint music or odd noises coming from upstairs. On October 30, prompted in part by mystery-mongering books and television shows, the pair investigated on their own, placing a tape recorder at the top of the stairs to record any ghostly sounds. They waited downstairs and did not hear anything at that time. But, upon listening to the tape, they heard one or more faint but distinct voices, various sounds, and what could have been a dog barking. I asked if it could be their dog, but Tom pointed out that theirs is a small breed (Pomeranian) that sounds different. He also dismissed the idea that passersby on the sidewalk might have made the noise, saying that people don't walk their dogs in his neighborhood at night, and besides, Monica and Tom couldn't hear street sounds from downstairs. "What I heard on the [audio] tape scared the shit out of me," Tom said—the conclusive evidence that convinced the couple (and Monica's parents) that the house was indeed haunted. That night, they left the house and refused to return. Many ghost investigators claim that they can record voices of the dead using microphones in empty (but supposedly haunted) areas.

6. Tom claimed that the family pets were afraid to enter his room and sometimes acted strangely. "The cat won't come in my bedroom. . . . He will stay right here [at the doorway], look, and walk right back out."

7. The couple complained of strange "cold spots" in the house and claimed that the upstairs would often be cold. "If you go upstairs, it's cold," Monica said. "My room is like the master bedroom of the house, and it's always like there's no heat up there. We always thought heat rises and the upstairs is always warmer than the downstairs."

INVESTIGATING THE HOUSE

After I got the basic facts of this case, Tom led me on a tour of their house. My first impression of the house was that, ghost or no, it was a strange and dark place. The house was not in great shape even though it had been extensively remodeled; some places had peeling paint, baseboards that were not flush, and minor cracks in the walls and ceilings. The upstairs, in particular, had unusual features, such as half-doorways leading to irregularly shaped storage spaces. The stairway was remarkably squeaky, straight out of a haunted-house film. Many of the rooms, especially those downstairs, seemed strangely dark even during the daytime.

We entered Monica's bedroom, and I asked for quiet so I could get a sense of the house's ambient sounds. There was a mild wind outside, and a few normal snaps and creaks could be heard. As I listened, I heard two muffled, mysterious scraping sounds coming from just below us, or just outside the window. Ghostly moans beyond the grave? Bloody fingernails on a rusty screen door? I looked through the window, but it was covered with opaque, plastic sheeting for insulation: I didn't see any figures moving. I stepped away from the window, and Tom went to look. "There's nobody there," he said soberly. Both of them looked spooked. I quickly walked out of the room, down the squeaky stairs, through the living room, down the porch, and into the front yard. Monica and Tom followed behind me. I asked them which side of the house we had just been on. Tom pointed to the left side of the house, and I peered back between the houses.

A maroon truck and a black car were parked in the driveway, with a low picket fence just beyond. I didn't see anything moving—no tree branches brushing against the roof, nothing that might have made the scraping sound. It seemed to be a genuine mystery, but a few moments' more patience paid off.

As I watched, an old woman's head and shoulders suddenly popped into view from behind the car. I then saw a rake in her hand and heard the distinct scraping again: a neighbor was raking leaves. When the "ghost" noticed me, I took a photo and gave a polite wave.

We went back inside, and I told them I'd like to experience the haunting and asked when they would be available to be in the house after dark. They said I could come over later that night. I headed home; in the twilight, a few people wandered the sidewalks and streets, including a young man walking a Dalmatian.

I returned to the house around 10:30 p.m. The night was unusually blustery, definitely appropriate for a haunted-house stakeout. (In fact, the windstorms that night knocked out power to many areas of Buffalo and the Eastern Seaboard.) Monica and her daughter had refused to return to their haunted home, but Tom was there to greet me. I told Tom to do all the things he usually does: turn on or off any lights that are usually on or off, close whatever doors are normally closed, etc. While we waited for any ghostly phenomena, I asked him more about his job. He'd been at the same factory job for thirteen years, driving a forklift at a nearby Ford plant. He often worked from 3 p.m.

until 11 p.m., though that had changed around June or July. At that time, he had adopted a new work schedule, 7 a.m. to 3 p.m. He had occasionally felt tapping on his feet at night but ignored it. Tom said that "It got a lot worse when I got the 7 to 3 shift."

Tom looked over at me between draws on his cigarette. "Can they follow you?" he asked. I told him that if there was something in his home, I didn't think that it would pursue him anywhere else; most likely it would remain confined to the house. If worse comes to worst, his family could just move out. I asked Tom what he thought was going on. "I think there's something going on, a spirit or something—a pissed-off spirit."

I suggested we head upstairs to listen quietly for sounds, if they were to come. "Oh, they will. They usually do," he said. We watched and waited as minutes turned to hours, but nothing happened: no ghostly voices, no creaking, vacant stairs, no unusual sounds or events. Tom seemed puzzled by the silence but offered no explanation. Finally, at just past 1 a.m., I left.

In the following days, I struggled with what to tell the couple. I had not seen any evidence of any paranormal activity, so far, but I knew they were genuinely scared. My presence helped put them at ease, but I hadn't explained the audiotape of the ghostly voices, the ghost's kick attack, or much else. I wanted the family back in their home by Thanksgiving, and the clock was ticking.

I had an idea for an experiment, but it would have to be done on a calm day, without the howling winds which had recently battered Buffalo.

MORE EXPERIMENTS AND INVESTIGATION

I returned to the house on the following Saturday. When I arrived, I had an audience; in addition to the family, Monica's mother was there, as were two of her friends. As I set up, I realized why the downstairs interior was often dim: the walls were painted light brown. Even with two lamps on, the place was quite dark, because the walls absorbed much of the ambient light. The walls had originally been white, but, because they showed dirt and imperfections, Tom had repainted with tan-colored paint he had found in the basement.

While the family sat around the living room, sometimes watching me curiously, I took measurements of the stairs and unpacked my camera gear. It was time for an experiment. I asked everyone to leave the house except for Monica, who sat in the living room. I instructed Tom to put the small tape recorder, in exactly the same position he did the evening of October 30. I turned it on, and we walked down the stairs and out into the driveway by the sidewalk below the windows that faced the staircase. We talked about the weather, and Tom commented that the conditions were exactly as they had been the night he recorded the audiotape: no wind and very calm. We spoke for about five minutes, then went back up to get the tape. I wanted to see if normal voices could be recorded from across the staircase, through the plastic-coated windows, and out toward the sidewalk.

I rewound it and played it in front of Tom and Monica. Indeed, our conversation was muffled but clearly audible. Though I hadn't analyzed the audiotape yet, the sounds and voices they recorded upstairs were almost certainly from beyond the house instead of beyond the grave. Aside from that, if it was a ghost or spirit, why would it create such mundane sounds as a barking dog or faint conversation, exactly mimicking ambient street noise?

The most likely explanation is that these normal sounds had been seen as strange in the context of other odd noises, feelings, and fear. The couple agreed with my explanation, but Tom reminded me that I hadn't yet explained the tapping he felt on his foot at night. "I want to see if you can explain that." I nodded as I scribbled notes. "I'm working on it," I said. "One thing at a time."

For my next experiment, I scooped up Scrappy, a pliant, orange calico cat, and took him with me upstairs to Tom's room. Tom, taking a cue from the many books on ghosts that he and his wife had consulted in the past weeks, said that the animals would refuse to go in the room or acted strangely if they did. With a camcorder running, I placed the cat in the middle of the room to see if he would hiss, scamper out, or react in any unusual way. Instead, the cat sat there looking at me and soon came over to be petted. If he sensed any supernatural spirits, Scrappy didn't show it as he purred and enjoyed a scratch behind the ears. Another mystery evaporated.

As I packed up, I told Tom and Monica that I would be willing to come back that night for one more ghost stakeout. Since I had been told that the steps and squeaking happen "constantly" throughout the night, it seemed that surely a second night in the haunted house should provide some evidence. Both said they'd like that, so, about seven hours later, I returned.

I arrived just before 11 p.m.; Tom met me at the door and led me into the living room where their daughter was asleep on a sofa, near Monica. Tom had slept there the night before without incident. This was the first time Monica had been in the house at night in the last two weeks. Both felt comfortable enough downstairs (on the sofa and floor) while I was there, but neither would sleep upstairs. Tom showed me a new light fixture he was installing at the base of the stairs, shining a bright light not only up the stairs but also into the hallway. I told him I thought that was a very good idea, as was contacting a carpenter or contractor to get rid of the stairway's noisy creaks and squeaks.

I sat on the couch, my camera and tape recorder at the ready. We all listened intently to the house sounds: the clock in the kitchen, the songbird clock chime, the fish tank bubbles and chirps. We sat in silence for an hour, until just past midnight. We made small talk, and the pair chain-smoked as the minutes ticked by. Tom asked me if I had looked more closely at the photos they had taken. I had taken the negatives to a professional developer and asked for a print of the image with the demonic face all the way to the edge of the frame. (When prints are made, the operators typically center the image, which can leave out important information.)

I explained that, if a person takes enough photographs, eventually he or she will find a few that apparently have odd lights or reflections. The photographs taken in the house did not show anything unusual or paranormal. There were a few anomalies, such as a few white specks and dots, but they were clearly common artifacts of the photography (such as flash reflections) and processing (such as specks of dirt). It is important to note that the camera was not of good quality, the photographer was an amateur, and the developing adequate but not of high quality. Some self-styled ghost hunters try to claim that the "orbs" are images of ghosts, but this is due to photographic inexperience and ignorance. (For more on orbs, see "The [Non]Mysterious Orbs," *Skeptical Inquirer*, September/October 2007.) The demonic face Tom and Monica had seen was clearly a simulacrum, as I showed him with an enlargement. People see such faces and images in everything from stains to dirty windows to clouds. I showed Tom that the demon's face was simply a lamp reflected off a glass tabletop.

"That could make sense," he admitted. "But I want you to make sense of the tapping I felt. . . . Everything else you can find an explanation for, but you're not going to find out about the tapping. You're not gonna find out [about] the kick in my bed." Monica sleepily pulled a blanket over herself.

I told them that I wasn't done yet, but that as far as I was concerned, they should not be worried about staying in the house. Though the incidents had been alarming and scary, no one had been harmed, and they apparently did not feel personally threatened. I walked to the top of the stairs and sat at the top landing, waiting for something unusual or paranormal to happen. After about fifteen minutes, the only thing I was feeling was chilly. I realized that there was indeed a cold spot near the top of the stairs, and I could feel cold air drifting past my waist. I announced I was coming down (so as not to startle them with the stairway sounds) and asked if they had any incense. Tom got off the floor and brought me a lit vanilla stick from the kitchen. I took it upstairs and held the incense at the bottom of the doorways. I shined my flashlight on the smoke stream, and it was clear that the cold was not a "cold spot" at all but instead simply a cold draft coming from the poorly insulated bedrooms. I then asked

Tom to join me and showed him my findings. He agreed that the "cold spot" was, instead, a draft, and marveled that a ten-cent stick of incense could be as good an investigative tool as a $10,000 infrared camera. As always, the value is in the methods, not the tools.

At one point, I left my chair at the top of the steps to retrieve a camera tripod. When I returned, Tom said he heard tapping right after I left. "A tap [on the wall] like letting you know, 'Yeah, I'm here.'" I had left my tape recorder on while I was gone, but when I listened to it, I didn't hear anything strange or unusual. Yet, even if something had been recorded, the sound could be anything; short of Morse Code, I'm not sure what would distinguish a "regular" tapping sound from one that seemed to be a message indicating, "Yeah, I'm here." Tom's mind was clearly interpreting ordinary sounds in extraordinary ways.

I waited another hour or so, finally leaving after 1 a.m. I had been ready and willing to experience and record the ghostly footsteps, but for the second time, the ghost activity had failed to materialize. Perhaps the spirits were just shy with a skeptic around, but without any evidence, there was little I could do. The creaking stairway and footsteps seemed to be the house settling, normal noises, products of imagination, or sleep-induced hallucinations. There was no evidence of anything else.

With mother and daughter asleep in their home for the first time in weeks, I packed up my gear and headed out. I told Tom I'd meet with them the following week, but for the time being, I thought everything would be fine. As I left, Tom suggested that maybe the spirits had gone, that it was over. I heartily agreed, suggesting that, if they did hear anything, they should just try to ignore it. If something was there at one point, it certainly didn't seem to be around anymore.

I called the following Monday and was told that no further activity had occurred. They felt the haunting might be over. I went to the house one last time to discuss the results of my investigation point by point. The kicking and the tapping were among the most puzzling phenomena and first on the agenda. The tapping on Tom's feet was most likely his imagination, a medical condition, or both. The fact that it invariably happened when he was going to sleep—or, by Tom's own account, actually asleep—is significant. This might be considered a hypnagogic hallucination (a sensory illusion that occurs in the transition to sleep), a fairly common phenomenon that can easily lead to misperceptions.

Tom was particularly disturbed when his bed was "kicked" by an unseen entity. I examined the metal-framed twin bed and found it to be fairly light-weight (Tom had carried it into his wife's room by himself). It is significant that the bed only jerked at night when Tom was in the bed and going to sleep. Several days into my investigation, I discussed the case with a police officer, who said that, when he is drifting off to sleep, his arm or leg will often twitch or jerk. In fact, his wife told me that, just the previous evening, her husband's leg had twitched as he was going to sleep and had not only shaken the bed but woken her up. It is not unusual for people to do that in bed; I have done it myself. It seems likely that Tom, as he drifted to sleep, had simply twitched or jerked, causing the bed he was on to shake. Tom, at over 250 pounds, was a big man, and a leg spasm could easily shake the lightweight bed. The fact that Tom has sleep apnea is even further evidence for this explanation; restless legs (restless leg syndrome) is actually one of the most common symptoms of apnea.

The ghostly activity coincided with a change in Tom's work (and, therefore, sleep) schedule. The strange events were infrequent and minor until Tom told his wife about what he had been experiencing. "When we admitted it to each other, it got tremendously worse," Monica said. With my background in psychology, this struck me as very strong evidence for a psychological explanation. It seems a near-classic case of each person reinforcing the other's expectations and interpretations of strange events. The fact that all this occurred right before Halloween might also be significant; it is the season when ghosts, spooks, witches, and spirits are in the public's consciousness. Tom said that the haunting had stopped, and they would be spending Thanksgiving in their own home.

People misunderstand and misperceive things all the time; it doesn't mean that they are stupid or crazy, just that they do not necessarily know what to look for. Much of this case was simply a collection of unrelated and mundane phenomena that, taken together and in the context of a possible haunting, seemed to be evidence for supernatural activity. Monica and Tom tried their best to understand and explain them. When that failed, they tried to record the events—but misinterpreted their findings. Normal ambient noise became mysterious ghostly voices; normal photographic glitches and artifacts became unexplained lights and faces.

In a very real way, Tom and Monica haunted their own home. As the weeks passed, Monica sought out books about ghosts and hauntings, none of which were skeptical or science-based. These served as a haunting blueprint for the couple and told them what signs to look for in a haunted home, and sure enough, they found them.

This terrified family and their (non)haunted house is a result of a society in which those seeking answers to strange phenomena are directed to the local bookstore or library shelf, only to find mystery-mongering misinformation by the ream and few skeptical resources. Popular books, television shows, and even a local "psychic" all made the problem worse.

In a Halloween 2007 interview on ABC's *Nightline,* Grant Wilson of the TV show *Ghost Hunters* said, "I don't care at all what the skeptics think of what we're doing because they don't need help. There are people who need help in their homes. Who's helping them? Are the skeptics going to help them? No!"

In fact, it's often skeptics who help families terrified by misinformation from "experts" such as the Ghost Hunters. In this case, and in many others, it is science and skepticism that solved the mystery and made a real difference in these people's lives.

NOTE

1. Respecting a request for anonymity, names have been changed except for the cat. A more detailed examination of this case is available online at www.csicop .org/specialarticles/.

THE DEVIOUS ART OF IMPROVISING,
Lessons One and Two

Massimo Polidoro

The great fake psychics are great improvisationists. This means that a really good pseudo-psychic is able to produce phenomena under almost any circumstance. A quick mind and a good knowledge of the techniques and psychology of deception are all that is needed. Sometimes, only a quick mind is enough.

In one early test of telepathy, in 1882, pseudo-psychic G. A. Smith and his accomplice, Douglas Blackburn, were able to fool researchers of the Society for Psychical Research. In a later confession, Blackburn described how they had to think fast and frequently invent new ways of faking telepathy demonstrations. Once, for example, Smith had been swathed in blankets to prevent him from signaling Blackburn. Smith had to guess the content of a drawing that Blackburn had secretly made on a cigarette paper. When Smith exclaimed, "I have it," and projected his right hand from beneath the blanket, Blackburn was ready. He had transferred the cigarette paper to the tube of the brass projector on the pencil he was using, and when Smith asked for a pencil, he gave him his. Under the blanket, Smith had concealed a slate coated with luminous paint, which in the dense darkness gave sufficient light to show the figure on the cigarette paper. Thus he only needed to copy the drawing.

I was lucky enough to learn the art of improvising from one of the greatest "teachers" on the subject, the Amazing Randi. I had met him only a few hours before, nearly twenty years ago, and he was already teaching me how to conduct a perfect swindle!

A PUZZLE

Randi had come to Italy to help us promote CICAP, the Italian Committee for the Investigation of Claims of the Paranormal, and he was expected to be on a talk show in Rome to discuss his work and talk about the Committee. The host, an actress called Marisa Laurito, asked him what he was going to do in front of the cameras, and he said that he planned to duplicate a drawing made by her in secret. She agreed and asked what was needed.

"Just some paper and some envelopes," said Randi.

"Chiara," said Marisa, addressing her secretary, "please, go and get those things from the office."

Randi shot a glance at me and said, "Massimo, maybe you should accompany her, to see if they are the right size."

The right size? I did not know what the right size was; I had never seen him perform up close, and I could not imagine what was needed. But, as soon as I was going to open my mouth, Randi smiled and said, "Go, Massimo, please," pushing me ahead.

I went out the door, following the woman, and a moment later, Randi came out as well and shouted at me: "Massimo! I am sorry, while you go, please throw away this junk I had in my coat."

There was a wastebasket in Marisa's dressing room; why did he need me to throw things away? However, he was my hero, and I was glad to help. I immediately went back to collect some scraps of paper and used train tickets, and, under his breath, Randi said to me, "When you are in the office, just get a few envelopes and sheets of paper without her seeing you, and then come back here. Now, go!"

I was quite confused, but I did as he asked. I made some gracious comments about the woman's blue eyes while we were in the office, and a little chitchat later, I had some sheets and envelopes hidden under my jacket without her knowing it.

When we got back to Marisa's dressing room, she and Randi were laughing at something he was telling her.

"Good, now here you are," Randi said to the secretary. "I do not want to touch anything," he stated, raising his hands up in the air, like a surgeon ready to operate. "Please give this stuff to Marisa."

The girl obeyed, and Randi continued: "Now, Marisa, please go to another room, the bathroom will be fine, close yourself inside and draw whatever you like on that piece of paper. When you are done, fold the paper and seal it in one of the envelopes."

As soon as she closed the bathroom door, Randi addressed her assistant. "Er...Chiara, I am sorry, could I have a glass of water? I need to take my medication."

"Sure," she said, and went out of the room. Now we were alone.

"Quick!" said Randi. "Give me the things you took in the office."

Randi took one of the blank sheets of paper, folded it in thirds, and placed it in an envelope, which he sealed and then put in his jacket's inner pocket.

I was more and more confused. "Mr. Randi," I said, "would you please tell me what this is all about?"

"Later, now she's coming."

Sure enough, the door of the bathroom opened up and Marisa was out, waving her envelope. "Here it is! Now, what do we do?"

Randi looked puzzled. "Hmmm . . . you know, that envelope doesn't convince me . . . excuse me." He took the envelope from her, even though he had said that he was not going to touch anything. Holding it quite high, with just two fingers, as if it might be contaminated, Randi approached the window.

"Please, I won't look. Tell me if you can see through the envelope, Marisa."

"No, it's quite thick. You can't see anything."

"Good! I do not want anyone to think that I merely saw your drawing through it. Well then, keep your envelope with you all the time now. Don't tell anyone what you drew and then, when we are on stage, keep hold of the envelope until I have made my guess. You will agree that there is no way for me to know what's in it."

"Quite right!"

"Good. If you think so, just say it when we are on air. Then, if I am able to correctly guess your drawing, this will mean that I have some extraordinary ESP ability—or I will have showed you that what I do is quite indistinguishable from real ESP. Ergo, the public should always doubt these kinds of demonstrations, unless there is someone really expert in this kind of thing checking everything out." I noticed that he did not use the word *tricks.* "Now, if you will excuse us, I need to get some rest before we start. I have arrived only a few hours ago from Miami and I am still jet-lagged."

It wasn't true, he had arrived about a week earlier, but as I would discover, he needed some time in private.

THE SWITCH

"Mr. Randi," I said as we were walking down the empty hall, "can you explain to me what we are doing? Why did you place an envelope in your pocket?"

"You mean this one?" he said, taking it out of his jacket.

"Yes. How can a blank sheet of paper be of some help in. . . ?"

The words died in my mouth as he opened up the envelope and I saw a drawing on it: a very simple pencil drawing showing a house and a cat!

"What the—?"

"Later. Now get in our dressing room."

When the door closed, Randi took a good look at the drawing.

"Quite simple, isn't it?"

"Do you mean that this one—"

"—is the drawing that Marisa drew, yes," he finished. He was clearly amused by the look on my face. "You wonder when I took it, right? Well, there was really no need to check if the envelope could be seen through, it's really thick and I probably looked a bit dumb to her by asking that question. But I needed to have the envelope with her drawing in my hands for just a few seconds, in order to do the switch."

"You mean. . . ?"

"Yes, when I approached the window, I turned my back to you all for just an instant, but that was enough for me to throw her envelope inside my jacket and take out the dummy one."

"But I didn't see you do it!"

"Well, thank you. That was the point."

Quite ingenious, I thought. I was soon going to learn that the best way to duplicate a drawing sealed in an envelope (and so far nobody has shown that another way exists) is to "somehow" secretly get a look at the drawing. That's all there is to it. It doesn't matter how: switching envelopes, looking at a reflection in a mirror, watching the pencil move on paper, have an accomplice take a peek. What matters is that you know beforehand what's inside that envelope. Well, most of the time: Randi has been able to go even further than this, but that is another lesson.

CONFUSION

Now, the problem was that Marisa had in her possession an envelope containing a blank sheet of paper: what was Randi going to do?

"Well, now that I know what she drew, I need to give this back to her . . . without her realizing it."

So he placed the drawing in another similar envelope—that's why he had asked me to get "a few" of those—and put it back again in his jacket pocket.

"Now, we only need to wait."

"Wait? Wait for what?"

"For the show to begin."

"Do you mean that you are going to do the switch live on camera?"

"Of course not, but I need her to be distracted a bit more now, and so we will wait just five minutes before the show starts. She will have a thousand things to think about, and will not have much time for me."

And that's what he did. Five minutes before showtime, Randi knocked on Marisa's dressing-room door just as she was coming out with all her assistants, producers, writers, coiffeur, and make-up artist all buzzing around her like she was the queen bee.

"I am sorry, Marisa" he said with a smile. "But while I was resting, I had this great idea. Let's put your drawing in one of the bigger envelopes there on the table. This way, we can show that your drawing was really impossible to see and the effect will be much stronger."

She had many people and distractions around her. "Yes . . . well, whatever you say. Here is my drawing, where should I put it?"

"Here," said Randi, taking her envelope and placing it in a bigger envelope. "Now we can seal it and you can put your signature on it. This will really shock the viewers!"

"Okay, just let's move on, we are about to start."

She signed her name on the envelope and then took it along with her.

RESOLUTION

We remained alone, again, in the dressing room. I stared at Randi, and said, "Now that that didn't work, what will you do?"

"What do you mean it didn't work?"

I was silent for a minute. "But you never had a chance. . . . When did you do the switch? It was impossible."

Randi chuckled. "Okay, okay, I will tell you. When we got in, with all those people and the confusion, I quickly put the envelope with her drawing inside one of the bigger ones that were resting on the table. Then, when she gave me the envelope with the blank sheet—and, of course, she thought that it contained her drawing—I acted as if I was placing it inside the envelope, but, actually, I was putting it behind it. So, when I placed the whole thing on the table to have her seal and sign it, on top of all the other envelopes lying there, the thing was done: the drawing was already inside, and the envelope with the blank one was mixed with all the other envelopes. In fact, here it is."

He took a sealed envelope from the lot and put it back in his jacket. It was later destroyed to avoid even the remotest risk of someone discovering the trick.

Of course, later on, when the show started and Randi joined Marisa on stage, all went perfectly. Marisa told the viewers how she had kept hold of the envelope the whole time, and when Randi—after much concentration, grinning, and sweating—duplicated her drawing, she was flabbergasted.

For me, that was the first important lesson I got from Randi: real tricksters rarely read magic books and magazines, they just invent their methods along the way, quickly improvising something on the spur of the moment.

I was going to find that out at my expense very soon, as we shall see in the next "lesson."

Harry Houdini, discussing fraud detection, once wrote in his book *A Magician Among the Spirits* that, "It is manifestly impossible to detect and duplicate all the feats attributed to fraudulent mediums who do not scruple at outraging propriety and even decency to gain their ends. . . . Again, many of the effects produced by them are impulsive, spasmodic, done on the spur of the moment, inspired or promoted by attending circumstances, and could not be duplicated by themselves."

In the first half of this chapter, I detailed my initiation, through James Randi, to the art of creating irreproducible feats by "impulsive, spasmodic" and "on the spur of the moment" techniques.

The first time I put to practice what I had learned came very soon after that episode. I had been invited on a late night talk show to discuss psychic phenomena and paranormal claims. The host of the show, Italian political commentator Giuliano Ferrara, asked me if I could do something to show how it is possible to obtain seemingly paranormal phenomena through trickery.

"However," he specified, "I am not going to be a stooge or help you with any trick. If you say you are able to demonstrate something that is indistinguishable from a psychic demonstration, then you should do it without my help."

I said that was fair enough, and suggested he make a drawing on a piece of paper and then I would try to duplicate it live on the show.

"Great!" he said. I was ready.

Or so I thought.

TAKING THE CHANCE

I had learned through various books on magic and mentalism many ways of duplicating a drawing done by someone who is hiding it. You can look at the movement of the pencil above the paper and guess the shape he is creating. In some cases, you can even guess a shape just by listening to the sound of the pen on the paper. Or you can look at the impression left by the tip of the pen on the piece of paper below the one being drawn. Or you could try to take a peek between your fingers while you pretend to cover your eyes. Or you could have an accomplice strolling around and taking a look at the drawing on your behalf. Or, as Randi did in Rome, you can switch an envelope . . . and so on.

Ferrara, however, took a single sheet of paper, asked me to turn my back to him and made his drawing very quickly, leaving me no chance to use some of the ruses I was ready to put in practice.

When I turned around, he had already folded his sheet of paper and placed it inside his jacket pocket. He used no envelopes.

"Good" I said, while I was actually thinking: "Bad. Really bad!" I had not been able to stay in charge of the situation and had let him take charge and make the rules.

A fatal mistake.

"Well, I will see you later on the show" he added, in fact, waving me goodbye.

"Oh no!" I thought. "This can't happen! He can't leave: I have no idea what he drew!"

And this is where Randi's lesson on improvising suddenly struck me. An idea popped into my mind.

"Right . . . ehm, Mr. Ferrara, I just thought something. . . ."

"What is it?"

"Well, did you make your drawing with a pen or a pencil?"

"With a pencil, why?"

"Ah! You see, I think that if you hold up a white sheet of paper to the camera with a drawing done in pencil, it will be very hard for the viewers to see what you drew. . . ."

He thought about it for a few seconds, and then agreed. "You are right! Here . . .," he said, taking another piece of paper. This time, he used a felt tip marker and did another drawing.

When he was done, he put the new drawing in his pocket and *took the other drawing, crumpled it up and threw it in a waste paper basket!*

And then he left.

Well, I thought. Now, this is very interesting. . . .

For some inexplicable reason, the papers I was holding suddenly fell on the floor and I had to kneel down in order to collect them. When I stood up I had a crumpled up piece of paper hidden between my documents.

I immediately went to my dressing room and opened up the paper. There was a real risk in all this: Ferrara may have well decided to do a different drawing the second time. However, when I saw what he drew I knew for sure that he had not changed his mind, for he had drawn a hammer and sickle. Since Ferrara had been a left-wing supporter for years but had just recently adopted a conservative position, I thought that by drawing the symbol of the Communist party he was planning to make a joke of some kind while on air.

And, in fact, this is exactly what happened. During the show, after some deep concentration, I drew a hammer and a sickle and he showed to the audience that it was exactly what he had drawn.

It is true that luck played an important part in all this, but as the saying goes, fortune helps the daring.

But this is not always the case, and this is exactly what happened to Randi in a memorable television appearance some years ago.

THE BELLY WRITER

Randi told me he had been invited by Barbara Walters to appear on her television show *Not for Women Only.* She had already had Uri Geller as a guest a few months earlier and he had worked wonders on her: she had become a believer.

When the time came for an episode of the show devoted to "Magic and Magicians," Walters invited three very prominent magicians of the time: Doug Henning, then star of Broadway's "The Magic Show," Mark Wilson, one of the first TV magicians, and Randi himself.

During the show, the discussion turned to Uri Geller and to an incredible telepathy test he had done for her. He had divined what drawing Barbara had drawn and had hidden inside three envelopes and then kept inside a book. She wanted to see if Randi was up to his claim that he could reproduce everything that Geller did, so she said: "Just before the program, I did a drawing. I put it in three envelopes in this book. We have not had this envelope out of our sight. Randi has in no way been close to it. The closest you've been to it, to my knowledge, is right now. Reproduce it."

"Now," Randi explained to me, "her words were totally true. I never ever had a chance to get close to the drawing, as hard as I tried. I doubt they used the same strict conditions on Geller, but the fact remains that that was the very first time I saw the envelope."

So, without blinking, Randi acted as if everything was going perfectly. "Please put it back in the book. I want to be sure it's under control."

She obeyed and then Randi turned to his fellow magicians. "Now watch this, it may surprise you."

Both Henning and Wilson were staring, quite curious to see how Randi would escape from such an impossible situation.

Randi took out a pen and started to draw something on a piece of paper, without showing it to anyone. When he was done he put the drawing, face down, on the table. Put the cap back on the pen and put the pen on the table. "Now let's see your drawing" he asked Walters.

She was already seeing her illusions go down the drain, because Randi's confidence was so powerful that it could mean only one thing: somehow he had reproduced the drawing correctly.

"Oh no, I can't believe it" she said even before she had seen Randi's guess.

Walters took the envelope out of the book. She opened one envelope, opened the second, and when she was ready to open the third, Randi stopped her to ask if her drawing could be seen through it. She checked, but it could not be seen.

The final envelope was opened and the drawing shown: it was a house with a little girl standing outside, a chimney, smoke coming therefrom, and a doorway. Randi turned his drawing to the camera. It was a bit crude in style, but it showed exactly the same thing, the house, the smoke and the little girl (standing inside, rather than outside).

Both the magicians and Walters agreed that Randi truly was amazing.

Now, how did he do it? Randi had described this demonstration in his book *The Truth About Uri Geller* but had never revealed it in print. So, when he showed me the tape a few years ago, I immediately asked him about it.

"During the show, while Barbara was talking or we were watching some footage on Geller, I surreptitiously opened the pen and took out the cartridge from it. Then, I closed the pen again and put it on the table. I then inserted the cartridge in my belt, with the writing side upward. Now it was just a question of acting. I mimed the action of drawing, with the non-writing pen, and put the blank paper face down. When she started to open the envelopes, I could see through the last one a bit of the drawing, I saw a house and a stylized girl. So I took the paper again with both hands, and asked her to check if the envelope was transparent. It was just an excuse to gain some time for me to draw, as best as I could given the situation, a replica of her drawing. I simply placed the paper against the tip of the cartridge and started to slowly move my belly in order to make the drawing. Yes, the final product was not perfect, but the result I got from her reaction was fantastic, and that is all that matters. To the viewers, and to my fellow magicians, I was able to reproduce exactly what Geller had done, even though the conditions had been quite bad for me!"

Randi had invented this ingenious system on the spur of the moment! He has done this time again throughout his career.

A PATENTLY FALSE PATENT MYTH—STILL!

Samuel Sass

For close to a century there has periodically appeared in print the story about an official of the US Patent Office who resigned his post because he believed that all possible inventions had already been invented. Some years ago, before I retired as librarian of a General Electric Company division, I was asked by a skeptical scientist to find out what there was to this recurring tale. My research proved to be easier than I had expected. I found that this matter had been investigated as a project of the D.C. Historical Records Survey under the Works Projects Administration. The investigator, Dr. Eber Jeffery, published his findings in the July 1940 *Journal of the Patent Office Society*.

Jeffery found no evidence that any official or employee of the US Patent Office had ever resigned because he thought there was nothing left to invent. However, Jeffery may have found a clue to the origin of the myth. In his 1843 report to Congress, the then-commissioner of the Patent Office, Henry L. Ellsworth, included the following comment: "The advancement of the arts, from year to year, taxes our credulity and seems to presage the arrival of that period when human improvement must end." As Jeffery shows, it's evident from the rest of that report that Commissioner Ellsworth was simply using a bit of rhetorical flourish to emphasize that the number of patents was growing at a great rate. Far from considering inventions at an end, he outlined areas in which he expected patent activity to increase, and it is clear that he was making plans for the future.

When Commissioner Ellsworth did resign in 1845, his letter of resignation certainly gave no indication that he was resigning because he thought there was nothing left for the Patent Office to do. He gave as his reason the pressure of private affairs, and stated, "I wish to express a willingness that others may share public favors and have an opportunity to make greater improvements." He indicated that he would have resigned earlier if it had not been for the need to rebuild after the fire of 1836, which had destroyed the Patent Office building. In any case, the letter of resignation should have put an end to any notion that his comment in the 1843 report was to be taken literally.

Unfortunately, the only words of Commissioner Ellsworth that have lived on are those about the advancement of the arts taxing credulity and presaging the period when human improvement must end. For example, the December 1979 *Saturday Review* contained an article by Paul Dickson titled "It'll Never Fly, Orville: Two Centuries of Embarrassing Predictions." He lists a page full of "some of the worst wrong-headed predictions." Ellsworth's rhetorical sentence is included with such laughable statements as that said by Napoleon to Robert Fulton: "What sir, you would make a ship sail against the wind and currents by lighting a bonfire under her decks? I pray excuse me. I have no time to listen to such nonsense."

If in the case of Commissioner Ellsworth there was at least a quotation out of context on which the "nothing left to invent" story was based, a more recent myth attributing a similar statement to a commissioner who served a half-century later is totally baseless. This news story surfaced in the fall of 1985, when full-page advertisements sponsored by the TRW Corporation appeared in a number of leading periodicals, including *Harper's* and *Business Week*.

These ads had as their theme "The Future Isn't What It Used to Be." They contained photographs of six individuals, ranging from a baseball player to a president of the United States, who had allegedly made wrong predictions. Along with such statements as "Sensible and responsible women do not want to vote," attributed to President Cleveland, and "There is no likelihood man can ever tap the power of the atom," attributed to physicist Robert Millikan, there is a prediction that was supposedly made by Commissioner of the US Patent Office Charles H. Duell. The words attributed to him were: "Everything that can be invented has been invented." The date given was 1899.

Since I was certain that the quotation was spurious, I wrote to the TRW advertising manager to ask its source. In response to my inquiry, I received a letter referring me to two books, although I had specifically asked for the *primary* and not secondary sources. The books were *The Experts Speak*, by Christopher Cerf and Victor Navasky, published in 1984 by Pantheon, and *The Book of Facts and Fallacies*, by Chris Morgan and David Langford, published in 1981 by St. Martin's Press.

When I examined these two volumes I found that the 1981 Morgan and Langford work contained Commissioner Ellsworth's sentence about the advancement of the arts taxing our credulity, although the quote was somewhat garbled. It also contained the following comment by the authors: "We suppose that at just about any period in history one can imagine, the average dim-witted official will have doubted that anything new can be produced; the attitude cropped up again in 1899, when the director of the US Patent Office urged President McKinley to abolish the office, and even the post of director," since "everything that can be invented has been invented." The authors do not give the name of the commissioner whom they call "director," but it was Charles H. Duell who held that office in 1899. They don't offer any documentation to support that alleged statement, and they would have had a tough time finding any.

It's easy enough to prove that Duell was not the "dim-witted official" so glibly referred to. One need only examine his 1899 report, a document of only a few pages, available in any depository library. Far from suggesting to the president that he abolish the Patent Office, Duell quotes the following from McKinley's annual message: "Our future progress and prosperity depend upon our ability to equal, if not surpass, other nations in the enlargement and advance of science, industry and commerce. To invention we must turn as one of the most powerful aids to the accomplishment of such a result." Duell then adds, "May not our inventors hopefully look to the Fifty-sixth Congress for aid and effectual encouragement in improving the American patent system?" Surely these words are not those of some kind of idiot who believes that everything has already been invented. Other information in that report also definitely refutes any such notion. Duell presents statistics showing the growth in the number of patents from 435 in 1837 to 25,527 in 1899. In the one year between 1898 and 1899 there was an increase of about 3,000. It's hardly likely that he would expect a sudden and abrupt ending to patent applications.

The other book cited by the advertising manager of TRW, Inc., *The Experts Speak*, by Cerf and Navasky, offers a key to how myths are perpetuated. This volume, published three years after the Morgan and Langford work, contains the spurious Duell quote, "Everything that can be invented has been invented," and prints it as though it had formed part of the commissioner's 1899 report to President McKinley. However, unlike the earlier work, *The Experts Speak* contains source notes in the back. The source given reads as follows: "Charles H. Duell, quoted from Chris Morgan and David Langford, *Facts and Fallacies* (Exeter, England, Webb & Bower, 1981), p. 64." Unlikely as it is for the head of the US Patent Office to have said something so silly, evidently it did not occur to Cerf and Navasky to question that statement. They simply copied it from the earlier book. One can expect that in the future there will be more copying because it is easier than checking the facts.

The irony is that the subtitle of *The Experts Speak* is "The Definitive Compendium of Authoritative Misinformation." One can only wonder how much more misinformation is contained in this nearly 400-page compendium. On the title page the book is described as a "joint project of the *Nation* magazine and the Institute of Expertology." Whatever this institute may be, on the theory that the *Nation* is a responsible publication, I wrote to Mr. Navasky, who is editor of that magazine and coauthor of the book, to ask if he could tell me where and when Commissioner Duell made the stupid statement attributed to him. I did not receive a reply.

ADDENDUM TO ORIGINAL ARTICLE

The earliest appearance of the patent myth in print that I am aware of is the October 16, 1915, issue of the *Scientific American*. It contains the following item: "Someone poring over the old files in the United States Patent Office in Washington the other day found a letter written in 1833 that illustrates the limitations of the human imagination. It was from an old employee of the Patent Office, offering his resignation to the head of the department. His reason was that as everything inventable had been invented the Patent Office would soon be discontinued and there would be no further need of his services...."

As in all "urban legends," the details are vague. Neither "the old employee" who is supposed to have resigned nor the "someone poring over the old files" is identified. The fact is that when the Patent Office burned to the ground in 1836, all records were destroyed so that even if that 1833 letter had ever existed it could not have been seen "the other day" in 1915.

With the imaginative addition later of the names of Commissioners Ellsworth and Duell, the myth kept cropping up sporadically for decades but then received a major boost twenty years ago by the publication of the Morgan and Langford book in 1981 and the Cerf and Navasky book three years later. The 1985 TRW ad, which made use of the misinformation in these volumes, helped spread it more widely.

A particularly discouraging aspect of the repetition of this fable is that it is repeated by individuals who have research and fact-checking facilities at their disposal. For instance, in his book *The Road Ahead*, published in 1995, Bill Gates relates the myth as fact. He was chided for that in the February 1996 issue of *Scientific American* in these words: "If Bill Gates's grasp of the past is any guide, readers should take his visions of the future with a dose of skepticism." In view of the fact that eighty years earlier the *Scientific American* had made the same mistake it's encouraging to know that the editors felt compelled to correct Gates.

Two other prominent individuals who failed to take a second look are Carol Browner, former Administrator of the Environmental Protection Agency, and Hugh Downs, radio and television journalist. The former repeated the myth in an October 2000 talk to the National Press Club, and Downs stated in his July 17, 1996, "Perspective" radio commentary spiel that in "a handwritten note" Commissioner Duell had urged President McKinley to abolish the Patent Office because "Everything that can be invented has been invented." He was evidently so pleased with himself for being a purveyor of that news that he ended by telling his audience, "Remember that you heard it here first, on radio."

It would be the height of optimism to believe that efforts to debunk this myth will cause it to disappear. It's too good a story and lends itself too readily to those who are eager to make a point and to whom facts and truth are secondary.

CONTRIBUTORS

MARIO BUNGE, a physicist and philosopher, is the Frothingham Professor of Logic and Metaphysics at McGill University and author of more than 50 books and 500 papers in physics and philosophy. Among them are *The Methodological Unity of Science, Understanding the World, Treatise on Basic Philosophy, Philosophy of Science* (two volumes), *The Mind-Body Problem, Causality and Modern Science, Scientific Research,* and *Chasing Reality: Strife Over Realism.*

THOMAS R. CASTEN, an energy entrepreneur and energy policy analyst, is founding chairman of Recycled Energy Development LLC, of Westmont, Illinois, whose mission is to profitably reduce greenhouse gas emissions by recycling wasted energy. He is author of *Turning Off The Heat: Why America Must Double Energy Efficiency to Save Money and Reduce Global Warming* (Prometheus, 1998). Companies he has led have invested more than $2 billion in 200-plus energy recycling projects that save users $500 million per year while reducing CO_2 by 5 million tons per year—$100 per ton of savings of avoided CO_2.

CLARK CHAPMAN is a planetary scientist at the Southwest Research Institute in Boulder, Colorado. As he and Alan Harris study the hazards of asteroids crashing into the Earth, they have also looked at other societal threats, large and small, including that from terrorists, and have analyzed how society responds to perceived threats and fears. They believe that their early controversial analysis of 9/11 in this chapter has been bolstered by subsequent events.

P. MICHAEL CONN is a senior scientist of the Oregon National Primate Research Center, as well as a professor of physiology and pharmacology and of cell biology and development at Oregon Health Sciences University. He and James V. Parker wrote about their experience as targets of animal rights terrorism in *The Animal Research War* (Macmillan/Palgrave, 2008), a book that also traces the evolution of the animal rights movement, profiles its leadership, and examine the effects of violent protest on science and scientists.

BRENNAN DOWNES is a senior project development engineer with EPCOR Power USA.

ANN DRUYAN is a co-writer with the late Carl Sagan of the Emmy and Peabody award-winning series *Cosmos*. Their twenty-year professional collaboration included NASA's Voyager Interstellar Message and many speeches, articles, and books, including *Shadows of Forgotten Ancestors* and *Comet*. She was co-creator with Sagan of the motion picture *Contact*, and a credited contributor to his *Pale Blue Dot*, *The Demon-Haunted World*, and *Billions and Billions*. She is co-founder and CEO of Cosmos Studios, as well as Program Director of the first solar sail spacecraft mission. She and Carl Sagan were married until his death in 1996. They have two children.

BRUCE L. FLAMM, MD, is a clinical professor of obstetrics and gynecology at the University of California, Irvine. He is the author of several medical books, book chapters, and research articles. In addition to his work in the medical field he is an expert in the history of calculating devices and has co-authored a book on the subject.

BARBARA FORREST is the co-author with Paul R. Gross of *Creationism's Trojan Horse: The Wedge of Intelligent Design* (Oxford University Press, 2004, 2007), which details the political and religious aims of the intelligent design creationist movement. She served as an expert witness for the plaintiffs in the first legal case involving intelligent design, *Kitzmiller et al. v. Dover Area School District*, which was decided in favor of the plaintiffs in December 2005. She is a Professor of Philosophy in the Department of History and Political Science at Southeastern Louisiana University.

KENDRICK FRAZIER, the editor of this volume, is the longtime editor of the *Skeptical Inquirer* and a fellow of the American Association for the Advancement of Science. He is a former editor of *Science News*, and has also been an editor at Sandia National Laboratories and the National Academy of Sciences. He is author of four science-related books and editor of five previous *SI* anthologies. He is a fellow of the Committee for Skeptical Inquiry, a member of its Executive Council, and a recipient of its In Praise of Reason Award.

MARTIN GARDNER is author of numerous books about mathematics, science, pseudoscience, philosophy, and literature. Among them are the classic *Fads and Fallacies in the Name of Science, The Ambidextrous Universe, Order and Surprise, Science: Good, Bad, and Bogus, Weird Water and Fuzzy Logic, The Night Is Large, Did Adam and Eve Have Navels?, Are Universes Thicker Than Blackberries?*, and *The Jinn from Hyperspace*. His latest is *The Fantastic Fiction of Gilbert Chesterton*.

HARRIET A. HALL, MD, Colonel, USAF (Ret.), also known as The SkepDoc, is a retired family physician living in Puyallup, Washington. She is the author of *Women Aren't Supposed to Fly: The Memoirs of a Female Flight Surgeon*. A contributing editor to both *Skeptical Inquirer* and *Skeptic* magazines, an editor of *The Scientific Review of Alternative Medicine*, and a medical advisor to the Quackwatch website, she also writes a weekly article for the Science-Based Medicine website.

ALAN W. HARRIS is a planetary scientist for the Space Science Institute, Boulder, Colorado, although he lives in La Cañada, California. As he and Clark Chapman study the hazards of asteroids crashing into the Earth, they have also looked at other societal threats, large and small, including that from terrorists, and have analyzed how society responds to perceived threats and fears. They believe that their early controversial analysis of 9/11 in their chapter in this volume has been bolstered by subsequent events.

RAY HYMAN is professor emeritus of psychology at the University of Oregon. His published research has been in such areas as pattern recognition, perception, problem solving, creativity, and related areas of cognition. He has written and published extensively on the psychology of deception and critiques of paranormal and other fringe claims.

STANLEY JEFFERS is an Associate Professor in the Department of Physics and Astronomy of York University in Toronto. His graduate work was in

Optoelectronics at Imperial College. After a year as a Post Doctoral Fellow in the Department of Astronomy at the University of Toronto, he joined the faculty at York University. His interests have included astronomical instrumentation, observational astronomy, and some aspects of theoretical physics.

JOHN E. JONES III is a federal judge on the United States District Court for the Middle District of Pennsylvania. He presided in the landmark *Kitzmiller v. Dover Area School District* case, in which he ruled the teaching of intelligent design in public school science classes was unconstitutional. A Republican lawyer, Jones was appointed to his post by President George W. Bush in 2002.

STUART JORDAN is a Senior Staff Scientist (retired emeritus) at the NASA Goddard Space Flight Center. He holds a PhD in physics and astrophysics from the University of Colorado. He is a Rhodes scholar.

RICHARD JUDELSOHN, MD, is a pediatrician in urban/suburban practice for 30 years; Medical Director, Erie County Department of Health; and Associate Professor of Pediatrics, University of Buffalo School of Medicine.

PAUL KURTZ is professor emeritus of philosophy at the State University of New York at Buffalo; chairman and founder of the Center for Inquiry; chairman of the Committee for Skeptical Inquiry; and editor of *Science and Religion, Science and Ethics,* and numerous other works.

CHRIS MOONEY is science journalist who writes on the intersection of science and politics and a contributing editor of *Seed* magazine. He is author of *The Republican War* on *Science* and *Storm World: Hurricanes, Politics, and the Battle Over Global Warming.*

NICOLI NATTRASS is Director of the AIDS and Society Research Unit at the University of Cape Town. She has published books and articles on AIDS denialism, the scientific governance of medicine, and the political-economy of South African AIDS policy.

JOE NICKELL is senior research fellow of the Committee for Skeptical Inquiry and Skeptical Inquirer's "Investigative Files" columnist. A former stage magician, private investigator, and academic, he is author of numerous books, among them *Adventures in Paranormal Investigation, Real-Life X-Files, Relics of the Christ,* and *Looking for a Miracle.* His Web site is at www.joenickell.com.

STEVEN NOVELLA, MD, is Assistant Professor of Neurology, Yale University School of Medicine. He is host of The Skeptics' Guide to the Universe, a weekly science podcast (www.theskepticsguide.org), and President of The New England Skeptical Society (www.theness.com).

JAMES OBERG (www.jamesoberg.com) worked twenty-two years at NASA Mission Control in Houston, specializing in orbital rendezvous, and is now a full-time freelance consultant, author, speaker, and examiner of space folklore. He is a founding Fellow of CSICOP and a *Skeptical Inquirer* consulting editor.

JAMES V. PARKER, now retired, studied and wrote about the animal-rights movement for two decades from his position as public information officer of the Oregon National Primate Research Center. He holds a PhD in religious studies from the University of Louvain, Belgium, and taught theology at Mt. Angel Seminary in Oregon prior to his career as a science writer. He and P. Michael Conn are authors of *The Animal Research War* (Macmillan/Palgrave, 2008).

MASSIMO PIGLIUCCI is a professor of evolutionary biology and philosophy at Stony Brook University in New York. He has published *Denying Evolution: Creationism, Scientism, and the Nature of Science* and writes regularly for *Skeptical Inquirer.* His blog can be found at rationallyspeaking.blogspot.com

STEVEN PINKER is a Harvard College Professor at Harvard University and the author of *The Stuff of Thought, How the Mind Works, The Blank Slate,* and four other books.

BENJAMIN RADFORD has investigated ghosts, psychics, lake monsters, UFOs, mass hysterias, and many other paranormal phenomena for more than a decade. He is managing editor of the *Skeptical Inquirer* and the author or coauthor of three books; his latest (with fellow investigator Joe Nickell) is *Lake Monster Mysteries: Investigating the World's Most Elusive Creatures.* His Web site is at www.RadfordBooks.com.

AMIR RAZ holds the Canada Research Chair in the Cognitive Neuroscience of Attention at McGill University and the SMBD Jewish General Hospital, where he heads the Cognitive Neuroscience Laboratory and the Clinical Neuroscience and Applied Cognition Laboratory, respectively.

CARL SAGAN was the David Duncan Professor of Astronomy and Space Sciences and director of the Laboratory for Planetary Studies at Cornell University. He was author of thirty books including the *Cosmic Connection, Cosmos, Broca's Brain, Shadows of Forgotten Ancestors, Comet* (the latter two with Ann Druyan), *Pale Blue Dot, The Demon-Haunted World, Billions and Billions,* and the Pulitzer-Prize Winning *The Dragons of Eden.* His Peabody Award-winning public television series, *Cosmos,* was seen by more than 500 million people in more than sixty countries.

SAMUEL SASS was librarian of the General Electric Company's transformer division for thirty-one years before his retirement in 1976.

ROBERT SHEAFFER worked for thirty years as a real-time software engineer in the Silicon Valley and elsewhere. He is a longtime skeptical UFO investigator, a columnist and contributing editor for the *Skeptical Inquirer,* and the author of *UFO Sightings* (Prometheus, 1998). His website is www.debunker.com.

CAMERON M. SMITH teaches and conducts research at Portland State University's Department of Anthropology. He has also published articles in *Scientific American MIND* and other magazines. His article and chapter in this volume, with Charles Sullivan, was a precursor to their book *The Top 10 Myths about Evolution* (Prometheus, 2007).

CHARLES SULLIVAN teaches philosophy and writing at Portland Community College in Portland, Oregon. He's currently working on a book debunking superstitious beliefs. His article and chapter in this volume, with Charles Sullivan, was a precursor to their book The Top 10 Myths about Evolution (Prometheus, 2007). His website is www.CharlesSullivan.org

REFERENCES

1. "Science and the Public: Summing Up Thirty Years of the *Skeptical Inquirer*," by Paul Kurtz, in *Skeptical Inquirer* 30, no. 5 (September/October 2006). Reprinted with Permission.
2. "In Defense of the Higher Values," by Kendrick Frazier, in *Skeptical Inquirer* 30, no. 4 (July/August 2006). Reprinted with Permission.
3. "Ann Druyan Talks about Science, Religion, Wonder, Awe…and Carl Sagan," by Ann Druyan, in *Skeptical Inquirer* (November/December 2003). Reprinted with Permission.
4. "Carl Sagan's last Q & A on Science and Skeptical Inquiry," and Ann Druyan's "The Great Turning Away," *Skeptical Inquirer* 29, no. 4 (July/August 2005): 29–37. Reprinted with Permission.
5. "Can the Sciences Help Us to Make Wise Ethical Judgments?" by Paul Kurtz, in *Skeptical Inquirer* (September/October 2004): 18–24. Reprinted with Permission.
6. "Science Wars: Science and the Bush Administration," by Chris Mooney, in *Skeptical Inquirer* 29, no. 6 (November/December 2005). Reprinted with Permission.
7. "Do Extraordinary Claims Really Require Extraordinary Evidence?" by Massimo Pigliucci, in *Skeptical Inquirer* (March/April 2005): 14, 43. Reprinted with Permission.
8. "The Fallacy of Misplaced Rationalism," by Robert Sheaffer, in Skeptical Inquirer 32, no. 4 (July/August 2008). Reprinted with Permission.
9. "We Find That Intelligent Design is not Science: Excerpts from the Dover Federal Court Ruling on Intelligent Design," with introduction by Kendrick Frazier, in *Skeptical Inquirer* 30, no. 2 (March/April 2006). Reprinted with Permission.

10. "The 'Vise Strategy' Undone: *Kitemuller et al. v. Dover Area School District*," by Barbara Forrest, in *Skeptical Inquirer* 31, no. 1 (January/February 2007). Reprinted with Permission.
11. "Four Common Myths About Evolution," by Charles Sullivan and Cameron McPherson Smith, in *Skeptical Inquirer* 29, no. 3 (May/June 2005). Reprinted with Permission.
12. "Only a Theory?: Framing the Evolution/Creation Issue," by David Morrison, in *Skeptical Inquirer* 29, no. 6 (November/December 2005). Reprinted with Permission.
13. "Is Intelligent Design Creationism?" by Massimo Pigliucci, in *Skeptical Inquirer* 32, no. 1 (January/February 2008). Reprinted with Permission.
14. "The Memory Wars," (Part 1 only) by Martin Gardner, in *Skeptical Inquirer* 30, no. 1 (January/February 2006) Reprinted with Permission.
15. "AIDS Denialism vs. .Science," by Nicoli Nattrass, in *Skeptical Inquirer* 31, no. 5 (September/October 2007) Reprinted with Permission.
16. "Global Climate Change Triggered by Global Warming," by Stuart O. Jordan, in *Skeptical Inquirer* 31, no. 3 (May/June and July/August 2007). Reprinted with Permission.
17. "A Skeptical Look at September 11th: How We Can Defeat Terrorism by Reacting to It More Rationally," by Clark R. Chapman and Alan W. Harris, in *Skeptical Inquirer* (September/October 2002): 29–34. Reprinted with Permission.
18. "Strong Response to Terrorism Not a Symptom of Fallacious Statistical Reasoning or Human Cognitive Limitations," by Steven Pinker, in *Skeptical Inquirer* (July/August 2003): 59–60. Reprinted with Permission.
19. "The Anti-Vaccination Movement," by Steven Novella, in *Skeptical Inquirer* 31, no. 6 (November/December 2007). Reprinted with Permission.
20. "Vaccine Safety: Vaccines Are One of Public Health's Great Accomplishments," by Richard G. Judelsohn, in *Skeptical Inquirer* 31, no. 6 (November/December 2007). Reprinted with Permission.
21. "Warning: Animal Extremists are Dangerous to Your Health," by P. Michael Conn and James V. Parker, in *Skeptical Inquirer* 32, no. 3 (May/June 2008). Reprinted with Permission.
22. "Predator Panic: A Closer Look," by Benjamin Radford, in *Skeptical Inquirer* 30, no. 5 (September/October 2006). Reprinted with Permission.
23. "The Case for Decentralized Generation of Electricity," by Thomas R. Casten and Brennan Downes, in *Skeptical Inquirer* (January/February 2005): 25–33. Reprinted with Permission.
24. "The Philosophy behind Pseudoscience," by Mario Bunge, in *Skeptical Inquirer* 30, no. 4 (July/August 2006). Reprinted with Permission.
25. "The Columbia University 'Miracle' Study: Flawed and Fraud," by Bruce Flamm, in *Skeptical Inquirer* (September/October 2004): 25–31; and "Third Strike for Columbia University Prayer Study: Author Plagiarism," by Bruce Flamm, in *Skeptical Inquirer* 31, no. 3 (May/June 2007). Reprinted with Permission.
26. "Anomalous Cognition: A Meeting of Minds" by Amir Raz, in *Skeptical Inquirer* 32, no. 4 (July/August 2008). Reprinted with Permission.

27. "Anomalous Cognition? A Second Perspective," by Ray Hyman, in *Skeptical Inquirer* 32, no. 4 (July/August 2008). Reprinted with Permission.

28. "The PEAR Proposition: Fact or Fallacy?" by Stanley Jeffers, in *Skeptical Inquirer* 30, no. 3 (May/June 2006); and "PEAR Lab Closes, Ending Decades of Psychic Research," by Stanley Jeffers, in *Skeptical Inquirer* 31, no. 3 (May/June 2007). Reprinted with Permission.

29. "Gary Schwartz's Energy Healing Experiments: The Emperor's New Clothes?" by Harriet Hall, in *Skeptical Inquirer* 32, no. 2 (March/April 2008). Reprinted with Permission.

30. "Testing Natasha," by Ray Hyman, in *Skeptical Inquirer* 29, no. 3 (May/June 2005). Reprinted with Permission.

31. "'Dr.' Bearden's Vacuum Energy," by Martin Gardner, in *Skeptical Inquirer* 31, no.1 (January/February 2007). Reprinted with Permission.

32. "Lessons of the 'Fake Moon Flight' Myth," by James Oberg, in *Skeptical Inquirer* (March/April 2003): 23, 30. Reprinted with Permission.

33. "Magnet Therapy: A Billion-dollar Boondoggle," by Bruce L. Flamm, in *Skeptical Inquirer* 30, no. 4 (July/August 2006). Reprinted with Permission.

34. "Oxygen Is Good—Even When It's Not There," by Harriet A. Hall, in *Skeptical Inquirer* (January/February 2004): 48–50, 55. reprinted with Permission.

35. "Undercover among the Spirits: Investigating Camp Chesterfield," by Joe Nickell, in *Skeptical Inquirer* (March/April 2002): 22–25. reprinted with Permission.

36. "How to 'Haunt' a House," by Benjamin Radford, in *Skeptical Inquirer* 32, no. 1 (January/February 2008). Reprinted with Permission.

37. "The Devious Art of Improvising, Lessons One and Two," by Massimo Polidoro, in *Skeptical Inquirer* 32, no. 1 (January/February 2008). Reprinted with Permission.

38. "A Patently False Patent Myth—Still!" by Samuel Sass, in *Skeptical Inquirer* (May/June 2003): 43–45, 48. Reprinted with Permission.